Introduction to Forestry
(Objective Questions)

THE AUTHOR

Prof. (Dr.) C.S. Bohra (b.1980) has more than 16 years of teaching and research experience in higher education at various universities and research organizations in the field of Forestry, Agroforestry, Forest Ecology and Biodiversity conservation, Soil Science, Plantation Forestry, Medicinal Herbs, Climate Change, Carbon sequestration, Environmental Science, Disaster management and Agriculture sciences. He has started his journey in 2004 from Kumaun University, Nainital (Uttarakhand) and worked with many other organizations of International repute in the last sixteen years of his career such as, Govind Ballabh Pant University of Agriculture and Technology, Pantnagar and other affiliated colleges from H. N. B. Garhwal Central University, Srinagar, Uttarakhand. He was associated with many funding agencies like University Grants Commission (UGC), New Delhi, National Oilseeds and Vegetable Oil Development Board (NOVOD), Gurgaon, Govt. of India, National Mission on Bamboo Application (NMBA), Govt. of India, New Delhi and Department of Science and Technology (DST), Govt. of India. In Agroforestry domain, Dr. Bohra is the only PDF Fellow of Dr D S Kothari Post Doctorate Fellowship in the country. Prof. Bohra has published five books and more than two dozen of research papers in the International/National journals of repute. He is also linked with many journals as member of editorial board and reviewer. In the past years, many conferences has been organized by Dr. Bohra and also presented his research in various National/International conferences/symposiums. Currently, Dr. Bohra is associated with Dev Bhoomi Group of Institutions, Dehradun (Uttarakhand) as Professor and Dean, School of Agriculture and Allied Sciences.

Introduction to Forestry
(Objective Questions)

Dr. C.S. Bohra

Professor and Dean
School of Agriculture and Allied Sciences
Dev Bhoomi Group of Institutions
(Affiliated to Sri Dev Suman University, Badshahithol, Uttarakhand)
Navgaon, Manduwala, Dehradun
(Uttarakhand)

2021
Astral Cracker
A Division of
Astral International Pvt. Ltd.
New Delhi – 110 002

ISBN 978-93-90384-61-7 (PB)

Published by : **Astral Cracker**
A Division of
Astral International Pvt. Ltd.
– ISO 9001:2015 Certified Company –
4736/23, Ansari Road, Darya Ganj
New Delhi-110 002
Ph. 011-43549197, 23278134
E-mail: info@astralint.com
Website: www.astralint.com

Printed at : **Replika Press Pvt Ltd**

— *Dedicated to* —

Bundle of Joy, My Nephews

Master Ojas Pratap Singh and
Master Shaurya Pratap Singh

Preface

The growing number of professionals from multifarious organizations has created significant competition among the graduates of various subjects. The success of any professional lies with the assured employment in government, NGOs and industrial sectors. Forestry subject is of no exception. Being a narrow science, but offered almost several SAUs and private universities distributed across the country, it has created large number of forestry graduates, which resulted in competition and compelled the forestry graduates to acquire up to date information in their subject for getting successful induction in public and private sectors. For being successful, one must be thorough in each and every aspects of the subject which demand a comprehensive book to resolve the issues.

Many competitive examinations conducted by state and central government organizations particularly UKPSC (Forest Ranger, ACFs, PCS), HPSC, UPSC (Preliminary exams), ARS (JRF, SRF and Scientist) and Bank exams, *etc.* are predominantly conducted through objective and subjective systems. For subjective examination wide range of books are available for the use by all competitions. However for objective type examinations, only very few books are available which again have not compiled the readymade information for each and every sector of the subject. The huge subject often pose and threat to the examinations which resulted in consuming more time and money in search of the actually relevant materials.

To mitigate this problem, the present book on forestry has been conceived and designed to benefit the forestry and agroforestry graduates who are preparing for the examinations like, UKPSC (Forest Ranger, ACFs, PCS), HPSC, UPSC (Preliminary exams), ARS (JRF, SRF and Scientist) and Bank exams., *etc.* This question bank on forestry provides the knowledge on the subject and extends adequate hands on experience through practice.

It encompasses several facets of forestry subject extensively which included Silviculture, Forest Mensuration, Forest Management, Forest Utilization, Wood Science and Technology, Forest Botany, Forest Genetics and Tree Breeding, Tree Physiology, Soil Science, Forest Microbiology, Forest Entomology, Forest Pathology, Wildlife, Ecology and Environment, Forest Laws and Policies and other additional and related informations on forestry.

This book covers wide range of questions with key answers which will be a ready reckoned and perfect workbook for wide range of competitive examinations. Adequate care has been taken to avoid errors. In forestry Sector, this is a maiden attempt and bound to happened with errors. Such errors if any, noticed by readers of this book can be informed to the author for necessary corrections and publications in all their future editions. Extending useful and constructive criticism and suggestions are most welcome for augmenting the contents in future.

Prof. (Dr.) C. S. Bohra

Contents

Chapter 1

Agroforestry

1. **The sustainable land use systems involving trees combined with crops and/or animals on the same unit of land is termed as:**
   (a) Agroforestry
   (b) Sustainable forest management
   (c) Social Forestry
   (d) Farm Forestry
2. **When the use of land is directed towards satisfying basic needs and is managed by the owner is said to be:**
   (a) Commercial Agroforestry systems
   (b) Intermediate Agroforestry systems
   (c) Subsistence Agroforestry systems
   (d) All of the above
3. **The practice of forestry ion and around the farms/village lands is called as:**
   (a) Social Forestry
   (b) Community forestry
   (c) Farm forestry
   (d) All of the above
4. **The practice of forestry in areas devoid of tree growth and other vegetation outside the conventional forest areas is:**
   (a) Social forestry
   (b) Extension forestry
   (c) Farm forestry
   (d) Afforestation
5. **The practice of forestry with the object of raising flowering trees is called as:**
   (a) Recreation forestry
   (b) Urban forestry
   (c) Social forestry
   (d) Farm forestry

6. **Plantations grow as liner strips of lands on the sides of public roads, canals and railway lines are called as:**
   (a) Linear strip plantations (b) Commercial forestry
   (c) Avenue plantations (d) Urban forestry

7. **Home Garden Agroforestry system is practiced in:**
   (a) Humid tropical climates (b) Semi arid lands
   (c) High lands (d) Sub humid low lands

8. **Shading effect of trees can be overcome by:**
   (a) Canopy management (b) Root management
   (c) Water (d) Fertilizer

9. **Alley cropping is also known as:**
   (a) Hedge row intercropping (b) Woody Hedges
   (c) Wood lots (d) Soil conservation hedges

10. **Multistory arrangement reduces:**
    (a) Yield (b) Nutrition
    (c) Competition (d) All of the above

11. **The oldest known agroforestry systems is:**
    (a) Alley cropping (b) Taungya systems
    (c) Home gardens (d) Shifting cultivation

12. **Social forestry programme was launched in the:**
    (a) Fifth five year plan (b) Sixth five year plan
    (c) Seventh five year plan (d) Eighth five year plan

13. **The collection period for neem *Azadiracta indica* seed is:**
    (a) January-March (b) March-May
    (c) May-June (d) June-July

14. **In the manufacture of Braille paper the _______ species is used:**
    (a) *Eucalyptus globules* (b) *Eucalyptus tereticornis*
    (c) *Eucalyptus camandulensis* (d) *Eucalyptus urophylla*

15. ***Indigofera tinctoria* is used in:**
    (a) Tannin industry (b) Dye industry
    (c) Resin (d) Gum

16. **Which classification of agroforestry is more important and commonly adopted?**
    (a) Socio-economic classification (b) Structural classification
    (c) Floristic classification (d) Ecological classification

17. In which agroforestry system, agricultural crops is of secondary importance?
(a) Silvipasture (b) Agrisilviculture
(c) Silviagriculture (d) Hortisilviculture

18. The most prevalent agroforestry systems in high rainfall areas of Kerala is:
(a) Silvipasture (b) Agrisilviculture
(c) Horti-silvi-pasture (d) Home gardens

19. Which among these tree species are best suited for alley cropping?
(a) *Eucalyptus tereticornis* (b) *Casuarinas equisetifolia*
(c) *Cassia siamea* (d) *Aianthus excels*

20. Which among these factors has degraded the Indian grasslands the most?
(a) Adverse site factors (b) Poor management
(c) Poor productivity (d) Over grazing

21. Carrying capacity is more related to:
(a) Agrisilviculture (b) Improvement of grasslands
(c) Improvement of nutritive value of grassland (d) None of the above

22. Trade/common name of *Dalbergia latifolia* is:
(a) Rose wood (b) Beef wood
(c) Soft wood (d) Hard wood

23. Animals feed in fenced pastures with control access is terms as:
(a) Paddoching (b) Pasture reseeding
(c) Grazing (d) Browsing

24. Indian grassland and fodder research institute is located at:
(a) Dehradun (b) Jhansi
(c) Jodhpur (d) Nigeria

25. Rotation of aerable crops with two or more years of sown pasture is called as:
(a) Mixed farming (b) Mixed cropping
(c) Land use systems (d) Lay farming

26. Forest products contribution to world gross domestic product (GDP) is:
(a) 1 per cent (b) 1.7 per cent
(c) 2 per cent (d) 2.7 per cent

27. Tree planted in silvi-agriculture should be oriented in the:
(a) East west direction (b) North south direction
(c) East south direction (d) West south direction

28. Most common agroforestry system practiced in arid regions are:
(a) Agrisilvicultural systems
(b) Silvopastural systems
(c) Agrosilvopastoral systems
(d) Other systems

29. Social forestry was first coined by:
(a) Westoby
(b) Shah
(c) Brandis
(d) Nair

30. Social forestry was first adopted successfully in:
(a) Gujarat
(b) Tamil Nadu
(c) Kerala
(d) Rajasthan

31. Van mahotasav is a festival of:
(a) Wild life
(b) Tree planting
(c) Timbers
(d) All of the above

32. Van mahotsav was started in:
(a) 1847
(b) 1947
(c) 1850
(d) 1950

33. National wasteland development board came into existence during the period:
(a) 1970
(b) 1975
(c) 1980
(d) 1985

34. The India, Plantation forestry was started in:
(a) 1942
(b) 1842
(c) 1947
(d) 1847

35. The major species used for planting on tank bed is:
(a) *Acacia leucophloea*
(b) *Prosopis juliflora*
(c) *Acacia nilotica*
(d) *Eucalyptus* spp.

36. Which type of tree is better for noise abatement?
(a) Evergreen trees
(b) Deciduous trees
(c) Semi evergreen trees
(d) All of the above

37. Heliotropic type of branching should be preferred for:
(a) Roadside planting
(b) Canal planting
(c) Railway planting
(d) All of the above

38. Neem and Mango are preferred for:
(a) Railway line planting
(b) Canal bank plantation
(c) Tank bed plantation
(d) Avenue planting

39. Social forestry project in Tamil Nadu was first launched with the assistance of:

(a) CIDA (b) SIDA

(c) JBIC (d) All of the above

40. SIDA Project phase – I was implemented during:

(a) 1981-86 (b) 1987-88

(c) 1988-89 (d) 1992-93

41. Protection and reforestation of degraded forests with the involvement of local people is:

(a) Sustainable forest management (b) Joint forest management

(c) Social forestry (d) Community forestry

42. The unit of management under JFM is:

(a) Hamlet (b) Division

(c) Beat (d) Village

43. The people's representation body on each of the identified management unit of JFM is:

(a) Sustainable forest management (b) Joint forest management

(c) Social forestry (d) Community forestry

44. TCIP stands:

(a) Tree cultivation intensive programme (b) Tree cultivation incentive programme

(c) Tree cultivation inducement programme (d) Tree cultivation income programme

45. SIDA project phase II was implemented during:

(a) 1981-86 (b) 1987-88

(c) 1988-89 (d) 1992-93

46. Tamil Nadu afforestation project was implemented with financial assistance from:

(a) Denmark (b) Japan

(c) Netherland (d) Sweden

47. When an immediate supply of a single necessary factor falls below the combined demand of the plant is called:

(a) Allelopathy (b) Allelospoly

(c) Both (a) and (b) (d) None of the above

48. The direct or indirect of one plant upon another through the production of chemical inhibitors is called:

(a) Allelopathy (b) Allelospoly

(c) Both (a) and (b) (d) None of the above

49. Allelopathy denotes:

(a) Positive interaction (b) Negative interaction

(c) Both (a) and (b) (d) None of the above

50. Allelochemicals produced by plants are:

(a) Primary plant metabolites (b) Secondary plant metabolities

(c) Both (a) and (b) (d) None of the above

51. Allelochemicals are released by means of:

(a) Volatilization (b) Leaching

(c) Exudation (d) All of the above

52. *Prosopis cineraria* and *Cenchrus ciliaris* is a good fodder tree plus grass combination for:

(a) Semi-arid climate with red soil (b) Temperate climate

(c) Arid climate with sandy soil (d) Semi-arid block cotton soil

53. Tree competes with the neighboring crops mainly for:

(a) Light (b) Moisture

(c) Nutrients (d) All of the above

54. Allelopathy is the best example for:

(a) Association (b) Ammensalism

(c) Symbiosis (d) Photo corporation

55. The broad and well known classification of agroforestry systems was given by:

(a) Chundawat (b) Gautam

(c) Nair (d) Dwivedi

56. A common example of the zonal pattern is:

(a) Hedgerow inter cropping (b) Home gardens

(c) Silvopastural systems (d) Agrosilvopastural systems

57. The fundamental attributes of all agroforestry systems are:

(a) Recover depleted soil nutrients (b) Increase crop yields

(c) Production and protection (d) Stabilize the soil structure

58. Integration of trees with grasses is known as:

(a) Apiculture (b) Silvipasture
(c) Aquaforestry (d) Multipurpose wood lots

59. Taungya was first introduced by:

(a) Dr. Brandis (b) Nair
(c) Clements (d) L.S. Khanna

60. Commercial agroforestry systems is used:

(a) Production of perennial cash crops (b) Satisfying basic needs
(c) Sale of the production of output (d) All of the above

61. The deliberated use of land for the production of crops including trees and crops is called as:

(a) Agri horticulture (b) Agripasture
(c) Agrisilvipasture (d) Agrisilviculture

62. The main aim of raising crops in departmental taungya is to:

(a) Reduce the weed growth (b) To provide additional income
(c) Maximum utilization of site (d) All of the above

63. The practice of managing rows of closely planted woody plants with annual crops planted in alleys in between hedges is called as:

(a) Hedge row intercropping (b) Woody hedgerows
(c) Soil conservation hedges (d) Multipurpose tree gardens

64. Shelter belts should possess a typical:

(a) Triangular shape (b) Linear Belt
(c) Square shape (d) Rectangular shape

65. Strips of trees planted to protect fields, homes, canal from wind is referred to:

(a) Shelter belts (b) Wind breaks
(c) Alley cropping (d) All of the above

66. Planting nectar producing trees on the boundary of the agricultural field is called as:

(a) Apiculture with trees (b) Aquaforestry
(c) Multipurpose wood lots (d) Woody hedgerows

67. The hydrological unit that drains at a common point is called:

(a) Watershed (b) Catchment
(c) Drainage basin (d) All of the above

68. A unit of landscape that contains all the drainage areas and channels contributing to a single stream/river system:

(a) Watershed (b) Dam

(c) Seashore (d) River beds

69. The drainage area of tertiary steam or river is:

(a) 400 to 40000 ha (b) 4000 to 40000 ha

(c) 40000 to 400000 ha (d) None of the above

70. The drainage area of catchments is:

(a) 400-200 ha (b) 2000-4000 ha

(c) >40000 ha (d) More than one lakh hectare

71. The drainage area of a sub catchment is:

(a) <40000 ha (b) >40000 ha

(c) >100000ha (d) 4000-40000ha

72. The continuous circulation of water among the hydrosphere, atmosphere and lithosphere is a called as:

(a) Nutrient cycle (b) Hydrological cycle

(c) Both (a) and (b) (d) None of the above

73. The drainage area of sub watershed is:

(a) 200-400 ha (b) 200-4000ha

(c) 200-2000ha (d) 2000-4000ha

74. The drainage area of a mini watershed is:

(a) 200-400 ha (b) 200-4000ha

(c) 200-2000ha (d) 400-2000ha

75. The drainage area of a micro water shed is:

(a) <400ha (b) >400 ha

(c) 400-2000 ha (d) 400-4000ha

76. The practice of rotational cropping helps ion:

(a) Improving soil structure (b) Maintaining soil fertility

(c) Soil and water conservation (d) All of the above

77. Erosion permitting and erosion resisting crops grows alternatively at right angle to the slope of the land is called as:

(a) Linear planting (b) Strip planting

(c) Cover planting (d) All of the above

78. The process of taking up nutrients from deeper soil profiles and depositing them on the surface layer is referred to:

(a) Nutrient uptake
(b) Nutrient absorption
(c) Nutrient pumping
(d) Nutrient cycling

79. The process of cycling of nutrients from soil to the plant and back to the soil is called as:

(a) Nutrient uptake
(b) Nutrient absorption
(c) Nutrient pumping
(d) Nutrient cycling

80. The process by which the complex organic substances are broken down to simple inorganic elements is called as:

(a) Mineralization
(b) Nitrification
(c) Ammonification
(d) All of the above

81. NWDB classified wastelands into:

(a) Two types
(b) Three types
(c) Four types
(d) Ten types

82. Extend of wastelands in India:

(a) 63.52 m ha
(b) 36.52 m ha
(c) 83.52 m ha
(d) 38.52 m ha

83. The land comprising deep gorges, gullies in deep alluvium and draining into a major river is known as:

(a) Sheet
(b) Drill
(c) Eroded land
(d) Gullied/Ravinous land

84. The last stage of water erosion is the formation of:

(a) Sheet
(b) Drill
(c) Gullies
(d) Ravines

85. The wastelands formed due to canal seepage and improper drainage is:

(a) Gullies and Ravines lands
(b) Water logged and marshy lands
(c) Mangroves
(d) Sand dunes

86. The lands which are rich in soluble salts and exchangeable sodium:

(a) Saline soil
(b) Alkaline soil
(c) Saline – Alkaline
(d) All of the above

87. The suitable species for reclamation of saline and sodic soil is:

(a) *Prosopis juliflora*
(b) *Albiiza lebbeck*
(c) *Azadirachta indica*
(d) *Casuarinas equisetifolia*

88. **The most suitable species for reclamation of sandy soils is:**
   (a) *Casuarinas equisetifolia* (b) *Acacia* spp.
   (c) *Melia dubia* (d) *Azadirachta indica*

89. **A family of procedure for the diagnosis of land management problems and potentials and design of agroforestry solution is:**
   (a) Sustainability (b) Productivity
   (c) Adoptability (d) Agroforestry D and D

90. **Agroforestry D and D covers entire ecological zone within a country is:**
   (a) Macro D and D (b) Micro D and D
   (c) Both (a) and (b) (d) None of the above

91. **Agroforestry D and D covers within the chosen ecological zone is:**
   (a) Macro D and D (b) Micro D and D
   (c) Both (a) and (b) (d) None of the above

92. **The main criteria for good agroforestry design are:**
   (a) Productivity (b) Sustainability
   (c) Adoptability (d) All of the above

93. **The term Social Forestry first time used by:**
   (a) Shah (b) Brandis
   (c) Nair (d) Jack Westoby

94. **Tree means water, water is bread and bread is life was first time quoted by:**
   (a) Westoby (b) Brandis
   (c) Sunderlal Baguguna (d) Nair

95. **The international donor agencies for social forestry in India:**
   (a) CIDA (b) USAID
   (c) SIDA (d) All of the above

96. **The cross sectional shape of any shelter belt is:**
   (a) Parallel (b) Neloid
   (c) Cylinder (d) Pyramid

97. **The practice of forestry with the object of obtaining maximum gain per unit area called as:**
   (a) Mixed forestry (b) Multiple use forestry
   (c) Intensive forestry (d) All of the above

98. The practice of forestry for the simultaneous use of a forest area for two or more purpose called as:

(a) Multiple use forestry
(b) Mixed forestry
(c) Multipurpose trees and shrubs
(d) Multispecies tree garden

99. Shelter belt have (Pick the odd man out):

(a) Single tree species
(b) Multiple tree species
(c) Used to arrest wind
(d) Prevent sand dune formation

100. Jhuming is the other name for:

(a) Shifting cultivation
(b) Taungya
(c) Both (a) and (b)
(d) None of the above

101. Components of social forestry:

(a) Farm forestry
(b) Extension forestry
(c) Recreational forestry
(d) All of the above

102. Hedge row intercropping is:

(a) Agricilvicultural systems
(b) Sivopastural systems
(c) Agrosilvopastural systems
(d) Other systems

103. Example for MPTs:

(a) Casuarinas
(b) Acacia
(c) Gliricidia
(d) All of the above

104. The ratio of height and width roughly in shelter belt:

(a) 1:10
(b) 1:15
(c) 1:20
(d) 1:25

105. The home garden is characterized by:

(a) 2-3 vertical strata
(b) 3-4 vertical strata
(c) 3-5 vertical strata
(d) 2-5 vertical strata

106. A well designed windbreak can offer protection up to _______ times of its height towards leeward side:

(a) 10
(b) 20
(c) 25
(d) 15

107. Essential attributes of agroforestry system are:

(a) Productivity
(b) Sustainability
(c) Adoptability
(d) All of the above

108. Based on the nature of components, Nair (1985) classified the agroforestry system into _______ groups:

(a) Four (b) Five

(c) Six (d) Sixteen

109. Home gardens found in tropics are one of the classic example of:

(a) Agrisilvicultural systems (b) Silvopastural systems

(c) Agrosilvopastural systems (d) Apiculture with trees

110. Taungya system was originated in:

(a) North Bengal (b) Kerala

(c) Tamil Nadu (d) Myanmar

111. Head quarters of international center for research in Agroforestry is located at:

(a) Ibadan (b) Nairobi

(c) Canada (d) Denmark

112. An example for NFTs:

(a) Teak (b) Leuceana

(c) Neem (d) All of the above

113. The first state in India, created a separate social Forest wing in the forest department was:

(a) Kerala (b) Gujarat

(c) Tamil Nadu (d) Karnataka

114. In Southern India, the taungya system is called as:

(a) Kumara (b) Jhum

(c) Podu (d) Dao or axe and fire

115. Taungya is a _______ word:

(a) Gujarat (b) Assamese

(c) Nepali (d) Burmese

116. Social forestry was first recommended by:

(a) National commission on Agriculture (b) Van mahotsav

(c) Ninth commonwealth forestry congress (d) All of the above

117. Joint forest management evolved in early:

(a) 1950 (b) 1960

(c) 1970 (d) 1980

118. The litteral meaning of agroforestry is:

(a) Agri+Forest (b) Agri+Horti

(c) Agri+Fish (d) Agri+Honeybees

119. Ninth commonwealth congress was held in:

(a) 1960 (b) 1968

(c) 1988 (d) 1952

120. Which one of the following states is having more arid areas?

(a) Gujarat (b) Haryana

(c) Andhra Pradesh (d) Rajasthan

121. The scientific name of pons cypress is:

(a) *Taxodium distichum* (b) *Taxodium scandens*

(c) *Salix alba* (d) *Celtis australis*

122. Characteristics of multipurpose tree suitable for agroforestry is:

(a) High adaptability (b) Nitrogen fixing

(c) Coppicing ability (d) All of the above

123. Social forestry project phase II was implemented during 1988-89 to 1992-93 with a financial outlay of:

(a) Rs.591 million (b) Rs. 584 million

(c) Rs.854 million (d) Rs. 951 million

124. The National Afforestation programme is being implemented through a _______ tier structure.

(a) 2 (b) 3

(c) 4 (d) 5

125. Institute of Tropical forestry is situated at:

(a) Pune (b) Puerto Rica

(c) Jhansi (d) Kenya

126. Basic set of elements or components of Agroforestry are:

(a) Tree (b) Herbs

(c) Animals (d) All of the above

127. Social forestry aids were received from:

(a) SIDA (b) CIDA

(c) USAID (d) All of the above

128. "The Journal of Agroforestry Today" is published from:

(a) Kenya (b) Denmark

(c) Japan (d) Canada

129. ICAR-Indian Grassland and Fodder Research Institute is located at:

(a) Jhansi (b) Solan

(c) Ranchi (d) Jodhpur

130. An example for live fence is:

(a) *Euphorbia* (b) *Gliricidia*

(c) *Acacia* (d) *Sesbania*

131. Nitrogen fixing ability of Casuarina is attributable to:

(a) *Azotobacter* (b) *Frankia*

(c) *Rhizobium* (d) *Azospirillum*

132. The system in which woody component is combined with agricultural crop is known as:

(a) Agri-silivi (b) Agri-horti

(c) Silvi pastoral (d) Apiculture

133. Shifting cultivation is most prevalent in:

(a) North East India (b) Eastern part of India

(c) Northern part of India (d) All of the above

134. Tropical Forest Research Institute is located in:

(a) Jhansi (b) Jabalpur

(c) Agra (d) Jodhpur

135. In north eastern hill region, shifting cultivation is called as:

(a) Jhum (b) Podu

(c) Kumara (d) Slash and burn

136. Most common agroforestry system practiced in arid regions are:

(a) Agrisilvicultural systems (b) Silvipastoral systems

(c) Agrosilvipastoral systems (d) Aquaforestry

137. Which forestry practice is not included in the scope of social forestry:

(a) Farm forestry (b) Extension forestry

(c) Recreational forestry (d) Protection forestry

138. Which social forestry programmes is purely managed by private efforts:

(a) Farm forestry (b) Community forestry

(c) Rehabilitation of degraded forest area (d) Urban forestry

139. Based on temporal arrangement of AF system, 'homegarden' is the example of:

(a) Coincident (b) Concomitant

(c) Interpolated (d) Relay cropping

140. Central Arid Research Institute is located at:

(a) Jabalpur (b) Jhansi
(c) Jodhpur (d) Dehradun

141. Protein bank is a sub system of:

(a) Agri silviculture (b) Silvi pastoral
(c) Agri horti (d) Agri silvi pastoral

142. The first Taungya plantation was established in:

(a) Maharashtra (b) Gujarat
(c) North Bengal (d) Rajasthan

143. ICFRE is located in:

(a) New Delhi (b) Dehradun
(c) Jhansi (d) Ibadan

144. Green gold is:

(a) Sal (b) Sandal
(c) Bamboo (d) Teak

145. Home garden is highly suitable for:

(a) Humid/sub humid region (b) Arid/Semi arid region
(c) Hilly Region (d) High lands

146. Generally Shelter belt assumes the shape of:

(a) Quadrangle (b) Rectangle
(c) Triangle (d) Tropizoidal

147. Energy plantation is established for:

(a) Pulp wood production (b) Fuel wood production
(c) Soil conservation (d) Wind break

148. TAP is funded by:

(a) SIDA (b) CIDA
(c) DANIDA (d) JBIC

149. When was JFM concept initiated in India:

(a) 1980 (b) 1970
(c) 1973 (d) 1972

150. Who initiated JFM concept?

(a) Sundar Lal (b) K.M. Munshi
(c) A.K. Banerjee (d) B.K. Nair

151. Who initiated Van Mahotsava?

(a) Brandis (b) K.M. Munshi
(c) A.K. Banerjee (d) Westoby

152. When was TAP programme initiated in Tamil Nadu?

(a) 1981-1986 (b) 1987-88
(c) 1988-89 (d) 1997-2001

153. Social forestry is a forestry which aims at producing flow of protection and recreation benefits for the community. This definition was give by:

(a) Westoby (b) Brandis
(c) Sundar lal (d) SIDA

154. Social Forestry project Phase – I launched in:

(a) 1981 (b) 1985
(c) 1988 (d) 1992

155. Social Forestry project Phase – II launched in:

(a) 1981 (b) 1985
(c) 1988 (d) 1992

156. The average size of tree patta land could be:

(a) 2.0 (b) 1.2
(c) 4.0 (d) 5.0

157. NRCAF is located in:

(a) Agra (b) Jhansi
(c) Jodhpur (d) Jabalpur

158. Benefit sharing concept is a core objective in:

(a) Joint forest management (b) Sustainable forest management
(c) Community forestry (d) Social forestry

159. Joint forest management concept was first implemented at:

(a) Jabalpur (b) Ayyalur
(c) Arabari (d) Jhansi

160. Wind breaks are:

(a) Strips of trees (b) Line of trees
(c) Both (a) and (b) (d) None of the above

161. Example of Allelo chemicals:

(a) Gallic acid (b) Vanilic acid
(c) Syringic acid (d) All of the above

162. The Diagnosis and Design (D and (d) survey in agroforestry was given by:

(a) PKR Nair
(b) Dwivedi
(c) Chundawat
(d) Gautam

163. In North India agroforestry the combination of _______ is common:

(a) Rice with poplar
(b) Rice with eucalyptus
(c) Rice with melia
(d) All of the above

164. The combination wheat with eucalyptus was common in:

(a) Punjab
(b) Rajasthan
(c) Karnataka
(d) Odisha

165. In Kerala the most commonly found agroforestry system is:

(a) Alley cropping
(b) Multispecies tree garden
(c) Protein bank
(d) Home garden

166. In Tamil Nadu in the Kangeyam region the famous silvipastoral system consists:

(a) *Cenchrus ciliaris* and *Acacia nilotica*
(b) *Cenchrus ciliaris* and *Acacia leucophlea*
(c) *Eulaliopsis binata* and *Acacia nilotica*
(d) *Eulaliopsis binata* and *Acacia leucophlea*

167. Most commonly used tree species for shade in tea estates is:

(a) *Grevillea robusta*
(b) *Eucalyptus* spp.
(c) *Melia dubia*
(d) All of the above

168. In joint forest management programme, the unit of management is:

(a) Village
(b) Division
(c) Range
(d) Beat

169. In joint forest management programme, the objectives are decided at:

(a) Division
(b) Village level
(c) Range
(d) Beat

170. Waste land:

(a) Land with <12 per cent productivity
(b) Land with >12 per cent productivity
(c) Land with <20 per cent productivity
(d) Land with >20 per cent productivity

171. ICRAF stands for:

(a) International Council for Research in Agroforestry

(b) Indian Council for Research in Agroforestry

(c) Indian Centre for Rainfall Assesment in Forest

(d) International Centre for Rainfall Assesment in Forest

172. National watershed development programme for rural areas (NWDPR(A) has been launches in the year:

(a) 1997 (b) 1995

(c) 1972 (d) 1982

173. Loyal timber of the poor is:

(a) Prosopis (b) Teak

(c) Acacia (d) Bamboo

174. Indian lilac is the common name for:

(a) *Azadirachta indica* (b) *Tectona grandis*

(c) *Casuarinas equisetifolia* (d) *Acacia nilotica*

175. The queen of timbers is:

(a) Rose wood (b) Vagai

(c) Teak (d) Eucalyptus

176. Which one of the following species is preferred for water logged areas:

(a) Coconut (b) Eucalyptus

(c) Neem (d) *Casuarinas* spp.

177. In *Populus deltoids* based Agrisilvicultural systems, the suitable intercrop is:

(a) Ginger (b) Paddy

(c) Wheat (d) Sugarcane

178. Calorific value (k cal/kg) of *Acacia catechu* is:

(a) 3124 (b) 4400

(c) 4700-4900 (d) 5142-5244

179. Calorific value (k cal/kg) of *Prosopis juliflora* is:

(a) 3800 (b) 5000-5500

(c) 5200-5400 (d) 5400-5600

180. Which one of the following is a specialized agroforestry systems?

(a) Agrisilviculture (b) Silvipasture

(c) Aquaforestry (d) Hortisilvicuture

181. ICAR had launched All India Coordinated Research Project on Agroforestry in the country in:

(a) 1974 (b) 1979

(c) 1981 (d) 1983

182. The characteristics of arid areas are:

(a) Salinity (b) Saline water

(c) Alkalinity (d) All of the above

183. The tree species having xerophytic characteristics are suitable for:

(a) Humid Bengal Assam basin (b) Arid zones

(c) Sub humid areas (d) Humid to semi arid areas

184. Which one of the following species is preferred for water logged areas:

(a) Coconut (b) Eucalyptus

(c) Neem (d) *Casuarinas* spp.

185. Which one of the following crop has high salt tolerant capacity:

(a) Wheat (b) Rice

(c) Sunhemp (d) Mustard

186. Important silvicultural operation in agroforestry is:

(a) Pruning (b) Thinning

(c) Weeding (d) Climber cutting

187. Which is not a factor in which the agroforestry systems are classified?

(a) Floristic (b) Function

(c) Agro climatic zones (d) None of the above

188. Classification of agroforestry based on arrangement of components in respect of time (Pick the odd out):

(a) Coincident (b) Mixed dense

(c) Concomitant (d) Interpolated

189. Agroforestry practice of growing *Albizia lebbeck* with Sorghum is called:

(a) Hydromorphic system (b) Xeromorphic system

(c) Mesophytic system (d) None of the above

190. Swidden is a synonym for one of the following agroforestry systems. Identify the same:

(a) Taungya (b) Home garden

(c) Parkland system (d) Shifting cultivation

191. An agroforestry system suitable for small orchards prone to wind damage is:

(a) Shelterbelt
(b) Swidden
(c) Windbreak
(d) Multistory cropping

192. A major thrust to "social forestry concept" was given by:

(a) NCA report
(b) Forest Conservation Report
(c) National forest policy
(d) Indian forest act

193. How may agroclimatic zones are there in Tamil Nadu?

(a) 15
(b) 07
(c) 09
(d) 08

194. Silvipasture means:

(a) Growing trees with grass
(b) Growing trees with annuals
(c) Growing trees with fruit crops
(d) Growing trees with medicinal plants

195. HDSR is applied for which plantation?

(a) Timber
(b) Pulpwood
(c) Fuel wood
(d) Agroforestry

196. Which coming there is not an advantage of agroforestry system?

(a) Maximum production
(b) Improve soil fertility
(c) Maximize adverse effect of climatic factors
(d) Aid industrial growth

197. In order to get maximum net return from agroforestry we have to:

(a) Choose suitable tree species
(b) Choose high yielding agricultural crops
(c) Plant clonally propagated tree seedlings
(d) Adopt suitable tree crop combination

198. Planting of forest trees in agroforestry systems (Pick the odd out):

(a) Reduces temperature
(b) Reduces evaporation
(c) Increase irrigation requirement
(d) Increase humidity

199. In which agroforestry system, agricultural crops is of secondary importance?

(a) Silvipasture
(b) Agrisilviculture
(c) Silviagriculture
(d) Hortisilviculture

200. Shelterbelt have (Pick the odd out):

(a) Single species
(b) Multiple tree species
(c) Prevent sand dune formation
(d) Used to arrest wind

Answer Sheet

1.	(a)	31.	(b)	61.	(d)	91.	(b)	121.	(b)
2.	(a)	32.	(d)	62.	(d)	92.	(d)	122.	(d)
3.	(d)	33.	(d)	63.	(a)	93.	(d)	123.	(c)
4.	(b)	34.	(b)	64.	(a)	94.	(a)	124.	(a)
5.	(a)	35.	(c)	65.	(b)	95.	(c)	125.	(b)
6.	(a)	36.	(a)	66.	(a)	96.	(b)	126.	(d)
7.	(a)	37.	(a)	67.	(a)	97.	(c)	127.	(d)
8.	(a)	38.	(d)	68.	(a)	98.	(a)	128.	(a)
9.	(a)	39.	(b)	69.	(b)	99.	(b)	129.	(a)
10.	(c)	40.	(a)	70.	(d)	100.	(a)	130.	(a)
11.	(d)	41.	(b)	71.	(b)	101.	(d)	131.	(b)
12.	(a)	42.	(a)	72.	(b)	102.	(a)	132.	(a)
13.	(c)	43.	(b)	73.	(d)	103.	(d)	133.	(a)
14.	(a)	44.	(b)	74.	(d)	104.	(a)	134.	(b)
15.	(b)	45.	(c)	75.	(a)	105.	(b)	135.	(a)
16.	(b)	46.	(b)	76.	(d)	106.	(b)	136.	(a)
17.	(c)	47.	(b)	77.	(b)	107.	(d)	137.	(d)
18.	(d)	48.	(a)	78.	(c)	108.	(a)	138.	(a)
19.	(c)	49.	(c)	79.	(d)	109.	(c)	139.	(c)
20.	(d)	50.	(b)	80.	(b)	110.	(d)	140.	(a)
21.	(b)	51.	(d)	81.	(a)	111.	(b)	141.	(b)
22.	(a)	52.	(c)	82.	(a)	112.	(b)	142.	(c)
23.	(a)	53.	(d)	83.	(d)	113.	(b)	143.	(b)
24.	(b)	54.	(c)	84.	(d)	114.	(a)	144.	(c)
25.	(d)	55.	(c)	85.	(b)	115.	(d)	145.	(a)
26.	(b)	56.	(a)	86.	(c)	116.	(a)	146.	(c)
27.	(a)	57.	(c)	87.	(a)	117.	(c)	147.	(b)
28.	(a)	58.	(b)	88.	(a)	118.	(a)	148.	(d)
29.	(a)	59.	(a)	89.	(d)	119.	(b)	149.	(b)
30.	(a)	60.	(c)	90.	(a)	120.	(d)	150.	(c)

151.	**(b)**	**161.**	**(d)**	**171.**	**(b)**	**181.**	**(d)**	**191.**	**(c)**
152.	**(d)**	**162.**	**(a)**	**172.**	**(b)**	**182.**	**(d)**	**192.**	**(a)**
153.	**(a)**	**163.**	**(a)**	**173.**	**(a)**	**183.**	**(b)**	**193.**	**(d)**
154.	**(a)**	**164.**	**(a)**	**174.**	**(a)**	**184.**	**(b)**	**194.**	**(a)**
155.	**(c)**	**165.**	**(d)**	**175.**	**(c)**	**185.**	**(b)**	**195.**	**(c)**
156.	**(a)**	**166.**	**(b)**	**176.**	**(b)**	**186.**	**(a)**	**196.**	**(c)**
157.	**(b)**	**167.**	**(a)**	**177.**	**(c)**	**187.**	**(c)**	**197.**	**(d)**
158.	**(a)**	**168.**	**(a)**	**178.**	**(d)**	**188.**	**(b)**	**198.**	**(c)**
159.	**(c)**	**169.**	**(b)**	**179.**	**(b)**	**189.**	**(b)**	**199.**	**(c)**
160.	**(a)**	**170.**	**(c)**	**180.**	**(c)**	**190.**	**(d)**	**200.**	**(a)**

Chapter 2

Silviculture

1. **The art of growing and reproducing a forest is called:**
   (a) Silviculture (b) Mensuration
   (c) Ecology (d) None of the above
2. **The scientific basis for the practice of silviculture is:**
   (a) Silvics (b) Forest Management
   (c) Forestry Biology (d) None of the above
3. **The word FOREST is derived from the Latin word:**
   (a) Foris (b) Fores
   (c) Forex (d) Forix
4. **Number of major forest types available in India is:**
   (a) 16 (b) 17
   (c) 18 (d) 19
5. **Food, Fuel, Fodder, Fibre obtained from the forests is its:**
   (a) Protective role (b) Productive role
   (c) Ameliorative role (d) Developmental role
6. **The forest products can be broadly classified into:**
   (a) Major Timber Products (b) Non Timber Forest Products
   (c) Both (a) and (b) (d) Secondary Timber Products

7. **Site index is a term used by foresters to explain the relationship of _______ in forest stands.**
   (a) Height to age (b) Diameter to age
   (c) Diameter to height (d) Volume to age

8. **Within the forest, trees that have tall, well formed crowns or tops that receive light from above and the sides are called:**
   (a) Suppressed (b) Co-dominant
   (c) Dominant (d) Inter mediate

9. **Which of the following is not an element of an ecosystem?**
   (a) Mineral matter (b) Producers and consumers
   (c) Decomposers (d) All of the above are elements of an ecosystem

10. **The ability of a tree to withstand shade is called:**
    (a) Intolerance (b) Tolerance
    (c) Hardiness (d) All of the above

11. **The most limiting element of all the environmental factors on the growth of trees is:**
    (a) Carbon dioxide (b) Moisture
    (c) Day length (d) All of the above

12. **Coarse textured soils such as sands are _______ in water holding capacity and nutrient content.**
    (a) High (b) Low
    (c) Moderate (d) Texture does not affect these

13. **A forest which has been reproduced through sprouting is called a _______ forest.**
    (a) Epicormic (b) Stocked
    (c) Coppice (d) All of the above

14. **A stand of trees produced after a clear cut is most likely to be:**
    (a) Uneven aged (b) Even aged
    (c) All aged (d) All of the above

15. **Based on crown classification which of the following is defined as a smaller tree with a crown below the upper level of the forest canopy and receiving only scattered sunlight from above?**
    (a) Intermediate (b) Co-dominant
    (c) Dominant (d) Overtopped

16. Stocking level is best described as:

(a) Closeness of tree crowns
(b) Numbers of trees/acre
(c) Percent utilization of the site potential
(d) Both (b) and (c)

17. _______ indicates the productive capacity of a specific area of forest land for a particular tree species.

(a) Stand density
(b) Site quality or index
(c) Tree hardiness
(d) All of the above

18. The periodic removal of individual trees in order to produce better yields is referred as:

(a) Clearcutting
(b) Seed-tree cutting
(c) Selective cutting
(d) Shelterwood cutting

19. Tree seeds may be dispersed by which of the following methods?

(a) Wind
(b) Birds and animals
(c) Water
(d) All of the above

20. Identification of trees according to species is called _______.

(a) Forestry
(b) Silvics
(c) Ecology
(d) Dendrology

21. The following method of timber harvesting often appears to be the most aesthetically drastic:

(a) Clear cutting
(b) Seed tree cutting
(c) Selection cutting
(d) Shelterwood cutting

22. Which of the following is not a function of tree roots?

(a) Serve as an anchor for the tree
(b) Absorb water from the soil
(c) Conduct photosynthesis
(d) All of the above

23. Which of the following are ways in which coniferous trees reproduce?

(a) Root suckers
(b) Stump sprouts
(c) Direct seeding
(d) All of the above

24. The most important characteristic(s) to use in identifying a tree is (are): q:

(a) Leaves
(b) Bark
(c) Twigs
(d) All of the above

25. The principal objectives of intermediate cuttings include:

(a) Improvement of the existing stand
(b) Manipulation or regulation of tree and stand growth
(c) Creation of stand conditions unfavorable to insects and diseases
(d) All of the above

26. Leaves do all of the following during the day except _______.

(a) Manufacture plant food
(b) Give beauty
(c) Give off oxygen
(d) Give off carbon dioxide

27. Which is NOT a common term when referring to coniferous trees?

(a) Softwoods
(b) Evergreens
(c) Hardwoods
(d) Pines

28. Deciduous trees are NOT referred to as _______.

(a) Broadleaf
(b) Softwood
(c) Hardwood
(d) All of the above

29. Which group of conifers has its needles in a cluster of from two to five?

(a) Pines
(b) Spruces
(c) Cedars
(d) Firs

30. Which tree leaves are least likely to turn red in fall?

(a) Sugar maple
(b) Silver maple
(c) Red maple
(d) Red oak

31. New terminal growth on pines is called _______.

(a) Pinettes
(b) Needlettes
(c) Spikes
(d) Candles

32. The proper depth to plant seedlings is _______.

(a) The same depth as they were in the nursery
(b) Slightly deeper than they were in the nursery
(c) Slightly shallower than they were in the nursery
(d) 12 inches

33. Basal area of a woodlot represents:

(a) Area occupied by the tree in square feet
(b) Crown density
(c) Stand density
(d) Taper of trunk

34. When referring to trees as tolerant or intolerant, a tolerant species is one that will withstand:

(a) Drought
(b) Cold
(c) Shade
(d) Flooding

35. A tree sprout arising from the root or stump is called a _______.

(a) Slash
(b) Bole
(c) Sucker
(d) Sapling

36. Another term for the trunk of a tree is the _______.

(a) Bole
(b) Club
(c) Skid
(d) Pike

37. The leaves and branches of a tree are referred to as the _______ of the tree.

(a) Dominant
(b) Bole
(c) Crown
(d) Diameter

38. Over crowding reduces the _______ of the tree.

(a) Stand density
(b) Height
(c) Weeding
(d) Diameter

39. The term _______is defined as the reproduction and management of the forest to obtain continuous high yields of forest products.

(a) Dendrology
(b) Ecology
(c) Silviculture
(d) Diameter breast height

40. Which would be the cheapest way to buy seedlings?

(a) Bare root
(b) Balled
(c) Container grown
(d) From a garden center

41. The best time of the year to plant seedlings is _______.

(a) Early spring
(b) Early summer
(c) Early fall
(d) Early winter

42. The most common cause for seedling death is _______.

(a) Poor seedling stock
(b) Roots have dried out before planting
(c) Diseases
(d) Insects

43. Temporary planting of several trees in long narrow rows is called _______.

(a) Hedging
(b) Trenching
(c) Heeling-in
(d) Stocking

44. Another name for the hand planting bar is the _______.

(a) Wedge prism
(b) Plunger
(c) Dibble
(d) Increment borer

45. The closeness of the tree tops is called _______.

(a) Closed canopy
(b) Open canopy
(c) Crown density
(d) Carrying capacity

46. The scientific management of forests for the continued output of products is called _______.

(a) Dendrology
(b) Forestry
(c) Ecology
(d) Forest mensuration

47. The art of producing and caring for a forest is called _______.

(a) Dendrology
(b) Silviculture
(c) Arboriculture
(d) Ecology

48. A method of harvesting that removes all of the trees in the stand is called _______.

(a) Clearing
(b) Clear cutting
(c) Total cutting
(d) Total removal

49. A cutting which removes trees of poor quality, undesirable species or those in poor condition is called.

(a) Improvement cutting
(b) Upgrading
(c) Thinning
(d) Defect clearing

50. Cutting of burnt or damaged trees before they become completely worthless is called _______.

(a) Improvement cutting
(b) Release cutting
(c) Salvage cutting
(d) Scalping

51. Cutting of large trees which throw shade on the young trees is called.

(a) Thinning
(b) Selection cutting
(c) Shade tree cutting
(d) Release cutting

52. Removal of mature trees which are the highest quality and most valuable is known as _______.

(a) Clear cutting
(b) Selection cutting and economic high grading
(c) Type cutting
(d) Mature cutting

53. Removing trees damaged by fire, insects or disease is called _______.

(a) Salvage cut
(b) Clean cutting
(c) Clear cut
(d) Sanitation cut

54. Removal of mature timber in a series of cutting so that the remaining trees might reseed the area is called:

(a) Selection cutting
(b) Natal cutting
(c) Shelterwood cutting
(d) Density cutting

55. Removal of the lower branches of a tree is called _______.

(a) Thinning
(b) Cutting back
(c) Pinching
(d) Pruning

56. Which is not considered in stand density?

(a) Basal area
(b) Tree height
(c) Crown density
(d) Number of stems per acre

57. Selling standing timber is called _______ sale:

(a) Live
(b) Green
(c) Stumpage
(d) Standing

58. A tree stand regeneration almost totally from sprouts is called a _______.

(a) Sprout forest
(b) Seedless forest
(c) Stump forest
(d) Coppice forest

59. Girdling is a method used to:

(a) Kill unwanted trees
(b) Harvest mature trees
(c) Properly space veneer quality trees
(d) Produce knot free lumber

60. Pruning is done to produce logs with _______.

(a) Knot free lumber
(b) A minimum of 24 inches in diameter
(c) No insects or diseases
(d) All of the above

61. The periodic removal of trees in order to produce better yields is:

(a) Clear cutting
(b) Seed-tree cutting
(c) Selection cutting
(d) Shelterwood cutting

62. Cutting made at the sapling stage for the purpose of improving the composition or the quality of the growing timber stand is called:

(a) Cleaning cuttings
(b) Improvement cutting
(c) Liberation cuttings
(d) Weeding cutting

63. A non grass like herbaceous plant is called:

(a) Forb
(b) Forage
(c) Forest
(d) Tree

64. The primary purpose of the flower is _______.

(a) Reproduction
(b) Respiration
(c) To attract beneficial insects
(d) Transpiration

65. Trees which are characterized by having flowers of both sexes on the same tree are:

(a) Coniferous (b) Deciduous

(c) Dioecious (d) Monoecious

66. The light colored part of a cross section of a tree that is responsible for conducting water and minerals to the crown is:

(a) Cambium (b) Heartwood

(c) Pith (d) Sapwood

67. The symbol for established regeneration whose height should be 2-5m or more and dbh 10cm is:

(a) e (b) u^+

(c) S^+ (d) W^+

68. A fire that has been started intentionally by a forester to be used as a management tool is a _______.

(a) Crown fire (b) Prescribed burn

(c) Surface fire (d) Wildlife

69. The part of a tree that grows into a seed is called as:

(a) Ovule (b) Stamen

(c) Ovary (d) Polyp

70. Flower part modified as a male reproductive structure:

(a) Stamen (b) Stigma

(c) Embryo sac (d) Sepal

71. The science of classification of trees or any other living matter is:

(a) Identification (b) Agronomy

(c) Horticulture (d) Taxonomy

72. In measuring the site index of a tree, which criteria must be measured?

(a) Age of tree and merchantable height

(b) Age of a tree and total tree height

(c) Dbh and total tree height

(d) Merchantable height and DBH

73. _______ is the weight of a complete tree including roots, stump, stem, limbs, branches and leaves.

(a) Basal area (b) Biomass

(c) Cubic volume (d) Stumpage

74. A conifer with scales rather than needles is _______.

(a) Cedar (b) Fir

(c) Spruce (d) Yew

75. The system of binomial nomenclature, is credited to:

(a) Carolus Linnaeus (b) Charles Darwin

(c) Charles E. Bessey (d) George Bentham

76. Silvics is the science of tree _______ and species distribution in the forest.

(a) Cutting (b) Disease

(c) Growth (d) Harvest

77. The most expensive timber management activity is _______.

(a) Estimating or cruising (b) Hand planting

(c) Marking (d) Site preparation

78. A timber harvesting method used to allow faster growth of remaining trees, to remove diseased trees or to remove undesirable trees before the final harvest is called _______.

(a) Initial forestation (b) Intermediate cuttings

(c) Selective cuttings (d) Shelter wood

79. Thinning is designed to _______.

(a) Concentrate growth on the best trees

(b) Reduce crown density

(c) Reduce the number of trees in an immature stand

(d) Reduce the time period to the final harvest

80. An economical way to remove excess logging debris and undesirable vegetation prior to planting is by.

(a) Bulldozer (b) Herbicide

(c) Prescribed fire (d) Wildlife

81. Tree injectors are used to:

(a) Stimulate growth (b) Mark trees

(c) Kill trees (d) Inoculate trees

82. An examples of a biotic factor in a forest is _______.

(a) Climate (b) Rainfall

(c) Soil depth (d) A tree

83. A cottonwood tree has a _______ leaf shape.

(a) Cordate (b) Deltoid

(c) Obovate (d) Oval

84. _______ is not a silvical characteristic.

(a) Growth habit
(b) Seeding habit
(c) Site and moisture requirements
(d) None of the above

85. The term commercial forest means.

(a) Land is capable of producing economically useful trees
(b) Land produces only commercial size harvestable trees
(c) Government owns a majority of the saleable timber
(d) Timber companies own a majority of the forestland

86. A tree with a small crown in the general canopy level and receives little direct light from the side is called.

(a) Codominant
(b) Dominant
(c) Intermediate
(d) Suppressed

87. The class of trees angiosperms is made up of _______ trees.

(a) Broadleaf
(b) Conifer
(c) Evergreen
(d) Hardwood

88. Roots which grow vertically downward and anchor the tree are _______.

(a) Feeder roots and hair roots
(b) Lateral roots
(c) Surface roots
(d) Tap roots

89. The trunk branches and roots grow in diameter by means of cell division in the:

(a) Cambium layer
(b) Heartwood
(c) Sapwood
(d) Outer bark

90. The inner core of a woody stem, wholly composed of non-living cells, darker in color than the outer layer of the stem is _______.

(a) Cambium layer
(b) Heartwood
(c) Sapwood
(d) Outer bark

91. Alternate, opposite and whorled describes.

(a) Leaf types
(b) Needle types
(c) Twig branching
(d) Tree classifications

92. A process which removes injured, disease or insect-infested trees is called _______.

(a) Liberation cutting
(b) Salvage cutting
(c) Sanitation cutting
(d) Thinnings

93. In the _______ harvest method, the entire stand of trees are cut except for a few which are selected for their quality and seed producing ability.

(a) Coppice (b) Clear cutting
(c) Selective cutting (d) Seed tree cutting

94. The purpose of this harvest method is to: Produce a large immediate income, provide for the natural reseeding of the area and provide partial shelter for the seedlings as they begin to germinate.

(a) Coppice (b) Selective cutting
(c) Seed tree cutting (d) Shelter wood cutting

95. Trees and forests are considered to be _______.

(a) Producers (b) First level consumers
(c) Second level consumers (d) Decomposers

96. A very complex community of associated trees, shrubs, other plants and animals is a/an:

(a) Arboretum (b) Forest
(c) Greenhouse (d) Park

97. Removing trees that have been or may be killed or damaged by insects, ice, fire and other agents is called:

(a) Liberation (b) Pruning
(c) Salvage (d) Sanitation

98. A tree that bears seeds in cones and that have long, narrow (needle like) leaves is a:

(a) Broadleaf (b) Conifer
(c) Deciduous tree (d) Foliage Plant

99. The class of trees, Gymnosperm, is made up of _______ trees.

(a) Broadleaf (b) Coniferous
(c) Deciduous (d) Hardwood

100. Roots which grow horizontally and help keep the tree upright are _______.

(a) Feeder roots and root hairs (b) Lateral roots
(c) Surface roots (d) Tap roots

101. The process by which excess water from the tree is allowed to evaporate through openings in the leaves is:

(a) Hydrolysis (b) Osmosis
(c) Photosynthesis (d) Transpiration

102. Applying tree seeds to the harvested area or prepared beds is called _______.

(a) Seed tree method
(b) Direct seeding
(c) Natural seeding
(d) All of the above

103. Neither natural seeding or _______ will result in trees being planted in rows unless a grain drill is used.

(a) Seed tree method
(b) Direct seedings
(c) Natural seeding
(d) Sheltherwood seeding

104. Dendrology is commonly defined as the _______.

(a) Study of forest measurement
(b) Study of tree growth and management
(c) Taxonomy of all woody plants
(d) Taxonomy of trees

105. As a general rule:

(a) Conifers are usually less tolerant of flooding than hardwoods
(b) Hardwood are usually less tolerant of flooding than conifers
(c) Most important forest species are flood tolerant
(d) No important forest species are flood tolerant

106. A silviculture prescription:

(a) Changes in forest density
(b) Is a type of forest inventory?
(c) Plans for pesticide application to control insects
(d) Usually contains cultural operations relating to objectives

107. Windrowing, herbicide application and prescribed burning are methods of:

(a) Artificial regeneration
(b) Natural regeneration
(c) Site preparation
(d) Uneven-aged management

108. A disadvantage to clearcutting is _______:

(a) Nothing can grow in clearcut area for 3 years
(b) Soil erosion can be a problem if t here is no advance tree regeneration
(c) It is the most costly harvest method
(d) There are no disadvantages to clearcutting

109. The term "alkaline" refers to a substance with a pH of _______.

(a) 7.0
(b) 7.1 or above
(c) 4.9 or below
(d) 3.0

110. A soil with a pH of 3.2 is _______.

(a) Neutral (b) Acidic

(c) Alkaline (d) All of the above

111. An herbicide that will not move or translocate within a plant is called.

(a) Surfactant (b) Contact

(c) Systemic (d) Cuticle

112. Stumpage refers to the _______.

(a) Number of stumps left after a timber harvest

(b) Number of stems per ha

(c) Not knowing what to do next

(d) The value of trees as they stand uncut

113. A forest fire which is out of control is called.

(a) Ground fire (b) Spot fire

(c) Suppression fire (d) Wild fire

114. A mixed stand is one in which less than _______ per cent of the trees are of same species.

(a) 75 to 80 (b) 60 to 70

(c) 50 (d) 30 to 40

115. Litter from leaves, especially from needles, at various stages of decomposition is called _______.

(a) Slash (b) Litter

(c) Trash (d) Duff

116. Which of the following factors would affect the success or failure of direct seeding?

(a) Climatic conditions (b) Consumption by animals

(c) Previous harvest method (d) Both (a) and (b)

117. The debris left after logging, such as limbs, tops, culls and knockdown trees is called:

(a) Conk (b) Kerf

(c) Stumpage (d) Slash

118. The most important characteristics to use in tree identification are _______.

(a) Bark, fruit, twigs and leaves (b) Bark, growth habit, leaves

(c) Fruit, twigs, leaves, color and taste (d) Twigs, bark, fruit, site

119. The function of the xylem in the tree is:

(a) Develop leaves from nutrients

(b) Conduct water and minerals to the crown

(c) Grow wood and bark cell

(d) Produce carbohydrates through photosynthesis

120. A totally occupied forest stand that allows plenty of room for development of crop trees is called a (an):

(a) Fully stocked stand
(b) Mixed stand
(c) Overstocked stand
(d) Pure stand

121. The best starter fertilizer for trees has a high concentration of:

(a) Nitrogen
(b) phosphorous
(c) Potassium
(d) All of the above

122. Which of the following is not a factor in using controlled burns?

(a) Area to be burned
(b) Intensity o the fire
(c) Humidity
(d) Site index

123. The harvest method that results in sustained production forestry by requiring foresters to choose trees for harvest based on maturity, size, species, growth rate or other factors is _______.

(a) Coppice cutting
(b) Clear cutting
(c) Selective cutting
(d) Thinning

124. Stumps should be cut _______.

(a) 1 foot above the ground

(b) At dbh

(c) At waist height

(d) Close to the ground to avoid waste

125. Thinning overstocked stands in order to manage or control pests is a _______ pest management strategy.

(a) Biological
(b) Chemical
(c) Regulatory
(d) Silvicultural

126. A snag is a _______.

(a) Blocked firebreak

(b) A branch from one tree caught up in branches from another tree

(c) A standing dead tree

(d) A pinched chain saw bar

127. A hypo-hatchet is used for _______.

(a) Applying a blaze to trees to mark a boundary

(b) Cutting small trees quickly

(c) Girdling a tree

(d) Injecting a tree with herbicide

128. The mid-summer water temperature of a stream flowing through a clear-cut with no stream buffer would be _______ the temperature of the water prior to the clearcut.

(a) Lower than (b) Higher than

(c) The same as (d) Clear cutting would

129. Which of the following would be considered mechanical tree damage?

(a) Frost cracks (b) Girdling roots

(c) Powdery mildew (d) Scorching

130. Which of the following processes uses shelterwood methods of harvesting to promote regeneration?

(a) Cuttings (b) Direct seeding

(c) Natural seeding (d) Seedlings

131. The practice most commonly used to produce high grade stands is _______.

(a) Cleaning (b) Improvement

(c) Release (d) Thinning

132. The process to remove damaging dominant trees in a all-aged stand is called_______.

(a) Liberations (b) Salvage cuttings

(c) Sanitation cuttings (d) Thinning

133. Tree cuttings made to allow faster growth of remaining trees, to remove diseased trees or to remove undesirable trees before the final harvest is called_______.

(a) Early forestation (b) Intermediate cuttings

(c) Seed tree cuttings (d) Shelterwood

134. The most accurate method of cruising is the _______.

(a) Average cruise (b) Ocular cruise

(c) One hundred percent cruise (d) Point sampling

135. Which of the following produces reproduction from stump sprouts?

(a) Coppice cutting (b) Seed tree cutting

(c) Selective cutting (d) Shelterwood cutting

136. Coppice shoots and root suckers are examples of _______.

(a) Sexual reproduction
(b) Vegetative reproduction
(c) Cloning
(d) Wood production

137. Fallen tree tops are generally referred to as _______.

(a) Biomass
(b) Brush piles
(c) Firewood
(d) Slash

138. Incendiary fires are _______.

(a) Ignited accidentally through use of power logging equipment
(b) Set maliciously
(c) Started by burning debris after logging
(d) The result of carelessly discarded cigarettes

139. The species which occurs under sclerophyll forests.

(a) *Quercus ilex*
(b) *Abies pindrow*
(c) *Cedrus deodara*
(d) *Populus tremuoides*

140. Contour lines show _______.

(a) Elevation and shape of the land
(b) Land quality
(c) Land use potential
(d) Resource distribution

141. On a topographic map which of the following is a dark, heavy line with an elevation figure?

(a) Contour index
(b) Contour interval
(c) Contour line
(d) Horizontal control station

142. Using the natural shape and texture of a tree to complement the architecture of a building or to enhance the vertical elements of a structure is an example of a _______ benefit of urban trees.

(a) Aesthetic
(b) Environmental
(c) Economic
(d) Wildlife

143. The function of root hairs is to _______.

(a) Anchor the tree
(b) Absorb water and nutrients
(c) Produce chlorophyll
(d) All of the above

144. _______ lack root hairs.

(a) Aspen and birch
(b) Casuarinas
(c) Birch
(d) None of the above

145. Instead of root hairs, pines have _______ attached to the rootlets which absob water and nutrients.

(a) Tap roots
(b) Branch roots
(c) Mycorrhizae (fungi)
(d) All of the above

146. Windfalls are most common on _______.

(a) Forest stands that have been heavily cut over
(b) On shallow soils
(c) On wet soils
(d) All of the above

147. The process whereby chloroplasts convert radiant energy into chemical energy is called _______.

(a) Transpiration and respiration
(b) Perspiration
(c) Chlorophyll
(d) Photosynthesis

148. In photosynthesis carbon dioxide from the air and water and minerals from the soil into _______ with oxygen as a by product.

(a) Fats and proteins
(b) Amino acids
(c) Sugars and starches
(d) Vitamins and minerals

149. The process of consuming carbohydrates and oxygen to obtain energy for the biological processes of life support, growth and reproduction is called_______.

(a) Transpiration
(b) Respiration
(c) Perspiration
(d) Chlorophyll

150. The loss of water vapor from the leaf of a tree is called _______.

(a) Transpiration
(b) Respiration
(c) Perspiration
(d) Chlorophyll

151. Most of the water vapor given off by a plant escapes through structures called _______.

(a) Pores
(b) Stomata
(c) Exit cells
(d) Cutin

152. Excess carbohydrates not used in respiration are converted to _______ and _______ and is then stored in the tree.

(a) Fats and protein
(b) Vitamins and minerals
(c) Cellulose and lignin
(d) Chlorophyll and chloroplasts

153. _______ and _______ are the main constituents of wood.

(a) Fats and protein

(b) Vitamins and minerals

(c) Cellulose and lignin

(d) Chlorophyll and chloroplasts

154. Sprouting, suckering and layering are examples of _______ reproduction in trees.

(a) Sexual (b) Asexual or vegetative

(c) Slow (d) Fast

155. During autumn and winter your best way to identify trees will be to compare _______ and _______ characteristics.

(a) Leaf and flower (b) Fruit and stem

(c) Twig and bark (d) All of the above

156. Major characteristics of _______used in identification are buds, leaf scars, lenticels and bark.

(a) Leaves (b) Bark

(c) Fruit (d) Twigs

157. On deciduous trees _______ is the most important identification feature in the dormant winter period.

(a) Fruit (b) Leaves

(c) Twigs (d) Bark

158. Poles and pilings are measured in _______.

(a) Linear feet (b) Cords

(c) Cubic feet (d) Weight

159. In a dormant condition trees can withstand _______ temperature extremes than during periods of active growth.

(a) Smaller (b) Greater

(c) The same (d) None of the above

160. In the temperate zone summer growth is often completed _______ high maximum summer temperatures are reached.

(a) Before (b) Long after

(c) At the same time as (d) At least 3 months before

161. Sunlight is important in regulating which of the following processes?

(a) Growth (b) Leaf fall and dormancy

(c) Flowering and reproduction (d) All of the above

162. Day length or _______ influences diameter growth in trees.

(a) Photoperiod (b) Perspiration

(c) Photosynthesis (d) Photochemical

163. Periods of ling daylight in spring and early summer result in the production of _______ cells.

(a) Large diameter, thick walled, spring wood

(b) Large diameter, thin walled, spring wood

(c) Small diameter, thick walled, summer wood

(d) Small diameter, thin walled, summer wood

164. Length of day light affects the time of flushing or _______ in the spring and the onset of dormancy in the autumn.

(a) Phototrophism (b) Defoliation

(c) New growth (d) Bud formation

165. Light distribution can affect tree form, resulting _______.

(a) Greater development of the tree on the side exposed to the most sunlight

(b) Less growth on the side of the tree ex posed to the most sunlight

(c) A tall, spindly tree because of exposure to the sun

(d) A short, overtopped tree because of exposure to the sun

166. Of all the factors t hat influence the growth of trees, _______ is the most limiting element.

(a) Soil fertility (b) Sunlight

(c) Moisture (d) Carbon dioxide

167. Seed formation depends on the amount of reserve _______ and nitrogenous materials in a tree.

(a) Fatty acids (b) Carbohydrates

(c) Proteins and amino acids (d) Oxygen

168. The term _______ is used to describe the physical expression of an inherited tree trait that can be measured and observed.

(a) Genotype (b) Genetic map

(c) Physical map (d) Phenotype

169. Winter-kill is the _______ of young trees and seedlings.

(a) Stunting (b) Deforming

(c) Killing (d) Weakening

170. In a forest ecosystem a tree with a small crown receiving sunlight from above and almost none from the sides would be considered a _______ tree.

(a) Co-dominant
(b) Dominant
(c) Inferior
(d) Intermediate

171. To determine site index the two measurements needed are _______:

(a) Age and diameter
(b) Age and height
(c) Diameter and distance from the closest tree
(d) Diameter and height

172. Growth at the end of the stem is called _______.

(a) Preliminary growth
(b) Primary growth
(c) Terminal growth
(d) Turgid growth

173. Tree species that have male and female flowers on separate trees are called _______.

(a) Bioecious
(b) Dioecious
(c) Monoecious
(d) Triecious

174. Trees are planted along city streets to help increase _______.

(a) Gas removal and particulate collection
(b) Summer air temperatures
(c) Temperatures of sidewalks and road surfaces
(d) Wind velocity

175. Using containerized seedlings rather than bareroot seedlings usually results in:

(a) A much shorter planting season
(b) Easier handling and transportation
(c) Improved rates o f survival and growth
(d) Lower initial cost for seedlings

176. The most widely known and used hardwood is _______.

(a) Ash
(b) Kickory
(c) Maple
(d) Oak

177. Of all the environmental factors that influence the growth of trees, which of the following is the most limiting?

(a) Topography
(b) Moisture
(c) Soil quality
(d) Temperature

178. Which of the following is NOT a common method of testing improved seeds or seedlings?

(a) Clone testing
(b) Progeny testing
(c) Provenance testing
(d) Regulation testing

179. Which of the following silvicultural systems is most likely to result in unevenaged forests?

(a) Clear cutting
(b) Shelterwood system
(c) Seed tree system
(d) Selection system

180. Beating up is a silvicultural operation that refers to:

(a) Pollarding
(b) Pruning
(c) Restocking
(d) Thinning

181. Stocking level is best described as:

(a) Closeness of tree crowns
(b) Numbers of trees/acre
(c) Percent utilization of ht site potential
(d) Both (b) and (c)

182. Which type of frost affects the forest the most?

(a) Radiation frost
(b) Pool frost
(c) Advective frost
(d) All of the above

183. Thinnings in young timber allow the timber grower to:

(a) Maintain high rates of growth
(b) Concentrate growth on the better trees
(c) Maintain a properly stocked forest
(d) All of the above

184. Removing trees that have been or may be killed or damaged by insects, ice, fire and other agents is called:

(a) Liberation
(b) Pruning
(c) Salvage
(d) Sanitation

185. Site index is a term used by foresters to explain the relationship of _______ in forest stands.

(a) Height to age
(b) Diameter to age
(c) Diameter to height
(d) Volume to age

186. In India Plantation forestry was started in:

(a) 1942
(b) 1842
(c) 1947
(d) 1847

187. Quantity of seed required for a nursery bed can be worked out by:

(a) W=(AXDX100)/(PXN)
(b) W=(PXDX100)/(AXN)
(c) A=(WXDX100)/(PXN)
(d) W=(AXNX100)/(PXD)

188. Tick the odd man out: Growth stages of tree:

(a) Pole
(b) Seedling stage
(c) Young timber
(d) Sapling stage

189. Choose the best tree species for coastal afforestation:

(a) *Albizia lebbeck*

(b) *Azadirachta indica*

(c) *Casuarinas equisetifolia*

(d) *Eucalyptus tereticornis*

190. CVP index refers to:

(a) Crop volume per cent index

(b) Cumulative volume per cent index

(c) Climate vegetation and productivity index

(d) Climate volume productivity index

191. The term provenance means:

(a) Land race

(b) Geographical source

(c) Seed source

(d) None of the above

192. Tropical dry evergreen forests are found in:

(a) Carnatic coast

(b) Coastal areas of Kerala

(c) Andaman and Nicobar island

(d) North east India

193. Shelterbelts have (pick the wrong statement):

(a) Single tree species

(b) Multiple tree species

(c) Prevent sand dune formation

(d) Used to arrest wind

194. Ecotype refers to:

(a) Variation in the phenotypic characters

(b) Phenotypic variation of different species

(c) Differential response of different species to environmental factors

(d) Differential response of the same species to environmental factors

195. Which is not a characteristic of gregarious flowering in Bamboo?

(a) Flowering occurs almost over the entire area

(b) It involves only few proportion of the culms

(c) Flowering is followed by the death of clump

(d) It follows a cycle which occur after long interval

196. High forest refers to:

(a) Forests regenerated through cutting

(b) Forests regenerated through seeds

(c) Forest regenerated through coppice

(d) Forest regenerated through tissue culture plants

197. Pure forest are those which cover:

(a) 40 per cent of the crop consist of single species

(b) 50 per cent of the crop consist of single species

(c) 60 per cent of the crop consist of single species

(d) 80 per cent of the crop consist of single species

198. Champion and Seth classified the Indian Forests into major:

(a) Six groups
(b) Sixteen groups
(c) Sixty groups
(d) None of the above

199. The leader of the 'Chipko' movement is:

(a) Westoby
(b) Bahuguna
(c) M.S. Swaminathan
(d) Tewari

200. Trees which do not form part of the upper most leaf canopy but the leading of shoots of which are not definitely overtopped by the neighbouring trees are called:

(a) Dominant
(b) Co-dominant
(c) Dominated
(d) Pre-dominants

201. The mean annual temperature of montane sub-tropical forest ranged between:

(a) 10°C and 18°C
(b) 7°C and 17°C
(c) 17°C and 24°C
(d) 18°C and 24°C

202. The mean annual temperature of montane temperate forest ranged between:

(a) 10°C and 18°C
(b) 7°C and 17°C
(c) 17°C and 24°C
(d) 18°C and 24°C

203. The term 'forest succession' was first used by:

(a) Clement
(b) Cawles
(c) Thoreau
(d) Dawson

204. An organism on or within which another organism feeds and develops is called:

(a) Parasites
(b) Epiphytes
(c) Host
(d) None of the above

205. ICFRE was established during the year:

(a) 1986
(b) 1980
(c) 1975
(d) 1990

206. The climax stage of estuarine succession:

(a) Mixed forests of spruce and fir (b) Mangrove scrub
(c) Evergreen forest (d) Dry deciduous forest

207. Tropical Forest Research Institute is located at:

(a) Jodhpur (b) Jorhat
(c) Jabalpur (d) Jhansi

208. Study of the biotic community and its abiotic environment is called:

(a) Population ecology (b) Community ecology
(c) Ecosystem ecology (d) None of the above

209. Population ecology is otherwise called as:

(a) Synecology (b) Autecology
(c) Ecosystem ecology (d) None of the above

210. The poly climax theory in plant succession was developed by:

(a) Cement (b) Tansley
(c) Fosberg (d) Whittaker

211. Fungal roots are otherwise called as:

(a) *Frankia* (b) *Mycorrhiza*
(c) *Rhizobium* (d) *Azospirillum*

212. The father of Indian Forestry is:

(a) O' Conolly (b) Dietrich brandis
(c) Voelckar (d) Gadgil

213. The Moist Deciduous and Rain Forest Institute is located in:

(a) Jorhat (b) Simla
(c) Jabalpur (d) Jhansi

214. The first Conservator of Forests.

(a) Trevor (b) Troupe
(c) Watson (d) Hartig

215. The first Forest Policy was enunciated during:

(a) 1864 (b) 1874
(c) 1884 (d) 1894

216. The per cent forest area in the country:

(a) 19.55 (b) 23.81
(c) 21.55 (d) 22.55

217. The OFI is located:

(a) London
(b) Edinburg
(c) Brussels
(d) Rome

218. Which of the following plants might be found in a primary state of succession?

(a) Beech
(b) White pine
(c) Aspen
(d) Lichen

219. Foresters recognize which of the following as the basic management unit in a forest?

(a) Stand
(b) Lot
(c) Section
(d) District

220. Secondary succession might be considered to start after which of the following?

(a) a landslide
(b) Back filling of a strip mine
(c) A glacier
(d) All of the above

221. Site index is a term used by foresters to explain the relationship of _______ in forest stands.

(a) Height to age
(b) Diameter to age
(c) Diameter to height
(d) Volume to age

222. Within the forest, trees that have tall, well formed crowns or tops that receive light from above and the sides are called:

(a) Suppressed
(b) Co-dominant
(c) Dominant
(d) Intermediate

223. Under which grade very ordinary thinning is carried out?

(a) B grade
(b) C grade
(c) D grade
(d) E grade

224. Which is the closest to a hectare?

(a) 1 acre
(b) 2.5 acres
(c) 5.2 acres
(d) 7 acres

225. The term stumpage refers to:

(a) Stumps left after harvesting
(b) The diameter of trees measured 1 foot off the ground
(c) The value of trees standing in a woodlot
(d) A measure of tree volume

226. An abandoned field changes into a mature forest over an extended period of time through a process called:

(a) Succession (b) Dominance

(c) Recycling (d) All of the above

227. Stump planting is very common in:

(a) Teak (b) Sal

(c) Deodar (d) Red sanders

228. Deodar is managed under:

(a) Clear felling system (b) Selection system

(c) Shelter wood system (d) None of the above

229. Scientific name of Toon:

(a) *Tectona grandis* (b) *Toona ciliate*

(c) *Shorea robusta* (d) *Cedrus deodara*

230. *Eucalyptus* is a:

(a) Bad coppice (b) Moderately goodcoppice

(c) Good coppice (d) Very badcoppicer

231. Example for Nitrogen fixing tree:

(a) Casuarinas (b) Acacia

(c) Prosopis (d) All of the them

232. Tendu is a source of:

(a) Tannin (b) Dye

(c) Gum (d) Beedi leaf

233. Sandal is commonly affected by:

(a) Spike disease (b) Pink disease

(c) Stem wilt (d) Leaf galls

234. *Terminalia* belong to the family:

(a) Dipterocarpaceae (b) Verbenaceae

(c) Papilionaceae (d) Combretaceae

235. Calculate number of plants/ha when teak is planted at 2m × 3m spacing:

(a) 666 (b) 1666

(c) 2000 (d) 2666

236. Ligno tuber is present in:

(a) Teak (b) Casuarina

(c) Rose wood (d) Eucalyptus

237. ***Prosopis juliflora*** **is native of:**

(a) Brazil (b) Mexico
(c) Thailand (d) Indonesia

238. Viability of the following tree seed slow down quickly:

(a) Sal (b) Teak
(c) Eucalyptus (d) Terminalia

239. "Mimosin" is present in:

(a) Leucaena (b) Acacia
(c) Neem (d) Sal

240. An important Multi-purpose tree in coastal area:

(a) Casuarinas (b) Neem
(c) Karanj (d) Albizia

241. Which is the family of Bamboo:

(a) Ebenaceae (b) Verbanaceae
(c) Combretaceae (d) Poaceae

242. Casuarina root nodules are formed in association with:

(a) Frankia (b) Rhizobium
(c) VAM fungi (d) AM fungi

243. ***Shorea robusta*** **is:**

(a) Susceptible to frost (b) Frost resistant
(c) Drought resistant (d) Susceptible to drought

244. ***Santalum album*** **is a:**

(a) Partial stem parasite (b) Complete stem parasite
(c) Partial root parasite (d) Complete root parasite

245. ***Meaua ferrea*** **belongs to:**

(a) Malvaceae (b) Combretaceae
(c) Guttiferae (d) Meliaceae

246. Most suitable Silvicultural system of deodar in hilly terrain for control of soil erosion is:

(a) Selection system (b) Clear felling
(c) Coppice system (d) All of the above

247. The Silviculture of Indian Trees was written by:

(a) L.S. Khanna (b) Brandis
(c) R.S. Troup (d) Banerjee

248. Neem is found to occur wild in India in:

(a) Irawadi valley (b) Western ghats

(c) Anamalai hills (d) All of the above

249. The scientific name of nagasampigo:

(a) *Hopea parviflora* (b) *Mesua ferrea*

(c) *Acacia catechu* (d) *Vateria indica*

250. Fragrant flowers characterized for:

(a) *Michelia champaca* (b) *Gmelina arborea*

(c) *Santalum album* (d) *Acacia catechu*

251. Very strong durable wood:

(a) *Casuarina equisetifolia* (b) *Anogeissue latifolia*

(c) *Mesua ferrea* (d) *Kydia calycina*

252. Commonest companion tree of Sal:

(a) *Eugenia jambolana* (b) *Buchanania latifolia*

(c) *Trevia nudiflora* (d) *Anogeissue latifolia*

253. Iron wood of Malabar:

(a) *Mesua faerrea* (b) *Hopea parviflora*

(c) *Pterocarpus marsupium* (d) *Tectona grandis*

254. Native of *Swietenia mahagoni:*

(a) West Indies (b) India

(c) Burma (d) Malaysia

255. *Dalbergia sissoo* grows most typically in:

(a) Alluvial (b) Laterite

(c) Loamy (d) Sandy loam

256. How many square feet are there in an acre?

(a) 40000 (b) 43522

(c) 43560 (d) 10000

257. Forest of *acacia Arabica* are frequently termed as:

(a) Babul meadow (b) *Mesquite bana*

(c) Sundra bana (d) None of the above

258. Calorific value of wood is highest in:

(a) *Casuarinas equisetifolla* (b) *Acacia nilotica*

(c) Teak (d) Betula

259. Artificially, *Terminalia belerica* can be raised by:

(a) Direct sowing (b) Entire plants

(c) Container seedlings (d) All of the above

260. Kail is:

(a) *Acacia nilotica* (b) *Acacia catechu*

(c) *Pinus wallichiana* (d) *Pinus excelsa*

261. Example for presence of root nodules in trees:

(a) *C. equistifolia* (b) *Terminalia tomentosa*

(c) *Adina cordifolia* (d) Teak

262. *Cedrus deodara* is a:

(a) Good coppice (b) Bad coppice

(c) Do not coppice (d) Moderate coppice

263. *Vitex negundo* is belongs to family:

(a) Verbenaceae (b) Rhizophoraceae

(c) Leguminsoe (d) Rubiaceae

264. *Hopea parviflora* belongs to the family:

(a) Dipterocarpaceae (b) Leguminosae

(c) Guttiferae (d) Mslvaceae

265. Scientific name of yellow bamboo is:

(a) *Bambusa tulda* (b) *Bambusa vulgaris*

(c) *Bambusa arundlnaria* (d) *Dendrocalamus strictus*

266. Sal is a:

(a) Poor coppice (b) Good coppice

(c) Moderate coppice (d) Non-coppicer

267. *Dalbergia sissoo* is:

(a) Frost hardy (b) Moderately frost hardy

(c) Frost tender (d) None of the above

268. *Atteva fabriciella* is a common pest of:

(a) *Ailanthus excelsa* (b) *Acacia auriculiformis*

(c) *Tectona grandis* (d) All of the above

269. Silk cotton tree belongs to the family:

(a) Malvaceae (b) Guttiferae

(c) Meliaceae (d) Caesalpinaceae

270. Good pollarding ability tree:

(a) *Mesua ferrea* (b) *Hopea parviflora*
(c) *Pterocarpus marsupium* (d) *Hardwckia binata*

271. Santalin is present in:

(a) *Santalum album* (b) *Pterocarpus marsupium*
(c) *Pterocarpus santalinus* (d) *Tectona grandis*

272. Mesquite bean is the common name for:

(a) *Prosopis spicigera* (b) *Prosopis juliflora*
(c) *Acacia tortilis* (d) None of the above

273. Azadirachta indica belongs to:

(a) Myrtaceae (b) Meliaceae
(c) Malvaceae (d) Mimoceae

274. *Eucalyptus tereticornis* belongs to family:

(a) Malavaceae (b) Mimoceae
(c) Myrtaceae (d) Meliaceae

275. Dry teak forest association:

(a) *Terminalia tomentosa* (b) *Acacia sundra*
(c) *Xylia xylocarpa* (d) All of the above

276. *Gmelina arborea* belongs to family:

(a) Verbenaceae (b) Combretaceae
(c) Myrtaceae (d) Meliaceae

277. *Pinus wallichlana* is a:

(a) 2 needle pine (b) 3 needle pine
(c) 5 needle pine (d) 7 needle pine

278. Origin of *Mangifera indica* is:

(a) Pakistan (b) India
(c) Srilanka (d) Burma

279. Indian Copal tree:

(a) *Vateria indica* (b) *Mesua ferrea*
(c) *Artocarpus hirsute* (d) *Michelia champaca*

280. *Terminalia arjuna* belongs to family:

(a) Verbenaceae (b) Combretaceae
(c) Malvaceae (d) Meliaceae

281. Karanj oil is extracted from:

(a) *Pongamia pinnata* (b) *Terminalia tomentosa*

(c) *Azadirachta indica* (d) *Santalum album*

282. *Pinus roxburghii* is a:

(a) 2 needle pine (b) 3 needle pine

(c) 5 needle pine (d) 7 needle pine

283. The art of growing and reproducing a forest is called:

(a) Ecology (b) Silviculture

(c) Mensuration (d) All of the above

284. A low forest is one that has developed from:

(a) Coppice (b) Cuttings

(c) Root sucker (d) Seed

285. A high forest is one that has developed from:

(a) Coppice (b) Cuttings

(c) Root sucker (d) Seed

286. The most limiting element of all the environmental factors on the growth of trees is:

(a) Carbon dioxide (b) Moisture

(c) Day length (d) All of the above

287. Which one of the following is not a stage in the growth of a tree:

(a) Sapling (b) Juvenile

(c) Pole (d) Saw timber

288. Canopy density is said to be dense when the density is:

(a) 1.0 (b) 0.75-1.0

(c) 0.50-0.75 (d) <0.50

289. Cane brakes in the Northern tropical wet evergreen forest is an example of:

(a) Climatic climax (b) Edaphic climax

(c) Pre climax (d) Post climax

290. A typical example of the southern montern wet temperate forest is:

(a) *Diperterocarpus turbinatus* (b) *Pertocarpus dalbergioides*

(c) *Lagerstroemia lanceolata* (d) *Rhododendron nilgiricum*

291. Seed dispersal in Dipterocarpus is by the agency of:

(a) Water (b) Wind
(c) Gravity (d) Birds

292. Tick the odd man out among the following species that coppice strongly:

(a) *Acacia catechu* (b) *Adina cordifolia*
(c) *Eucalyptus tereticornis* (d) *Albizia* species

293. Tick the odd man out among the following kinds of sowing:

(a) Continuous line sowing (b) Parallel sowing
(c) Strip sowing (d) Patch sowing

294. The number of plants required for 10 ha of plantation in which the plants are planted in a triangular pattern 2.5 m apart is:

(a) 4620 (b) 9840
(c) 13860 (d) 18480

295. Calculate the spacing of plants if there are 400 plants in a ha. Of plantation raised on square planting pattern:

(a) 4m (b) 5m
(c) 8m (d) 10m

296. Beating up is a silvicultural operation that refers to:

(a) Pollarding (b) Pruning
(c) Restocking (d) Thinning

297. Tick the odd man out of the following terms for ordinary thinning:

(a) Crown thinning (b) Low thinning
(c) Thinning from below (d) German thinning

298. Maximum thinning is a attributable to:

(a) Heck (b) Craib
(c) Gehrhardt (d) Glover

299. The rotation followed for deodar in U.P. is:

(a) 80 years (b) 100 years
(c) 120 years (d) 160 years

300. The practice of planting other valuable species of planting other valuable species under the main crop when the natural regeneration of the valuable species in the upper storey cannot be assured is called:

(a) Supplementary planting (b) Cover planting
(c) Enrichment planting (d) Under planting

301. Crop diameter:

(a) Diameter of a representative tree

(b) Diameter corresponding to the mean basal area of a uniform, generally pure crop

(c) Average of diameter measurements of 250 trees

(d) None of the above

302. A symbiotic association of fungus with the subterranean part of a plant is said to be:

(a) Mycorrhiza

(b) Endophyte

(c) Exophyte

(d) Parasite

303. Edaphic factor is related to:

(a) Termperature

(b) Wind

(c) Soil

(d) Rainfall

304. A pure forest:

(a) Consists of trees of exactly the same age

(b) Consists of only softwood species

(c) Consists of only hardwood species

(d) It in which over 80 per cent of the crop consists of a single species

305. A mixed crop is:

(a) One in which a number of species may be present

(b) In which trees of all ages are growing together

(c) An inventory of logs of various commercial timbers

(d) None of the above

306. Wind throw:

(a) Is trees which fall easily due to wind

(b) Is when all the pollen from trees is blow in one direction

(c) Is when all the dry leaves suddenly drop due to heavy wind

(d) None of the above

307. Epiphytes are:

(a) Plants severely affected by an epidemic

(b) Plants restricted to particular area

(c) Plants growing on the boundaries of a thick forest

(d) Plants growing on another plant, but making their own food

308. Dry nursery is:

(a) One which is established in a desert area

(b) One which is maintained on a rocky site

(c) One which is maintained without

(d) The planting stock stored under low temperature during winter

309. Sunken bed is required for _______ area:

(a) Dry (b) Wet

(c) Shallow sites (d) None of the above

310. Living fossil is:

(a) *Rhododendron* spp. (b) *Gingko biloba*

(c) *Sesquioia semeperverens* (d) None of the above

311. Total number of plants that can be accommodated in 1.0 ha area with the spacing of 3m × 2m spacing is:

(a) 1500 (b) 1800

(c) 1666 (d) 2500

312. Difference between western and eastern part of Himalayan region is:

(a) *Pinus* spp. is more (b) Denuded land

(c) Most warmer (d) None of the above

313. Phnematophore has function of:

(a) Respiration (b) Digestion

(c) Photosynthesis (d) All of the above

314. Tidal vegetation is found in:

(a) Coastal region (b) Gangetic plain

(c) Himalayan region (d) None of the above

315. For full stock crown density should be:

(a) >1 (b) <1

(c) 0.5 (d) 0.75

316. Forest types of India was given by:

(a) Khanna (b) Champion and Seth

(c) Pascal (d) None of the above

317. Tropical wet evergreen forests are found in:

(a) Leh and Ladakh area (b) Gangetic Plains

(c) Andamans and Southern India (d) Rajasthan

318. Pneumatophores are commonly seen in:

(a) Dry deciduous forests
(b) Mangrove forests
(c) Open forests
(d) Sub-tropical montane temperate forests

319. Non leguminous N_2 fixing tree is:

(a) *Alunus nepalensis*
(b) *Leuceanea leucocephala*
(c) *Acacia nilotica*
(d) *Calliandra callothyrsus*

320. Sal is naturally found in:

(a) Littoral moist deciduous forests
(b) Tropical moist deciduous forests
(c) Temperate forests
(d) Alpine forests

321. Teak is naturally found in:

(a) Littoral moist deciduous forests
(b) Tropical moist deciduous forests
(c) Tropical dry deciduous forests
(d) Alpine forests

322. The polyclimax theory of succession is attributable to:

(a) Whittaker
(b) Element
(c) Tansley
(d) Folsberg

323. Presence of chloroxylon in moist deciduous forest is:

(a) Pre climax
(b) Post climax
(c) Sub climax
(d) Zootic climax

324. _______ is the upper limit of the tree growth beyond which the growth of the trees is not seen:

(a) Tree line
(b) Tropic of cancer
(c) Tropic of Capricorn
(d) None of the above

325. Buttresses are commonly found in:

(a) Temperate evergreen forests
(b) Tropical evergreen forests
(c) Tropical Rain forests
(d) Tropical thorn forests

326. Rain forest occurring above the cloud level is known as:

(a) Cloud forests
(b) Heath forests
(c) Mequis
(d) Mangrove forests

327. In Mediterranean region _______ forests are seen:

(a) Sclerophyllous forests
(b) Tropical evergreen forests
(c) Mangrove forests
(d) None of the above

328. Tallest tree species of the world is seen at:

(a) Tropical seasonal forests
(b) Temperate forests
(c) Boreal Forests
(d) Maquis

329. Trees tend to more biomass for girth increment when they are planted at:

(a) Wider spacing
(b) Closer spacing
(c) Both
(d) None of the above

330. What does the type 1A/C4 indicate?

(a) West coast tropical evergreen forests
(b) Southern tropical semi evergreen forests
(c) Southern hill top tropical evergreen forests
(d) Giant evergreen forests

331. Practical application of silvics is:

(a) Silviculture
(b) Silvics
(c) All of the above
(d) None of the above

332. The study of life history and general characteristics of forest trees and crops with reference to environment is:

(a) Silviculture
(b) Silvics
(c) All of the above
(d) None of the above

333. Which among the following species is both deciduous and Evergreen species:

(a) *Melia azedararach*
(b) *Santalum album*
(c) *Mangifera indica*
(d) None of the above

334. Pollination by birds is called:

(a) Ornithophily
(b) Entomophily
(c) Zoophily
(d) All of the above

335. What poisonous alkaloid is extracted from tobacco leaves and widely used as an insecticide?

(a) Cinerin
(b) Nicotine
(c) Rotenone
(d) Pyrethrum

336. The study of phenomena at very low temperatures is called:

(a) Cryogenics
(b) Cytogenetics
(c) Both (a) and (b)
(d) None of the above

337. Pollination by wind is called:

(a) Anemophily
(b) Zoophily
(c) Entomophily
(d) All of the above

338. Pollination by birds is called:

(a) Anemophily
(b) Zoophily
(c) Entomophily
(d) All of the above

339. Pollination by insects is called:

(a) Anemophily
(b) Zoophily
(c) Entomophily
(d) All of the above

340. The portion of the solar energy which is visible to human eye is known as:

(a) Light
(b) Temperature
(c) Moisture
(d) Wind

341. In India the tree line is found around _______ from MSL:

(a) 3500M
(b) 3000M
(c) 2500M
(d) 5000M

342. Which is a topographic component of environment?

(a) Aspect
(b) Slope
(c) Exposure
(d) None of the above

343. The direction towards which the slope face is known as:

(a) Aspect
(b) Slope
(c) Exposure
(d) None of the above

344. Tropical rain forest is not found in:

(a) Central and South Australia
(b) North and Central South America
(c) Central and West Africa
(d) Indo-Malayan-Burmeo and New Guinea

345. Deodar is managed under:

(a) Clear felling systems
(b) Simple coppice systems
(c) Shelter wood system
(d) None of the above

346. Epicormic branches originate from:

(a) Adventitious buds
(b) Auxiliary buds
(c) Meristematic regions
(d) Internodal buds

347. Example for Nitrogen fixing tree:

(a) *Dalbergia*
(b) *Acacia*
(c) *Leucaena*
(d) All of the above

348. Vana Mahotsva was started in the year:

(a) 1952 (b) 1950

(c) 1988 (d) None of the above

349. Knee shaped or spike like projection of the roots of swamp tree called:

(a) Silt roots (b) Prop roots

(c) Pneumatophore (d) All of the above

350. Examples of pneumatophore:

(a) Heretiera (b) Bruguiera

(c) All of the above (d) None of the above

351. Examples of Lignotubers:

(a) Eucalyptus (b) Teak

(c) *Cassia* spp. (d) *Dalbergia* spp.

352. The science that deals with the time of appearance of characteristic periodic events:

(a) Silviculture (b) Phenology

(c) Vegetative propagation (d) None of the above

353. The forest composed of even age woods called:

(a) Even aged (b) Regular forest

(c) Both (a) and (b) (d) None of the above

354. If the male and female flower parts occur in separate flowers on the same tree, the species is said to be:

(a) Monoecious (b) Dioecious

(c) Polygamous (d) None of the above

355. The most important climate indices for estimating potential productivity of a site:

(a) CVP index (b) Ecological index

(c) Precipitation effective index (d) Dryness factor

356. Prop roots is a common phenomenon in:

(a) Ficus (b) Bamboo

(c) Rhizophora (d) Heritieras

357. A stand of trees produced after a clear cut is most likely to be:

(a) Even aged (b) Uneven aged

(c) Both (a) and (b) (d) None of the above

358. The profession dealing with maintenance of ornamental tress is called:

(a) Arboriculture
(b) Silviculture
(c) Floriculture
(d) None of the above

359. The only living part of the tree trunk is the:

(a) Cambium layer
(b) Pith
(c) Phloem
(d) Xylem

360. The leaf stem is called the:

(a) Pedicel
(b) Peduncle
(c) Petiole
(d) None of the above

361. Identification of trees according to species is called:

(a) Dendrology
(b) Wood Technology
(c) Silviculture
(d) All of the above

362. A tree sprout arising from the root or stump is called a:

(a) Coppice
(b) Sucker
(c) Cuttings
(d) All of the above

363. New terminal growth on pines is called:

(a) Candles
(b) Apical shoot
(c) Meristem
(d) None of the above

364. An examples of a trees having alternate leaves is:

(a) *Bombax ceiba*
(b) *Pinus wallichiana*
(c) Both (a) and (b)
(d) None of the above

365. Maximum forest area in India comes under:

(a) Tropical wet evergreen
(b) Tropical moist deciduous
(c) Tropical dry deciduous
(d) Tropical dry evergreen

366. The height of Tectonagrandis in the first quality site:

(a) 30-36m
(b) 24-30m
(c) 18-24m
(d) 12-18m

367. *Shorea robusta* is:

(a) Susceptible to drought
(b) Drought Hardy
(c) Moderately drought hardy
(d) None of the above

368. *Mesua ferrea* belongs to:

(a) Guttiferae
(b) Combertaceae
(c) Myrtaceae
(d) Verbanaceae

369. Most suitable Silvicultural system of deodar in hilly terrain for control of soil erosion is:

(a) Selection system
(b) Clear felling systems
(c) Shelter wood systems
(d) Coppicing systems

370. Forest of *Acacia Arabica* are frequently termed as:

(a) Babul meadow
(b) Carnatic
(c) Black thorn
(d) All of the above

371 Dry teak forest associated species.

(a) *Terminalia tomentosa*
(b) *Indigofera tinctoria*
(c) *Mallotus* spp.
(d) *Sterculia* spp.

372. *Melia dubia* belongs to family:

(a) Verbenaceae
(b) Melianceae
(c) Laminaceae
(d) Myrtaceae

373. The Silviculture of Indian Trees was written by:

(a) Banerjee
(b) L.S. Khanna
(c) Troups
(d) Negi

374. Site quality can be assured by:

(a) Indicative species
(b) Rate of growth of trees
(c) Total volume of trees
(d) All of the above

375. It is a measure of the relative productive capacity of a site for a particular species is called as:

(a) Site factors
(b) Site quality
(c) Both (a) and (b)
(d) None of the above

376. Site index is the relation between:

(a) Age and height
(b) Age and no of trees
(c) Number of stems/crop diameter
(d) Spacing and top height relationship

377. The first Inspector General of Forests of Government of India was:

(a) Dr. Dietrich Brandis
(b) Watson
(c) Chathu Menon
(d) Conolly

378. Size distribution of particles in the soil is called:

(a) Soil texture
(b) Soil structure
(c) Both (a) and (b)
(d) None of the above

379. Oxides of Iron and Aluminum is called:

(a) Sesquioxides (b) Silica

(c) Both (a) and (b) (d) None of the above

380. The plants which grow on the trunk and branches of other plants for nourishment is called:

(a) Epiphytes (b) Climbers

(c) Parasites (d) Abnoxious weeds

381. Rhizobium is _______fixing bacteria:

(a) Nitrogen (b) Phosphorus

(c) Potassium (d) All of the above

382. The best indicator for the site quality _______ Annual Increment:

(a) Mean (b) Periodic

(c) Total (d) None of the above

383. When regeneration is obtained from the seed, the forest may be named as:

(a) High forest (b) Low forest

(c) Coppice forest (d) All of the above

384. Removing dead, dying, diseased and suppressed stems during thinning process is:

(a) B grade (Moderate) thinning (b) C grade (Heavy) thinning

(c) A grade (Light) thinning (d) D grade (Very heavy) thinning

385. Central Board of Forestry was constituted in the year:

(a) 1947 (b) 1948

(c) 1952 (d) 1988

386. The effect of light on the form of trees is called:

(a) Phototropism (b) Photoperiod

(c) Photoperiodism (d) None of the above

387. The reflectivity of a surface or substance to the solar radiation is called:

(a) Albedo (b) Solar radiation

(c) Ultra violet (d) Infra red

388. Local climate of an area which differ from the general climate is:

(a) Microclimate (b) Weather

(c) Bio climate (d) None of the above

389. The study of an individual organism in relation to its environment is:

(a) Autecology
(b) Synecology
(c) Ecosystem ecology
(d) All of the above

390. The series of stages in succession is called:

(a) Climax
(b) Lithosere
(c) Psammosere
(d) Seres

391. When natural succession is pushed back to several stages due to natural or biotic factors is called:

(a) Retrogression
(b) Regression
(c) Both (a) and (b)
(d) None of the above

392. Succession false under the category of _______ ecology:

(a) Autecology
(b) Synecology
(c) Ecosystem ecology
(d) All of the above

393. The plants which grow on the trunk and branches of other plants for support is called:

(a) Epiphytes
(b) Climbers
(c) Parasites
(d) Abnoxious weeds

394. The negative logarithm of hydrogen ion concentration is called:

(a) pH
(b) EC
(c) Cation exchange capacity
(d) All of the above

395. Casuarinas root nodules are formed in association with:

(a) Mycorrhiza
(b) Frankia
(c) Both (a) and (b)
(d) None of the above

396. The largest family among the Indian flora is:

(a) Asteraceae
(b) Poaceae
(c) Fabaceae
(d) Myrtaceae

397. Sucking roots of parasites are:

(a) Haustoria
(b) Hypophysis
(c) Radical
(d) Transverse cell

398. The origin of teak is:

(a) Burma
(b) Nilambur
(c) West Bengal
(d) None of the above

399. In India, silvicultural systems are classified based on:

(a) Type of crop (b) Tending operations
(c) Mode of regeneration (d) All of the above

400. Low forests originated from:

(a) Regeneration through coppice (b) Regeneration through seeds
(c) Both (a) and (b) (d) None of the above

401. The Latin name for Andaman Padauk tree is:

(a) *Pterocarpus dalbergioides* (b) *Dipterocarpus macrocarpus*
(c) *Messua ferrea* (d) *Dipterocarpus tubinatus*

402. The habitat of deodar is:

(a) Evergreen forest (b) Temperate forest
(c) Deciduous forest (d) Alpine forest

403. The tree species belonging to the family Myrtaceae:

(a) Eucalyptus (b) Casuarinas
(c) Teak (d) Terminalia

404. Development of embryo without fertilization is termed:

(a) Apomixes (b) Parthnogensis
(c) Apogamy (d) Apospory

405. The latin name of golden bamboo is:

(a) *Bambusa vulgaris* (b) *Bambusa bamboo*
(c) *Bambusa balcooa* (d) *Bambusa lineate*

406. Nitrogen fixing non-leguminous t ree:

(a) *Dalbergia* (b) *Acacia*
(c) *Leucaena* (d) *Casuarina*

407. The upper branchy part of a tree above the bole called:

(a) Stem (b) Crown
(c) Root (d) None of the above

408. The expansion of crown measured as crown length and crown width called as:

(a) Crown development (b) Crown expansion
(c) Both (a) and (b) (d) None of the above

409. Example for Deciduous tree:

(a) *Acacia catechu* (b) *Melia azedarach*
(c) *Ailanthus excels* (d) All of the above

410. Example for Evergreen tree:

(a) *Abies pindrow*
(b) *Michelia champaca*
(c) *Hopea parviflora*
(d) All of the above

411. The decrease in diameter of the stem of a tree from the base upwards is known as:

(a) Taper
(b) Clean bole
(c) Commercial Bole
(d) None of the above

412. The part of the bole free of branches is called as:

(a) Stem
(b) Crown
(c) Root
(d) Both (a) and (b)

413. The irregular involutions and swelling on the bole called:

(a) Forked
(b) Fluting
(c) Buttressed
(d) None of the above

414. Examples of buttresses stem:

(a) *Bombax ceiba*
(b) *Terminalia myriocarpa*
(c) *Acrocarpus fraxinifolius*
(d) All of the above

415. The roots produced from parts of the plants other than the radical is called:

(a) Adventitious roots
(b) Taproot
(c) Lateral roots
(d) All of the above

416. Examples for prop roots:

(a) *Ficus bengalensis*
(b) Bamboo
(c) Rhizophora
(d) Heritieras

417. The tree appears on flying buttresses called as:

(a) Pneumatophore
(b) Silt roots
(c) Prop roots
(d) All of the above

418. A knee shaped or spike like projection of the roots:

(a) Pneumatophore
(b) Silt roots
(c) Prop roots
(d) All of the above

419. The structure produced from the combination of the modified rootlet with fungal tissue called as:

(a) Fungi
(b) Mycorrhiza
(c) Frankia
(d) None of the above

420. _______ increase the surface area of roots.

(a) Plasmodesmata
(b) Mycorrhizae
(c) Root hairs
(d) Root hairs and mycorrhizae

421. Flame of the forest:

(a) *Butea monosperma* (b) *Tectona grandis*
(c) *Shorea robusta* (d) *Cassia fistula*

422. The condition of the tree with dead upper branches projecting above the green portion of the crown is called:

(a) Thinning (b) Pruning
(c) Climber cutting (d) Stagheaded

423. Natural forest uninfluenced by human activity:

(a) Natural forest (b) Rain forest
(c) Virgin forest (d) All of the above

424. The male and female flowers are found on the same individual called:

(a) Monoecious (b) Dioecious
(c) Polygamous (d) None of the above

425. The male and female flowers are found on the different individual called:

(a) Monoecious (b) Dioecious
(c) Polygamous (d) None of the above

426. The unisexual and bisexual flowers are found on the same individual called:

(a) Monoecious (b) Dioecious
(c) Polygamous (d) None of the above

427. Examples of monoecious:

(a) Pinus (b) *Abies pindrow*
(c) Salix (d) Both (a) and (b)

428. Examples of dioecious:

(a) Salix (b) Diospyros
(c) *Cedrus deodara* (d) All of the above

429. Examples of polygamous flowering:

(a) Sterculia (b) Rhus
(c) Garuga (d) All of the above

430. The height corresponding to the mean diameter of the 250 biggest diameters per hectare:

(a) Crop Height (b) Top height
(c) Crop diameter (d) Crop volume

431. Average height of regular crop is called as:

(a) Crop diameter (b) Crop height

(c) Crop volume (d) Top height

432. Average height of regular crop is determined by:

(a) Lorey's formula (b) Smalian formula

(c) Heyer's formula (d) None of the above

433. The diameter corresponding to the mean basal area of a uniform called as:

(a) Crop diameter (b) Crop height

(c) Crop volume (d) Top height

434. The relative completeness of the stocking expressed as a decimal coefficient, taking normal number of trees, basal area or volume as unity is called as:

(a) Crop density (b) Canopy density

(c) Both (a) and (b) (d) None of the above

435. The relative completeness of canopy usually expressed as a decimal coefficient, taking closed as unity is called as:

(a) Crop density (b) Canopy density

(c) Both (a) and (b) (d) None of the above

436. The canopy density is 0.75 to 1.0:

(a) Closed (b) Thin

(c) Dense (d) Open

437. The canopy density is below 0.5:

(a) Closed (b) Thin

(c) Dense (d) Open

438. Examples of light demanders for Moist temperate forest:

(a) *Pinus wallichiana* (b) *Populus ciliata*

(c) Both (a) and (b) (d) None of the above

439. Examples of shade demanders for tropical Moist deciduous forest:

(a) *Syzygium cumini* (b) *Schleichera oleosa*

(c) *Mallotus philipinensis* (d) All of the above

440. Examples of shade bearers for tropical wet evergreen forest:

(a) *Artocarpus* spp. (b) *Dipterocarpus* spp.

(c) *Calophyllum* (d) *Mesua ferrea*

441. Based on the methods of regeneration, forests are classified as:

(a) Normal and abnormal forest
(b) Evenaged and unevenaged
(c) Pure and mixed forest
(d) High and low forest

442. The mean annual temperature of montane sub-tropical forest ranged between:

(a) 24°C
(b) 17°C and 24°C
(c) 7°C to 17°C
(d) < 7°C

443. The Forest Survey of India (FSI) conducts Forest Cover Survey once in every:

(a) Two years
(b) Five years
(c) Ten years
(d) None of the above

444. Climate Vegetation Productivity Index was given by:

(a) Thoreau
(b) Clements
(c) Dawson
(d) Paterson

445. First Teak Plantation in India:

(a) Nilampur
(b) West Bengal
(c) Siwalik
(d) Gorakhpur

446. Examples of frost hardy species:

(a) Acacia catechu
(b) Anogessue pendula
(c) Toonaciliate
(d) All of the above

447. Examples of frost tender species:

(a) Azadirachta india
(b) Tectona grandis
(c) Terminalia arjuna
(d) All of the above

448. Examples of drought hardy species:

(a) Acacia arabica
(b) Diospyros melanoxylon
(c) Syzygium cumini
(d) All of the above

449. A month which has <50mm of rainfall is referred to as:

(a) Dry month
(b) Rainy days
(c) Both (a) and (b)
(d) None of the above

450. A day which has minimum of 2.5 mm of rainfall is referred to as:

(a) Dry month
(b) Rainy days
(c) Both (a) and (b)
(d) None of the above

451. The commercial beedi leaves are plucked from:

(a) *Diospyros melanoxylon* (b) *Shorea robusta*

(c) *Pterosperma* spp. (d) *Gmelina arborea*

452. The dying back is a common disease in:

(a) Teak (b) Deodar

(c) Sal (d) None of the above

453. Example of indigenous populous species:

(a) *Populous ciliate* (b) *Populus robusta*

(c) *Populous deltoids* (d) *Populus rubrapoiret*

454. Agar wood tree is:

(a) *Aquilaria malaccensis* (b) *Teactona grandis*

(c) *Dalbergia sissoo* (d) *Acacia catechu*

455. The blowing of strong noisy wind up and down the valley locally called as:

(a) Dadu (b) Nor westers

(c) Mango showers (d) All of the above

456. A plant which grows in dry situation is called:

(a) Hydrophyte (b) Mesophyte

(c) None (d) Xerophytes

457. A plant which grows in moist situation is called:

(a) Hydrophyte (b) Hygrophyte

(c) Mesophyte (d) None of the above

458. A plant which grows wholly or partly immersed in water:

(a) Hydrophyte (b) Hygrophyte

(c) Mesophyte (d) None of the above

459. A plant whose normal habitat is neither very wet nor very dry:

(a) Hydrophyte (b) Hygrophyte

(c) Mesophyte (d) None of the above

460. Plants growing on, but not nourished by another pant is referred to as:

(a) Epiphytes (b) Climbers

(c) Parasites (d) Abnoxious weeds

461. Epiphytes commonly present in:

(a) *Ficus bengalensis* (b) *Ficus religiosa*

(c) Orchids (d) All of the above

462. Examples of Abnoxious weeds:

(a) *Lantana camara*
(b) *Eupatorium odoratum*
(c) Both (a) and (b)
(d) None of the above

463. Succession which have previously not born vegetation:

(a) Primary succession
(b) Secondary succession
(c) Both (a) and (b)
(d) None of the above

464. The series of plant communities resulting from processes of succession is called:

(a) Seres
(b) Climax
(c) Lithosere
(d) Psammosere

465. The term Succession was first used by:

(a) Thoreau
(b) Clement
(c) Cawles
(d) Sawson

466. Making high demands as regard soil and other factors is called:

(a) Exacting
(b) Reaction
(c) Competition
(d) None of the above

467. The succession initiated in extremely dry situation is called:

(a) Xerarch succession
(b) Hydrarch succession
(c) Both (a) and (b)
(d) None of the above

468. The successional stages of Xerarch succession:

(a) Xerarch succession
(b) Xeroseres
(c) Both (a) and (b)
(d) None of the above

469. Xerosers which originates on rock surface is called:

(a) Lithosere
(b) Psammosere
(c) Xerarch succession
(d) None of the above

470. Xerosere which originates on sand is called:

(a) Lithosere
(b) Psammosere
(c) Xerarch succession
(d) None of the above

471. What is the scientific term for a leaf blade?

(a) Lamina
(b) Pedicel
(c) Peduncle
(d) None of the above

472. An ash leaf generally contains _______ leaflets:

(a) Nine
(b) Five
(c) Seven
(d) None of the above

473. What is the scientific term for a leaf stalk?

(a) Petiole
(b) Pedicel
(c) Peduncle
(d) None of the above

474. Example of edible 'Fungi':

(a) Mucor
(b) Pencillium
(c) Agaricus
(d) Rhizopus

475. Energy transferred from one trophic level to another is:

(a) 15 per cent
(b) 42 per cent
(c) 54 per cent
(d) 85 per cent

476. Coir which is obtained from fruit of coconut is:

(a) Endocarp
(b) Periderm
(c) Epicarp
(d) Mesocarp

477. Solarisation is:

(a) Formation of chlorophyll
(b) Utilisation of sunlight
(c) Effects of solar light
(d) Destruction of chlorophyll

478. 'Pheumatophores' are present/common in:

(a) Xerophytes
(b) Hygrophytes
(c) Halophytes
(d) Mesophytes

479. Maximum absorption of rainfall water is done by:

(a) Tropical deciduous forest
(b) Tropical evergreen forest
(c) Tropical savanna
(d) Scrub forest

480. In which form does the food transported in plants:

(a) Sucrose
(b) Fructose
(c) Glucose
(d) Lactose

481. Largest flower is:

(a) Rafflesia
(b) Lotus
(c) Rose
(d) Lily

482. In angiosperms, the structures that produce male gametophyte are called the:

(a) Stigmas
(b) Anthers
(c) Pollen tubes
(d) Sepals

483. The outermost circle of flower parts consists of several:

(a) Sepals
(b) Petals
(c) Carpels
(d) Corollas

484. Pollination occurs when pollen lands on the:

(a) Stigma (b) Style
(c) Filament (d) Anther

485. The seed leaves of flowering plants are known as:

(a) Endosperm (b) Carpels
(c) Radicals (d) Cotyledons

486. Grasses have _______ meristems.

(a) Intercalary (b) Apical
(c) Lateral (d) None of the above

487. The scientist credited with being the Father of Botany is:

(a) Darwin (b) Theophrastus
(c) Hinterthuer (d) Linnaeus

488. The part of cotton producing pure cellulose is:

(a) Root hair (b) Leaf hair
(c) Seed hair (d) Stem hair

489. Which is considered to be the best chemical method of fixing atmospheric nitrogen?

(a) Fisher method (b) Decan method
(c) Haber-Bosch method (d) Paranas-Meyerhoff method

490. Plant length is increased by _______.

(a) Apical meristems (b) Lateral meristems
(c) Periblem (d) Parenchyma

491. The cell division is restricted to _______.

(a) Meristematic cells (b) Permanent cells
(c) Secretory cells (d) All of the above

492. Active division takes place in the cells of _______.

(a) Xylem (b) Phloem
(c) Sclerenchyma (d) Cambium

493. Collenchymas mainly forms _______.

(a) Hypodermis (b) Epidermis
(c) Phloem (d) Inner cortex

494. Plants growing in extremely cold soils are:

(a) Xerophytes (b) Psychrophyte
(c) Mesophyte (d) Hygrophyte

495. Xerophytes have long roots:

(a) Lower rate of transpiration than mesophytes
(b) Transpiration rate equal to mesophytes
(c) Transpiration rate higher than mesophytes
(d) No transpiration

496. The orchid plants, which are found growing on the trees, are:

(a) Epiphytes
(b) Climbers
(c) Parasites
(d) Abnoxious weeds

497. The first plants to reappear in a badly burned forest area will most probably be:

(a) Liverworts
(b) Ferns
(c) Grasses
(d) Mosses

498. The transfer of energy from organism to organism in a natural community establishes:

(a) Food chains
(b) Food web
(c) Ecosystems
(d) Population

499. Which is a primary consumer:

(a) Carnivore
(b) Herbivore
(c) Omnivore
(d) Detritivore

500. Trophic levels are formed by:

(a) Plants only
(b) Animals only
(c) Organisms linked in food chain
(d) Carnivores only

501. Driving force of ecosystem is:

(a) Carbohydrate in plant
(b) Biomass
(c) Solar energy
(d) Producer

502. Which group of plants possesses both animal and plant characteristics?

(a) Myxomycetes
(b) Basidiomycetes
(c) Both (a) and (b)
(d) None of the above

503. Cleistogamous flowers are always:

(a) Self pollinated
(b) Cross pollinated
(c) Both (a) and (b)
(d) None of the above

504. Flowers are small and unattractive in:

(a) Anemophily
(b) Ornithophily
(c) Entomophily
(d) None of the above

505. When birds are agents, the pollination is:

(a) Anemophily (b) Ornithophily
(c) Entomophily (d) None of the above

506. Palynology deals with study of:

(a) Pollen grains (b) Fossils
(c) Flowers (d) All of the above

507. The color of light least effective in driving photosynthesis is:

(a) Yellow (b) Green
(c) Blue (d) Orange

508. Sandal wood tree generally flowers in the month of:

(a) May to June (b) February to April
(c) September to February (d) March to June

509. The collection period for Neem (*Azadiracta indica*) seed is:

(a) February to April (b) September to February
(c) March to June (d) May to June

510. *Dendrocalamus strictus* is a good associate of:

(a) Trema (b) Callicarpa
(c) Macaranga (d) Teak

511. The beef wood is also known as:

(a) Casuarinas (b) Teak
(c) Dalbergia (d) Bamboo

512. Taiga form of vegetation are found in:

(a) Russia (b) Brazil
(c) Canada (d) Italy

513. The tropical and sub-tropical grassland with either sparsely scattered trees or bushes are called:

(a) Savannas (b) Prairies
(c) Steppe (d) Range land

514. Clear felling system has also been termed as:

(a) Coupe blanche (b) Coupe rare
(c) Clear cutting system (d) All of the above

515. Trade/common name of *Dalbergia latifolia* is:

(a) Rose wood (b) Beef wood
(c) Soft wood (d) Hard wood

516. The most limiting element of all the environmental factors on the growth of trees is:

(a) Moisture (b) Temperature
(c) Soil (d) None of the above

517. Course textured soils such as sands are _______ in water holding capacity and nutrient content.

(a) Low (b) High
(c) Medium (d) None of the above

518. A forest which has been reproduced through sprouting is called a _______ forest.

(a) Coppice (b) Low
(c) High forest (d) Both (a) and (b)

519. Based on crown classification which is defined as a smaller tree with a crown below the upper level of the forest canopy and receiving only scattered sunlight from above?

(a) Suppressed (b) Co-dominant
(c) Dominant (d) Intermediate

520. Stocking level is best described as:

(a) Percent utilization of the site potential (b) Numbers of trees/acre
(c) Both (a) and (b) (d) None of the above

521. The outgrowth from the base of the trees connecting it with the roots, especially common in tropical rain forest species is:

(a) Fluting (b) Forking
(c) Tapering (d) Buttressing

522. Natural pruning in trees can be accelerated by:

(a) Density of planting (b) Species composition
(c) Retention of understory trees (d) All of the above

523. Which one of the following is correctly matched?

(a) *Piceas mithiana* – Spruce (b) *Dysoxylumm alabaricum* – Elm
(c) *Ulmus wallichiana* – White cedar (d) *Tectona grandis* – Ven Teak

524. Which of the following is a moisture loving plant?

(a) *Tectona grandis* (b) *Prosopis juliflora*
(c) *Shorea robusa* (d) *Terminalia arjuna*

525. Which sub-type of forest will not be under Tidal Swamp Forests?

(a) Myristica swamp forest (b) Mangrove scrub

(c) Mangrove forest (d) Palm swamp

526. Dry red sanders forests occur in which state?

(a) Andhra Pradesh (b) Tamil Nadu

(c) Karanataka (d) Madhya Pradesh

527. The climate that modified by the adjusted climatic factors is called:

(a) Bioclimate (b) Climate

(c) Microclimate (d) All of the above

528. Koppen classified world climate into _______ groups:

(a) Five (b) Six

(c) Seven (d) Sixteen

529. Solar radiation, temperature and rainfall is mainly affected by:

(a) Altitude (b) Slope

(c) Exposure (d) All of the above

530. Moisture regime of the soil is highly influenced by the _______.

(a) Altitude (b) Slope

(c) Exposure (d) All of the above

531. The direction towards the slope faces is called:

(a) Altitude (b) Slope

(c) Exposure (d) Aspect

532. The relation of a site in relation to sun and wind is called:

(a) Altitude (b) Slope

(c) Exposure (d) Aspect

533. Southern aspect are _______than northern aspect:

(a) Drier (b) Wetter

(c) moisture (d) None of the above

534. The climate of small area that differ generally from the general climate is called:

(a) Microclimate (b) Bioclimate

(c) Both (a) and (b) (d) None of the above

535. Microclimate is important in:

(a) Artificial (b) Natural regeneration

(c) Both (a) and (b) (d) None of the above

536. The upper most weathered layer of the earth crust is called:

(a) Soil (b) Bed rock

(c) Regolith (d) Rock

537. The geological process through which the soil is formed is called:

(a) Weathering (b) Soil formation

(c) Soil tilth (d) None of the above

538. Physical disintegration of rock is called:

(a) Physical weathering (b) Chemical weathering

(c) Biological weathering (d) All of the above

539. Disintegration of rocks chemically is called:

(a) Physical weathering (b) Chemical weathering

(c) Biological weathering (d) All of the above

540. The structure of the surface layer of the soil is called:

(a) Weathering (b) Soil formation

(c) Soil tilth (d) None of the above

541. A sequence of soil derived from the same age and same parent material is called:

(a) Catena (b) Mature soils

(c) Immature soils (d) None of the above

542. Basis of soil farming factor, the soils are classified into:

(a) Endodynamorphic (b) Ectodynamorphic

(c) Endoectododynamorphic (d) All of the above

543. Soil host properties are influenced mainly by parental materials:

(a) Endodynamorphic (b) Ectodynamorphic

(c) Endoectododynamorphic (d) None of the above

544. Soil formed from parent material in situ:

(a) Sedentary soil (b) Alluvial soil

(c) Secondary soil (d) Aeolian soil

545. Soil derived from pre-existing soils by translocation and redeposition:

(a) Secondary soil (b) Alluvial soil

(c) Aeolian soil (d) Sedentary soil

546. Soil whose properties are influenced by the parent material is called:

(a) Endodynamorphic (b) Ectodynamorphic

(c) Endoectododynamorphic (d) None of the above

547. Soil whose properties are influenced by other than the parent material is called:

(a) Endodynamorphic (b) Ectodynamorphic

(c) Endoectododynamorphic (d) None of the above

548. A secondary immature soil derived from soil material of older origin deposited by local erosion on the lower slope of gentle hills is called:

(a) Colluvial soils (b) Alluvial soil

(c) Aeolian soil (d) Sedentary soil

549. The appearance of soils in vertical section with the particular reference to the sequence of layers that may be differentiated is termed as:

(a) Soil profile (b) Soil horizon

(c) Regolith (d) Bed rocks

550. Soils which contains an excess of soluble salts with low exchangeable Sodium is:

(a) Saline soils (b) Alkali soils

(c) Acid soils (d) All of the above

551. Kankar pan is commonly found in:

(a) Temperate region (b) Tropical region

(c) Alluvial soils (d) Red soils

552. Silican pan is formed in:

(a) Podsolic iron pan (b) Laterite and lateritic soils

(c) Kankar pan (d) Clay pan

553. Podsolic iron pan is formed in podsol soils of:

(a) Temperate region (b) Tropical region

(c) Alluvial soils (d) Red soils

554. A plant which tend to be restricted to calcarious soil is:

(a) Calciphilre (b) Calcifuges

(c) Calcigaurd (d) None of the above

555. Laterization is a process of accumulation in soil:

(a) Iron and Aluminium oxides (b) Silica

(c) $CaCO_3$ (d) $CaSO_4$

556. Podzolization is a process of accumulation in soil:

(a) Iron and Aluminium oxides (b) Silica

(c) $CaCO_3$ (d) $CaSO_4$

557. Concentration of nitrogen in soil air is:

(a) 70 per cent
(b) 74 per cent
(c) 78 per cent
(d) 82 per cent

558. Vertical section of a the soil is known as:

(a) Solum
(b) Regolith
(c) Soil profile
(d) A horizon

559. Unconsolidated material overlying rocks:

(a) Solum
(b) Regolith
(c) Bed rocks
(d) A horizon

560. The solum is comprised of:

(a) A, B and E horizon
(b) A and B horizon
(c) A, B and C horizon
(d) A+ regolith

561. The B horizon is generally absent in:

(a) Black soils
(b) Red soils
(c) Desert soils
(d) Brown hill soils

562. Passive factor of soil formation includes:

(a) Climate
(b) Organisms
(c) Parent material
(d) None of the above

563. The relative proposition of various size groups of individual soil particles is termed as:

(a) Soil texture
(b) Soil structure
(c) Soil class
(d) None of the above

564. Which type of structure is dominant in grasslands?

(a) Spheroidal
(b) Platy
(c) Columnar
(d) Blocky

565. White alkali soils are:

(a) Acid soils
(b) Saline soil
(c) Saline sodic soil
(d) Sodic soil

566. A group of soils having same range in particle size and physical properties:

(a) Soil class
(b) Soil structure
(c) Soil texture
(d) None of the above

567. Arrangement of individual soil particles into aggregates is called:

(a) Soil class
(b) Soil structure
(c) Soil texture
(d) None of the above

568. The root development of plant is affected by:

(a) Soil class (b) Soil structure

(c) Soil texture (d) None of the above

569. The percentage of finer particle governess _______ status of soil:

(a) Nutrient (b) Soil texture

(c) Soil structure (d) C: N ratio

570. In the nutrition of a plant, which element is classified as a macronutrient?

(a) Zinc (b) Chlorine

(c) Calcium (d) Molybdenum

571. The most suitable C:N ration is:

(a) 10:1 (b) 17:1

(c) 33:1 (d) 12:1

572. The influence of living organisms on the vegetation is called:

(a) Climatic factors (b) Edaphic factors

(c) Biotic factors (d) Topographic factors

573. An organism that draws part or whole of its nourishments from another living organisms is called:

(a) Parasite (b) Epiphytes

(c) Climbers (d) Abnoxious weeds

574. Viscum and Loranthus are the:

(a) Partial stem parasite (b) Total stem parasite

(c) Partial root parasite (d) Total root parasite

575. Cuscuta and Cassytha are the:

(a) Partial stem parasite (b) Total stem parasite

(c) Partial root parasite (d) Total root parasite

576. Sandal is a:

(a) Partial root parasite (b) Partial stem parasite

(c) Both (a) and (b) (d) None of the above

577. A plant growing on but not nourished by another plant is:

(a) Parasite (b) Epiphytes

(c) Climbers (d) Abnoxious weeds

578. Ficus, orichids and tinospora are:

(a) Parasite (b) Epiphytes

(c) Climbers (d) Abnoxious weeds

579. Large woody climbers in tropical areas are called:

(a) Liana or Liane
(b) Parasite
(c) Epiphytes
(d) Climbers

580. Dioscorea, Ipomaea are the example of:

(a) Parasite
(b) Epiphytes
(c) Climbers
(d) Lianas

581. Eupotorium and Lantana are classified as:

(a) Parasite
(b) Epiphytes
(c) Climbers
(d) Abnoxious weeds

582. A species that invades ina bare area such as newly exposed soil are rock surface is called:

(a) Pioneer species
(b) Succession
(c) Xerarch succession
(d) Lithosere

583. The gradual replacement of one community by another in the development of the vegetation towards a climax is called:

(a) Succession
(b) Climax
(c) Pioneer species
(d) None of the above

584. Culmination stage in plant succession for a given environment is called:

(a) Climax
(b) Climatic climax
(c) Biotic climax
(d) Pre climax

585. The series of plant communities resulting from process of succession is called:

(a) Primary succession
(b) Secondary succession
(c) Seres
(d) All of the above

586. What is haustoria?

(a) Sucking root of plants
(b) Creeping root of climbers
(c) Sucking root of parasite
(d) Climbing root of plants Agroforestry

587. Widely accepted succession theory was:

(a) Clements Theory
(b) Whittaker's theory
(c) Aubreville's theory
(d) All of the above

588. The Succession initiated in extremely dry situation is called:

(a) Xerarch succession
(b) Hydrarch succession
(c) Hydro succession
(d) None of the above

589. The different stages in a Xerarch Succession is called:

(a) Xeroseres
(b) Hydrosere
(c) Both (a) and (b)
(d) None of the above

590. Succession begins in water is defined:

(a) Xerarch succession
(b) Hydrarch succession
(c) Hydro succession
(d) None of the above

591. The different stages in a hydrarch sucession is called:

(a) Lithosere
(b) Psammosere
(c) Hydrosere
(d) All of the above

592. Succession take place where there is no previous vegetation is called:

(a) Primary succession
(b) Secondary succession
(c) Both (a) and (b)
(d) None of the above

593. A succession in which vegetation affects the ecosystems to bring about consequently changes is called:

(a) Allogenic succession
(b) Autogenic succession
(c) Both (a) and (b)
(d) None of the above

594. A succession which take place after the destruction of the whole or part of the original vegetation is called:

(a) Secondary succession
(b) Primary succession
(c) Both (a) and (b)
(d) None of the above

595. Secondary Succession otherwise termed as:

(a) Allogenic succession
(b) Autogenic succession
(c) Both (a) and (b)
(d) None of the above

596. The whole process where a plant establishes itself in a new area from germination to reproduction is called:

(a) Migration
(b) Ecesis
(c) Aggregation
(d) Competition

597. Grouping followed by establishment of scattered colonizing invaders as a result of propagation is called:

(a) Migration
(b) Ecesis
(c) Aggregation
(d) Competition

598. The effect of vegetation on the site is called:

(a) Reaction
(b) Migration
(c) Aggregation
(d) Competition

599. Riverine succession of Gangetic alluvium is _______ Succession:

(a) Primary succession (b) Secondary succession

(c) Both (a) and (b) (d) None of the above

600. A Major unit of vegetation comprising the climax communities of an area uniform in its major physiognomic features is termed as:

(a) Formation (b) Migration

(c) Establishment (d) Grouping

601. Monoclimax Theory was proposed by:

(a) Clements (b) Tansley

(c) Whittiker (d) Aubreville

602. Mosaic theory was proposed by:

(a) Clements (b) Tansley

(c) Whittiker (d) Aubreville

603. Theory of vegetation gradient and climax patterns was proposed by:

(a) Clements (b) Tansley

(c) Whittiker (d) Aubreville

604. The climax which owes distinctive characters due to climatic factors is referred as:

(a) Climatic climax (b) Edaphic climax

(c) Pre climax (d) Post climax

605. A plant which differs from the climatic climax due to the influence of special soil factors is defined:

(a) Edaphic climax (b) Pre climax

(c) Climatic climax (d) Post climax

606. The plant community immediately preceding the climatic climax is called:

(a) Edaphic climax (b) Pre climax

(c) Climatic climax (d) Post climax

607. A plant community more exacting that the climatic climax is defined:

(a) Edaphic climax (b) Pre climax

(c) Climatic climax (d) Post climax

608. A climax which differs from the climatic climax of the area due to biotic factors is referred as:

(a) Biotic Climax (b) Climatic climax

(c) Pre climax (d) Post climax

609. A vegetation development towards the climatic climax has been arrested due to some stage short of normal climax is called:

(a) Biotic climax (b) Sub climax
(c) Pre climax (d) Post climax

610. Study of plant succession is very important for classification of forests into:

(a) Forest types (b) Artificial regeneration
(c) Successional stage (d) All of the above

611. A category of forest defined with reference to its geographical location, climatic and edaphic feature, composition and condition is called:

(a) Forest type (b) Forestry
(c) Forest (d) None of the above

612. Indian Forest are classified into forest types based on _______ classification:

(a) Mayr (b) Koppen
(c) Seth and Khan (d) Champion and Seth

613. A general appearance of forest community is called:

(a) Physiognomy (b) Physiology
(c) Phenology (d) None of the above

614. Vegetation first classified according to temperate zones by:

(a) Schimper (b) Mayr
(c) Koppen (d) Seth and Khan

615. Classification of Vegetation according to temperate combined with latitude and altitude was done by:

(a) Schimper (b) Mayr
(c) Koppen (d) Seth and Khan

616. Thornthwait classified the veget5ation based on:

(a) Season of Rain fall (b) Temperature efficiency
(c) Bioclimatic (d) Eco systems

617. Classification of vegetation based on thermodynamic principles was done by:

(a) Shanbhag (b) Thornthwaite
(c) Kippen (d) Seth and Khan

618. Indian forest are classified into forest types in the year:

(a) 1947 (b) 1935
(c) 1952 (d) 1988

619. According to Champion and Seth Indian Forest have been divided into _______ groups:

(a) Five (b) Sixteen
(c) 221 (d) None of the above

620. According to Champion and Seth Indian Forest have been divided into _______ Sub groups:

(a) 5 (b) 16
(c) 221 (d) None of the above

621. According to champion and Seth Indian Forest have been divided into _______ types:

(a) 5 (b) 16
(c) 221 (d) None of the above

622. A renewal of forest crop by natural or artificial means:

(a) Regeneration (b) Natural regeneration
(c) Artificial regeneration (d) All of the above

623. The renewal of a forest crop by self-sown seed or by coppice is called:

(a) Regeneration (b) Natural regeneration
(c) Artificial regeneration (d) All of the above

624. The regeneration obtained from seed forms a crop called:

(a) Seedling crop (b) Croppice crop
(c) Both (a) and (b) (d) None of the above

625. The Seedling crop grows into a forest it is called:

(a) High forest (b) Low forest
(c) Coppice forest (d) All of the above

626. Regeneration obtained by coppice forms of a crop is called:

(a) Coppice forest (b) Low forest
(c) High forest (d) Both (a) and (b)

627. Coppice forest otherwise called as:

(a) Low forest (b) High forest
(c) Both (a) and (b) (d) None of the above

628. A year in which a given species bear seed abundantly called as:

(a) Seed year (b) Seedling year
(c) Germination (d) None of the above

629. Examples of seed dispersal by wind:

(a) *Pterocarpus marsupium*
(b) *Acacia catechu*
(c) *Casuarinas* spp.
(d) All of the above

630. Examples of seed dispersal by water:

(a) Mangrove species
(b) Dalbergia
(c) Teak
(d) All of the above

631. Examples of seed dispersal by gravity:

(a) Oaks
(b) *Juglans regia*
(c) Aesculus
(d) All of the above

632. Examples of seed dispersal by birds:

(a) Mulberry
(b) *Diospyros melanoxylon*
(c) *Prunus* spp.
(d) All of the above

633. Examples of seed dispersal by animals:

(a) *Anthocephalus cadamba*
(b) *Acacia arabica*
(c) *Prosopis juliflora*
(d) All of the above

634. The potential capacity of a seed to germinate is called:

(a) Viability
(b) Germination
(c) Germinative capacity
(d) None of the above

635. The percentage, by number of seeds in a given sample that actually germinate, irrespective of time is called:

(a) Germinative energy
(b) Germinative capacity
(c) Germination
(d) Plant percent

636. The percentage, by number of seeds in a given sample that have germinated up to the time when the rate of germination reaches its peak is called:

(a) Germinative energy
(b) Germinative capacity
(c) Germination
(d) Plant percent

637. The percentage of number of seeds in a sample that develop into seedlings at the first growing season is called:

(a) Germinative energy
(b) Germinative capacity
(c) Germination
(d) Plant percent

638. In a year which a given species produces abundant first year seedlings is called:

(a) Seedling year
(b) Seed year
(c) Both (a) and (b)
(d) None of the above

639. Area where plants are raised for eventual planting out is called:

(a) Nursery (b) Sowing
(c) Grading (d) All of the above

640. Salvage cutting is done primarily for:

(a) Improving the hygiene of the stand (b) Salvaging the timber
(c) Salvaging the stand from deterioration (d) Producing of large trees

641. Pick out the trees species with the correct spacing:

(a) Eucalyptus – 3 x 3m (b) Casuarinas – 2 x 3m
(c) Bamboo – 6 x 6m (d) Ceiba pentandra – 2.5 x 2.5m

642. Wide Nursery bed is called:

(a) Permanent nursery (b) Temporary nursery
(c) Kamara (d) None of the above

643. Standard soil mixture used for the nursery is (Soil: Sand: FYM):

(a) 2:1:1 (b) 3:2:1
(c) 3:2:2 (d) 2:2:1

644. Standard size of nursery bed in forestry is:

(a) 12.5m x 1.2m (b) 10m x 1.0m
(c) 12.5m x 1.5m (d) 10m x 2.0m

645. The technique of joining together parts of two plants to form one composite plant is:

(a) Budding (b) Cutting
(c) Grafting (d) Sowing

646. Grafting can be used only in plants having:

(a) Soft system (b) Hard tissue
(c) Buds (d) Cambium

647. In grafting scion forms:

(a) Root system (b) Shoot system
(c) Buds (d) Cambium

648. Scion is:

(a) Graft of shoot (b) Graft of root
(c) A bud (d) Clone

649. In which of following, diameter of stock is larger than scion:

(a) Side grafting (b) Tongue grafting
(c) Wedge grafting (d) Crown grafting

650. More than one scion are grafted on single stock, the grafting is:

(a) Crown grafting
(b) Wedge grafting
(c) Layering
(d) Bud grafting

651. Special type of grafting, in which scion unite with the stock while still attached to parent plant:

(a) Approach grafting
(b) Wedge grafting
(c) Side grafting
(d) Tongue grafting

652. Grafting technique is successful between two plants having:

(a) Viable seed
(b) Parenchymatous tissue
(c) Active buds
(d) Active vascular cambium

653. Which is the most common pattern of planting in forestry?

(a) Rectangular planting
(b) Line planting
(c) Triangular planting
(d) Quincunx planting

654. Planting of cuttings is common in:

(a) Ficus spp.
(b) Bambax
(c) Salix
(d) All of the above

655. Naked root planting has been successful in:

(a) *Casuarinas equisetifolia*
(b) *Alnus nitida*
(c) *Juglans regia*
(d) All of the above

656. Which is not a tending operation?

(a) Weeding
(b) Soil working
(c) Thinning
(d) Cleaning

657. Which of the following is not a harvesting system?

(a) Coppice cutting
(b) Clear cutting
(c) Selective cutting
(d) Thinning

658. Removing a part of the stand to promote faster growth of the remaining stand is called _______.

(a) Thinning
(b) Release cutting
(c) Improvement cutting
(d) Under story cutting

659. French thinning is otherwise called as:

(a) Ordinary thinning
(b) Crown thinning
(c) Free thinning
(d) Advance thinning

660. Maximum thinning was developed by:

(a) Heck (b) Gerhardt

(c) Craib (d) Glover

661. Germen thinning is otherwise called as:

(a) Ordinary thinning (b) Crown thinning

(c) Free thinning (d) Advance thinning

662. The whole set of thinning carried out in a crop for the earliest to maturity:

(a) Thinning cycle (b) Thinning regime

(c) Thinning intensity (d) Thinning

663. Thinning in young timber allow th timber grower to:

(a) Maintain high rates of growth (b) Concentrate growth on the better trees

(c) Maintain a properly stocked forest (d) All of the above

664. The thumb rule method adopted for deodar as per Gloven formula is:

(a) d=aD+b (b) d=D

(c) d=1.5D (d) d=3D

665. Laurie's formula for mechanical thinning for irregular sal is:

(a) d=aD+b (b) d=D

(c) d=1.5D (d) d=3D

666. Free Thinning is otherwise called as:

(a) Single stem silviculture (b) Elite thinning

(c) Both (a) and (b) (d) None of the above

667. In which of these system the regeneration fellings are distributed over the whole felling series:

(a) Clear felling systems (b) Group systems

(c) Shelter wood systems (d) Selection systems

668. In which of the tree species uniform systems is followed:

(a) Teak (b) Bamboo

(c) Eucalyptus (d) Sal

669. Salix in Kashmir valley is worked under:

(a) Coppice with standard (b) Coppice with two rotation

(c) Pollard systems (d) Selection systems

670. Silvicultural systems followed for Bamboo is:

(a) Clear felling systems
(b) Shelter wood systems
(c) Group selection systems
(d) Culm selection systems

671. Silviculture system for road side plantation is:

(a) Simple coppice systems
(b) Shelter wood systems
(c) Clear strip systems
(d) None of the above

672. Teak was managed under:

(a) Selection systems
(b) Clear felling systems
(c) Shelter wood systems
(d) Group selection systems

673. Natural selection sal is best achieved under:

(a) Clear felling systems
(b) Coppice systems
(c) Shelter wood systems
(d) None of the above

674. Poor regeneration of sal can be attributed to:

(a) Seed dormancy
(b) Dying back of seedlings
(c) Orthodox nature of the seeds
(d) None of the above

675. A patch usually 20x20 m in size burnt and artificially regenerated in a felled coupe to supplement natural regeneration is known as:

(a) Rab
(b) Block
(c) Compartment
(d) None of the above

676. Teak is generally:

(a) Shade bearer
(b) Light demander
(c) Shade demander
(d) None of the above

677. Silvicultural system prevalent during early part of 20th century was:

(a) Shelter wood systems
(b) Clear felling systems
(c) Selection systems
(d) Coppice systems

678. The clear felling or clear strip system is suitable for:

(a) Shade demanders
(b) Shade bearer
(c) Light demander
(d) Light bearer

679. Example of clear felling system:

(a) Clear felling system
(b) Clear strip system
(c) Alternate strip system
(d) All of the above

680. Example of shelter wood system:

(a) Clear felling system
(b) Uniform system
(c) Selection system
(d) None of the above

681. The first stage of regeneration felling under a shelterwood system:

(a) Seeding felling
(b) Final felling
(c) Secondary felling
(d) Felling series

682. Example of clear-felling followed by artificial regeneration by taungyas:

(a) Sukna technique for sal
(b) Allapalli technique for teak
(c) Cryptomeria plantations of WB
(d) None of the above

683. The shelterwood coppice system is applicable where:

(a) Frost is common occurrence
(b) Locality is poor
(c) There is frequent fire
(d) None of the above

684. Coppice with standards system is applicable in:

(a) Sal forests
(b) Teak forests
(c) Moist deciduous forests
(d) None of the above

685. Reservation of old crops in C.W.R. is done:

(a) By area
(b) By species
(c) By trees
(d) All of the above

686. Pollard system is applied for:

(a) Jamun
(b) Hardwickia
(c) Teak
(d) None of the above

687. Example of accessory system:

(a) Alternate strip system
(b) The group system
(c) Selection system
(d) Improvement fellings

688. Example of coppice system:

(a) Clear strip system
(b) Group system
(c) Selection system
(d) Pollard system

689. Regeneration felling includes:

(a) Seeding felling
(b) Secondary felling
(c) Final felling
(d) All of the above

690. Important consideration for application of clear felling system is:

(a) Factors of locality
(b) Potential productivity of site
(c) Species and composition of crop
(d) All of the above

691. Example of clear felling followed by natural regeneration:

(a) Natural regeneration by seed stored in the area
(b) Natural regeneration by seed received from outside

(c) Natural regeneration by seed from advance growth
(d) All of the above

692. Factors affect the length of regeneration period in uniform system:

(a) Soil conditions
(b) Fire
(c) Grazing and browsing
(d) All of the above

693. Operations carried out in Indian irregular shelterwood system in addition to prescribed cultural operations:

(a) Burning
(b) Soil working
(c) Weeding and shrub cutting
(d) All of the above

694. The major advantage of selection system over other systems are:

(a) More growing stock
(b) Producing large sized trees
(c) Produces of forest of aesthetic and scenic values
(d) All of the above

695. The height of stumps in simple coppice system for Eucalyptus hybrid is:

(a) 5-10cm
(b) 10-15cm
(c) 15-20cm
(d) 25-35cm

696. The main objective of the coppice of two rotation system:

(a) Large sized timber
(b) Soil conservation
(c) To protect young regeneration from frost
(d) None of the above

697. The shelterwood coppice system is applicable where there is:

(a) Good locality
(b) High rainfall areas
(c) Poor soil
(d) None of the above

698. Coppice with standards system is applicable in:

(a) Jamun belts
(b) Bamboo plantations
(c) Teak plantations
(d) None of the above

699. Example of Reservation of old crop by species in C.W.R. is:

(a) Teak
(b) Jamun
(c) Eucalypts
(d) None of the above

700. Pollard system is applied for:

(a) Salix
(b) Teak
(c) Jamun
(d) None of the above

701. Length of conversion period depends on:

(a) Sacrifice of immature crop

(b) Proportion of over mature growing stock with negative increment

(c) Both (a) and (b)

(d) None of the above

702. The silvicultural stem in which regeneration fellings are distributed over the whole felling series and the crop is always irregular is known as:

(a) Systems of diffused regeneration (b) Accessory system

(c) System of concentrated regeneration (d) Coppice system

703. The alternate strip system was experimentally applied to:

(a) *Cedrus deodara* (b) *Shorea robusta*

(c) *Acacia auriculiformis* (d) *Pinus roxburghii*

704. The parts of forest set aside to be regenerated or otherwise treated during a specified period is kknown as:

(a) Regeneration period (b) Periodic block

(c) Felling series (d) Annual coupe

705. The shelterwood coppice system is a modification of:

(a) Simple coppice system (b) The coppice of two rotation system

(c) Coppice selection system (d) The coppice with reserve system

706. Most suitable Silvicultural system o deodar in hilly terrain for control of soil erosion is:

(a) Selection system (b) Clear felling

(c) Coppice system (d) All of the above

707. Artificially, *Terminalia belerica* can be raised by:

(a) Direct sowing (b) Entire plants

(c) Container seedlings (d) All of the above

708. The pattern of felling to be adopted in harvesting a matured crop to be regenerated is depends of _______ requirements of the species:

(a) Mode of regeneration (b) Type of crop

(c) Tending operation (d) Character of crop

709. The form or character of the new crop produced in any Silvicultural systems depends on _______.

(a) Spacing (b) Age

(c) Light (d) Moisture

710. High forest systems are those Silvicultural systems in which the regeneration is normally of _______.

(a) Coppice (b) Cutting

(c) Seed origin (d) Root suckers

711. In _______ systems equal or equi productive of matured crop are successively clear felled in one operation:

(a) Clear felling system (b) Uniform system

(c) Group system (d) Shelter wood strip systems

712. High forest with reserves system is categorized under _______ systems:

(a) Clear felling system (b) Shelter wood systems

(c) Selection systems (d) Accessory systems

713. Opening the canopy of a matured stand to provide conditions for securing regeneration from the retained seed trees is called _______.

(a) Seeding felling (b) Secondary felling

(c) Final felling (d) Felling cycle

714. Wagner's Blender Saumschlag is otherwise called as _______ systems.

(a) Strip selection system (b) Pujab shelter wood system

(c) Progressive clear strip system (d) Compartment system

715. In_______ system tree stem has been cut off in order to obtain a flush of shoots:

(a) Coppice systems (b) Clear felling system

(c) Selection systems (d) Shelter wood systems

716. Change from one Silvicultural system or one species to another is called _______.

(a) Coversion (b) Regeneration

(c) Rotation (d) Periodic block

717. Enucalyptus hybrid is worked with _______ system:

(a) Simple coppice systems (b) Coppice of two rotations

(c) Shelter wood coppice (d) Coppice with standard systems

718. Deodar forest is commonly worked with _______ system in Himachal Pradesh:

(a) Eberhard's wedge (b) Shelter wood strip

(c) Irregular shelter wood systems (d) Indian irregular shelter wood systems

719. Shelter wood/selection system is suitable for _______ bearer species:

(a) Shade (b) Light

(c) Temperature (d) All of the above

720. Forest growing on rocky and precipitous slopes are worked with _______ system.

(a) Shelter wood (b) Selection

(c) Clear felling (d) Uniform

721. *Dendrocalamus strictus* forest in Uttar Pradesh and Madhya Pradesh are worked with _______ system:

(a) Selection cutting combine with cleaning and cultural operations (b) Selection

(c) Shelter wood (d) Coppice

722. In two storied high forest system, the crop in each storey is approximately _______ aged.

(a) Even (b) Uneven

(c) Both (a) and (b) (d) None of the above

723. Group selection system is categorized under system of _______ regeneration:

(a) Diffused (b) Concentrated

(c) Both (a) and (b) (d) None of the above

724. The interval between seeding felling and the final felling is called _______ interval:

(a) Seeding felling (b) Secondary felling

(c) Final felling (d) Regeneration interval

725. The time which elapses between successive main fellings on the same are is called _______.

(a) Felling cycle (b) Final felling

(c) Seeding felling (d) Secondary felling

726. In culm Selection System, the felling cycle will be _______.

(a) 4 years (b) 6 years

(c) 8 years (d) 10 years

727. *Acacia catechu* forests of Himachal Pradesh in worked with _______ system:

(a) Simple coppice (b) Shelter wood coppice

(c) Coppice of two rotation (d) Coppice selection

728. Chipko movement is an example of forest conservation through:

(a) Tehri-Garhwal district (b) Uttaranchal

(c) Public awareness and participation (d) Political issue

729. Chipko movement is related to:

(a) Forest conservation
(b) Soil conservation
(c) Water conservation
(d) Wetland conservation

730. Chipko movement was launched for protection of:

(a) Forests
(b) Grass lands
(c) Wet lands
(d) Live stocks

731. Conservation of species in its natural habitat is:

(a) *In situ*
(b) *Ex situ*
(c) *In vitro*
(d) Both (b) and (c)

732. Deforestation brings above:

(a) Soil erosion
(b) Weed control
(c) Decreases drought
(d) Increased sunlight

733. Deforestation is caused due to:

(a) Silviculture
(b) Construction of roads
(c) Rainfall
(d) Plantation of trees

734. Deforestation is the major causal agent of:

(a) Depletion of natural resources
(b) Environmental pollution
(c) Desertification of habitat
(d) Genetic erosion

735. Deforestation will decrease:

(a) Soil erosion
(b) Land slides
(c) Soil fertility
(d) Rainfall

736. Development of botanical garden is what type of conservation:

(a) *Ex situ*
(b) A common
(c) An easy
(d) *In situ*

737. Extensive planting of trees to increase forest cover is called:

(a) Deforestation
(b) Agroforestry
(c) Social forestry
(d) Afforestation

738. For mapping, Remote sensing equipment is mounted on:

(a) Aircraft and satellite
(b) Ship and spacecraft
(c) Satellite and spacecraft
(d) All of the above

739. Forests control drought by:

(a) Preventing soil erosion
(b) Increasing oxygen
(c) Increasing humidity and rainfall
(d) Preventing floods

740. Fresh water present on earth is:

(a) 97.5 per cent
(b) 0.01 per cent
(c) 2.5 per cent
(d) 1.97 per cent

741. Fresh water problem arises due to:

(a) Globally available stocks are insufficient
(b) Uneven distribution on earth
(c) No regeneration of required quantity through natural hydrological cycle
(d) Any one of above

742. In India, common type of forest is:

(a) Tropical thon forests
(b) Sal and Teak forests
(c) Tropical most deciduous forest
(d) Tropical dry deciduous forest

743. The silvicultural system, group system was developed by:

(a) Gayer
(b) Van rich cotta
(c) Emberhard
(d) Thevor

744. The silvicultural system in which regeneration fellings instead of being done uniformly all over the compartment are carried out in scattered group is defined as:

(a) Uniform systems
(b) Selection systems
(c) Shelter wood strip systems
(d) Group systems

745. Which system is called as Andaman's canopy Lifting Shelther wood system?

(a) Selection systems
(b) Indian irregular shelter wood systems
(c) Group systems
(d) Uniform systems

746. Sivicultural systems in which clear felling are done in the form of strips which progress successively in one direction is known as:

(a) Alternate strip system
(b) Clear strip system
(c) Shelter wood strip systems
(d) Clear felling with standards

747. Deodar forests of Uttar Pradesh and Himachal Pradesh are generally worked under:

(a) Clear strip system
(b) Indian Irregular Shelthwood system
(c) Coppice system
(d) Selection system

748. Center for International Forestry Research (CIFR) is in:

(a) Nairobi, Kenya (b) Vienna, Austria

(c) Bogor, Indonesia (d) Ibadan, Nigeria

749. Which of the silvicultural system is used for the regeneration and management of high forests:

(a) Clear felling (b) Shelter wood

(c) Selection felling (d) All of the above

750. The silvicultural system followed for black wattle (*Acacia mearnsii*) in the Nilgiris is:

(a) Clear felling followed by natural regeneration (b) Clear felling followed by artificial regeneration

(c) Uniform system (d) Simple coppice system

751. Blue colour in the regeneration map represents:

(a) Excellent regeneration (b) Good regeneration

(c) Moderate regeneration (d) Fair regeneration

752. Red colour in the regeneration map represents:

(a) Excellent regeneration (b) Good regeneration

(c) Moderate regeneration (d) Fair regeneration

753. Sandal wood forest is worked under:

(a) Group selection (b) Natural selective systems

(c) Group system (d) Indian Irregular Shelthwood system

754. Silvicultural systems prevalent during early part of 20th century was:

(a) Shelthwood system (b) Clear felling system

(c) Selection systems (d) Coppice systems

755. Clear felling or clear strip system is suitable for:

(a) Shade demanders (b) Shade bearers

(c) Light demanders (d) Light bearers

756. Rose wood tree includes:

(a) Blackwood rosewood (b) Indian rorewood

(c) Malabar black wood (d) All of the above

Answer Sheet

1.	(a)	31.	(d)	61.	(c)	91.	(c)	121.	(a)
2.	(a)	32.	(a)	62	(b)	92.	(c)	122.	(d)
3.	(a)	33.	(c)	63.	(a)	93.	(d)	123.	(c)
4.	(a)	34.	(c)	64.	(a)	94.	(d)	124.	(d)
5.	(b)	35.	(c)	65.	(c)	95.	(a)	125.	(d)
6.	(c)	36.	(a)	66.	(d)	96.	(b)	126.	(c)
7.	(a)	37.	(c)	67.	(c)	97.	(c)	127.	(d)
8.	(c)	38.	(d)	68.	(b)	98.	(b)	128.	(b)
9.	(a)	39.	(c)	69.	(a)	99.	(b)	129.	(b)
10.	(b)	40.	(a)	70.	(a)	100.	(b)	130.	(c)
11.	(b)	41.	(a)	71.	(d)	101.	(d)	131.	(b)
12.	(b)	42.	(b)	72.	(b)	102.	(b)	132.	(a)
13.	(c)	43.	(c)	73.	(b)	103.	(b)	133.	(b)
14.	(b)	44.	(c)	74.	(a)	104.	(c)	134.	(c)
15.	(a)	45.	(c)	75.	(a)	105.	(a)	135.	(a)
16.	(d)	46.	(b)	76.	(c)	106.	(d)	136.	(b)
17.	(b)	47.	(b)	77.	(d)	107.	(c)	137.	(d)
18.	(c)	48.	(a)	78.	(b)	108.	(b)	138.	(b)
19.	(d)	49.	(a)	79.	(d)	109.	(b)	139.	(a)
20.	(d)	50.	(c)	80.	(c)	110.	(b)	140.	(a)
21.	(a)	51.	(d)	81.	(c)	111.	(b)	141.	(a)
22.	(c)	52.	(b)	82.	(d)	112.	(d)	142.	(a)
23.	(c)	53.	(a)	83.	(b)	113.	(d)	143.	(b)
24.	(d)	54.	(c)	84.	(d)	114.	(a)	144.	(d)
25.	(d)	55.	(d)	85.	(a)	115.	(d)	145.	(c)
26.	(d)	56.	(b)	86.	(c)	116.	(d)	146.	(d)
27.	(c)	57.	(c)	87.	(d)	117.	(d)	147.	(d)
28.	(b)	58.	(d)	88.	(d)	118.	(a)	148.	(c)
29.	(a)	59.	(a)	89.	(a)	119.	(b)	149.	(b)
30.	(a)	60.	(a)	90.	(b)	120.	(a)	150.	(a)

151.	(b)	185.	(a)	219.	(a)	253.	(b)	287.	(d)
152.	(c)	186.	(b)	220.	(d)	254.	(a)	288.	(b)
153.	(c)	187.	(a)	221.	(a)	255.	(a)	289.	(b)
154.	(b)	188.	(c)	222.	(c)	256.	(c)	290.	(d)
155.	(c)	189.	(c)	223.	(d)	257.	(a)	291.	(b)
156.	(d)	190.	(c)	224.	(b)	258.	(a)	292.	(c)
157.	(d)	191.	(b)	225.	(c)	259.	(b)	293.	(b)
158.	(a)	192.	(a)	226.	(a)	260.	(c)	294.	(d)
159.	(b)	193.	(a)	227.	(a)	261.	(a)	295.	(b)
160.	(a)	194.	(d)	228.	(c)	262.	(c)	296.	(c)
161.	(d)	195.	(b)	229.	(b)	263.	(a)	297.	(a)
162.	(a)	196.	(b)	230.	(c)	264.	(a)	298.	(c)
163.	(b)	197.	(d)	231.	(d)	265.	(b)	299.	(c)
164.	(c)	198.	(b)	232.	(d)	266.	(b)	300.	(c)
165.	(a)	199.	(b)	233.	(a)	267.	(a)	301.	(b)
166.	(c)	200.	(c)	234.	(d)	268.	(a)	302.	(a)
167.	(b)	201.	(c)	235.	(b)	269.	(a)	303.	(c)
168.	(d)	202.	(b)	236.	(d)	270.	(d)	304.	(d)
169.	(c)	203.	(c)	237.	(b)	271.	(c)	305.	(b)
170.	(d)	204.	(c)	238.	(a)	272.	(b)	306.	(a)
171.	(b)	205.	(a)	239.	(a)	273.	(b)	307.	(d)
172.	(c)	206.	(c)	240.	(a)	274.	(c)	308.	(c)
173.	(b)	207.	(c)	241.	(d)	275.	(a)	309.	(a)
174.	(a)	208.	(c)	242.	(a)	276.	(a)	310.	(b)
175.	(c)	209.	(a)	243.	(d)	277.	(c)	311.	(c)
176.	(d)	210.	(b)	244.	(c)	278.	(b)	312.	(b)
177.	(b)	211.	(b)	245.	(c)	279.	(a)	313.	(a)
178.	(d)	212.	(b)	246.	(a)	280.	(b)	314.	(a)
179.	(d)	213.	(a)	247.	(d)	281.	(a)	315.	(a)
180.	(c)	214.	(c)	248.	(b)	282.	(b)	316.	(b)
181.	(d)	215.	(d)	249.	(b)	283.	(b)	317.	(b)
182.	(b)	216.	(b)	250.	(a)	284.	(a)	318.	(b)
183.	(d)	217.	(a)	251.	(c)	285.	(d)	319.	(a)
184.	(c)	218.	(d)	252.	(b)	286.	(b)	320.	(b)

321.	(c)	355.	(a)	389.	(a)	423.	(c)	457.	(b)
322.	(c)	356.	(a)	390.	(d)	424.	(a)	458.	(a)
323.	(a)	357.	(a)	391.	(c)	425.	(b)	459.	(c)
324.	(a)	358.	(a)	392.	(a)	426.	(c)	460.	(a)
325.	(c)	359.	(a)	393.	(a)	427.	(d)	461.	(d)
326.	(a)	360.	(c)	394.	(a)	428.	(d)	462.	(c)
327.	(a)	361.	(a)	395.	(b)	429.	(d)	463.	(a)
328.	(b)	362.	(a)	396.	(a)	430.	(b)	464.	(a)
329.	(a)	363.	(a)	397.	(a)	431.	(b)	465.	(a)
330.	(a)	364.	(c)	398.	(a)	432.	(a)	466.	(a)
331.	(a)	365.	(b)	399.	(c)	433.	(a)	467.	(a)
332.	(b)	366.	(a)	400.	(a)	434.	(a)	468.	(b)
333.	(b)	367.	(a)	401.	(a)	435.	(b)	469.	(a)
334.	(a)	368.	(a)	402.	(a)	436.	(c)	470.	(b)
335.	(c)	369.	(a)	403.	(a)	437.	(d)	471.	(a)
336.	(a)	370.	(a)	404.	(a)	438.	(c)	472.	(a)
337.	(a)	371.	(a)	405.	(a)	439.	(d)	473.	(a)
338.	(b)	372.	(b)	406.	(d)	440.	(a)	474.	(c)
339.	(c)	373.	(a)	407.	(b)	441.	(b)	475.	(a)
340.	(a)	374.	(d)	408.	(a)	442.	(b)	476.	(d)
341.	(a)	375.	(b)	409.	(d)	443.	(a)	477.	(d)
342.	(b)	376.	(a)	410.	(d)	444.	(d)	478.	(c)
343.	(a)	377.	(a)	411.	(a)	445.	(a)	479.	(b)
344.	(a)	378.	(a)	412.	(a)	446.	(d)	480.	(a)
345.	(c)	379.	(a)	413.	(b)	447.	(d)	481.	(a)
346.	(a)	380.	(c)	414.	(d)	448.	(d)	482.	(a)
347.	(d)	381.	(a)	415.	(a)	449.	(a)	483.	(b)
348.	(b)	382.	(a)	416.	(a)	450.	(b)	484.	(a)
349.	(c)	383.	(a)	417.	(b)	451.	(a)	485.	(c)
350.	(c)	384.	(c)	418.	(a)	452.	(c)	486.	(a)
351.	(a)	385.	(b)	419.	(b)	453.	(a)	487.	(b)
352.	(c)	386.	(a)	420.	(d)	454.	(a)	488.	(c)
353.	(c)	387.	(a)	421.	(a)	455.	(a)	489.	(c)
354.	(a)	388.	(a)	422.	(d)	456.	(d)	490.	(a)

491.	(a)	525.	(a)	559.	(b)	593.	(b)	627.	(a)
492.	(d)	526.	(a)	560.	(a)	594.	(a)	628.	(a)
493.	(a)	527.	(a)	561.	(a)	595.	(a)	629.	(d)
494.	(b)	528.	(a)	562.	(c)	596.	(b)	630.	(d)
495.	(a)	529.	(a)	563.	(a)	597.	(c)	631.	(d)
496.	(a)	530.	(b)	564.	(a)	598.	(a)	632.	(d)
497.	(b)	531.	(d)	565.	(b)	599.	(a)	633.	(d)
498.	(a)	532.	(c)	566.	(a)	600.	(a)	634.	(a)
499.	(b)	533.	(a)	567.	(b)	601.	(a)	635.	(b)
500.	(c)	534.	(a)	568.	(c)	602.	(d)	636.	(a)
501.	(c)	535.	(c)	569.	(a)	603.	(c)	637.	(d)
502.	(a)	536.	(a)	570.	(c)	604.	(a)	638.	(a)
503.	(a)	537.	(a)	571.	(a)	605.	(a)	639.	(a)
504.	(a)	538.	(a)	572.	(c)	606.	(b)	640.	(b)
505.	(b)	539.	(b)	573.	(a)	607.	(d)	641.	(c)
506.	(a)	540.	(c)	574.	(a)	608.	(a)	642.	(c)
507.	(b)	541.	(a)	575.	(b)	609.	(b)	643.	(a)
508.	(a)	542.	(d)	576.	(c)	610.	(d)	644.	(a)
509.	(d)	543.	(a)	577.	(b)	611.	(a)	645.	(c)
510.	(d)	544.	(a)	578.	(b)	612.	(d)	646.	(d)
511.	(a)	545.	(a)	579.	(a)	613.	(a)	647.	(b)
512.	(a)	546.	(a)	580.	(d)	614.	(a)	648.	(a)
513.	(a)	547.	(b)	581.	(d)	615.	(b)	649.	(d)
514.	(d)	548.	(a)	582.	(a)	616.	(b)	650.	(a)
515.	(a)	549.	(a)	583.	(a)	617.	(a)	651.	(a)
516.	(a)	550.	(a)	584.	(a)	618.	(b)	652.	(d)
517.	(a)	551.	(c)	585.	(c)	619.	(a)	653.	(a)
518.	(d)	552.	(b)	586.	(a)	620.	(b)	654.	(d)
519.	(d)	553.	(a)	587.	(a)	621.	(c)	655.	(d)
520.	(c)	554.	(a)	588.	(a)	622.	(d)	656.	(b)
521.	(d)	555.	(a)	589.	(a)	623.	(b)	657.	(d)
522.	(d)	556.	(b)	590.	(b)	624.	(a)	658.	(a)
523.	(a)	557.	(c)	591.	(c)	625.	(a)	659.	(b)
524.	(d)	558.	(c)	592.	(a)	626.	(d)	660.	(b)

661.	(a)	681.	(a)	701.	(c)	721.	(a)	741.	(b)
662.	(b)	682.	(a)	702.	(a)	722.	(a)	742.	(d)
663.	(d)	683.	(a)	703.	(b)	723.	(a)	743.	(a)
664.	(b)	684.	(a)	704.	(b)	724.	(b)	744.	(d)
665.	(c)	685.	(d)	705.	(a)	725.	(a)	745.	(b)
666.	(c)	686.	(b)	706.	(a)	726.	(a)	746.	(b)
667.	(d)	687.	(d)	707.	(b)	727.	(d)	747.	(b)
668.	(d)	688.	(d)	708.	(a)	728.	(c)	748.	(c)
669.	(c)	689.	(d)	709.	(a)	729.	(a)	749.	(d)
670.	(d)	690.	(d)	710.	(c)	730.	(a)	750.	(a)
671.	(c)	691.	(d)	711.	(a)	731.	(a)	751.	(b)
672.	(d)	692.	(d)	712.	(d)	732.	(a)	752.	(d)
673.	(c)	693.	(d)	713.	(a)	733.	(b)	753.	(b)
674.	(b)	694.	(d)	714.	(a)	734.	(c)	754.	(b)
675.	(a)	695.	(c)	715.	(a)	735.	(d)	755.	(c)
676.	(b)	696.	(a)	716.	(a)	736.	(a)	756.	(d)
677.	(b)	697.	(a)	717.	(a)	737.	(d)		
678.	(c)	698.	(a)	718.	(d)	738.	(a)		
679.	(d)	699.	(b)	719.	(a)	739.	(c)		
680.	(b)	700.	(a)	720.	(b)	740.	(c)		

Chapter 3

Forest Mensuration

1. **Mensuration is derived from _______:**
   (a) Russia
   (b) Latin
   (c) Greek
   (d) Both (a) and (b)

2. **One hectare _______:**
   (a) 2.47 acres
   (b) 2.54 acres
   (c) 2.49 acres
   (d) 2.59 acres

3. **In metric system girth of a tree is always taken in:**
   (a) Metres and to the nearest centimeter
   (b) Foot and to the nearest inches.
   (c) Centimeters to the nearest millimeter
   (d) Only in foot

4. **The FPS systems enacted during the year:**
   (a) 1956
   (b) 1966
   (c) 1962
   (d) 1967

5. **The CGS or metric systems enacted during the year:**
   (a) 1956
   (b) 1966
   (c) 1962
   (d) 1967

6. **The diameter class (cm) followed in India:**
   (a) 1,2,5
   (b) 2,5,10
   (c) 1,3,5,
   (d) 3,5,10

7. The diameter class (cm) followed in India:

(a) 1,2,5 (b) 2,5,10

(c) 5,15,30 (d) 3,5,10

8. The height class (m) following in India are:

(a) 1,3,5 (b) 1,2,5

(c) 1,3,10 (d) 1,2,10

9. Height class followed for tree attaining a height of 25 m:

(a) 5 (b) 10

(c) 2 (d) 1

10. Which of the following is not correctly matched:

(a) Mana – Measure of volume (b) Tulamana – Measure of weight

(c) Avamana – Measure of diameter (d) Kalapramana – Measure of time

11. The country, that adopts 1.3 m as breast height:

(a) Burma (b) America

(c) Europe (d) India

12. Basal Area of a tree refers to:

(a) Cross sectional area at stem height (b) Cross sectional area at base

(c) Cross sectional area at top height (d) Cross sectional area at breast height

13. On sloping ground, the dbh should be measured on:

(a) Down hill side (b) Accessible side

(c) Up hill side (d) None of the above

14. The maximum spread of the crown along its widest diameter is:

(a) Crown length (b) Crown height

(c) Crown width (d) Crown diameter

15. Relationship between the diameter and girth of trees is:

(a) 1/______ (b) 0.3182

(c) 7/22 (d) All of the above

16. Volume of the log calculated using the formula:

(a) g/4x1 (b) $(g^2/_4)$x1

(c) $(g/4)^2$x1 (d) None of the above

17. The instrument designed based on the principal of similar triangle:

(a) Christen Hypsometer
(b) Brandis Hypsometer
(c) Altimeter
(d) Abney level

18. Instrument used for measuring the height of the tree based on trigonometrical principles.

(a) Christen's hypsometer
(b) Smithies hypsometer
(c) Improvised calipers
(d) Brandis hypsometer

19. Abneys level is used for:

(a) Measurement of height
(b) Measurement of volume
(c) Measurements of hight and contour survey
(d) Past rate of growth of a tree

20. _______ instrument used for measure the basal area of the crop:

(a) Reloskop
(b) Abneys level
(c) Brandis hypsometer
(d) Improvised caliper

21. Pick the Newton's formula:

(a) $(S1+4Sm+S^2/6)x1$
(b) Smx1
(c) S1
(d) $S1+S^2/2x1$

22. Factors affecting the overall weight of a quantity of wood are:

(a) Density and moisture content
(b) Bank and foreign material
(c) Moisture content, Bank and foreign material
(d) All of the above

23. Pressler's increment borer is used to determine:

(a) Increment of a tree
(b) Height of a tree
(c) Volume of a tree
(d) Age of a tree

24. Metzer's theory is related to:

(a) Stem form
(b) Form height
(c) Sampling
(d) None of the above

25. In slopping ground diameter of tree is measured on:

(a) Down hill side
(b) Uphill side
(c) Leaning side
(d) None of the above

26. Stump analysis is done to determine:

(a) Age and diameter relation
(b) Age and Height relation
(c) Age and Volume relation
(d) All of the above

27. The method of sampling adopted in Bamboo enumeration is:

(a) Multistage sampling (b) Multiphase sampling
(c) Random sampling (d) List sampling

28. The breast height adopted by most of the European countries is:

(a) 1.3m (b) 1.37m
(c) 1.27m (d) 3.0m

29. The normal size of the sample plot is:

(a) 0.1 to 0.2 ha (b) 0.1 to 0.5 ha
(c) 0.1 to 1.0 ha (d) 0.1 to 2.0 ha

30. The Instrument, which is used to measure angle of slope, is:

(a) Altimeter (b) Clinometers
(c) Hypsometer (d) Abney's level

31. Artificial form factor is also known as:

(a) Absolute form factor (b) Breast height form factor
(c) Normal form factor (d) Tree form factor

32. True form factor is also known as:

(a) Absolute form factor (b) Breast height form factor
(c) Normal form factor (d) Tree form factor

33. Crop diameter refers to:

(a) Mean diameter of tree in a stand (b) Mean diameter of Dominant trees
(c) Mean diameter of suppressed trees (d) None of the above

34. The standard of weights and measures act enacted during the year:

(a) 1956 (b) 1966
(c) 1957 (d) 1967

35. The product of form factor and total height of tree:

(a) Form quotient (b) Form class
(c) Tree Height (d) Form Height

36. Crop diameter is:

(a) Mean diameter of tree in a stand
(b) Diameter corresponding to the mean basal area of uniform crop
(c) Mean diameter of dominant trees
(d) None of the above

37. Basal area of tree implies:

(a) Cross sectional area at stump height
(b) Cross sectional area at breast height
(c) Cross sectional area at base.
(d) None of the above

38. Age of trees with annual rings determined by:

(a) Swedish bark gauge
(b) Wheeler penta prism
(c) Pressler's increment borer
(d) Blume-leiss hypsometer

39. Average annual increment for any short period is called as:

(a) CAI
(b) PAI
(c) MAI
(d) IP

40. The relationship between CAI/MAI (Pick the odd out):

(a) CAI is greater than MAI, the MAI is raising
(b) To begin MAI keeps below CAI
(c) CAI attains its maximum before MAI
(d) MAI is fully under CAI

41. Which instrument is used to determine the age of a tree:

(a) Abney's level
(b) Pressler's borer
(c) Relascope
(d) Altimeter

42. Effective variables for volume table preparation:

(a) Single variable
(b) Double variable
(c) Trible variable
(d) Multi variable

43. The volume tables are combined fom DBH alone is called:

(a) General volume tables
(b) Regional volume tables
(c) Local volume tables
(d) Standard volume tables.

44. Crop height is determined by:

(a) Andre's formula
(b) Heyer's formula
(c) Gumpel's formula
(d) Lorey's formula

45. Proportionate distribution of trees in selection forests is attributable to:

(a) Mayers law
(b) Faustman law
(c) De Liocourtslaw
(d) None of the above

46. Optimal number of sample plots for each thinning regime in yield table preparation is:

(a) 500
(b) 600
(c) 700
(d) 800

47. The formula for converting d.o.b. into d.u.b. is:

(a) G=g′-2_______t (b) G′=g-2_______h

(c) G=d′-2_______t (d) G=g′-2_______h

48. Smalian's formula for calculating volume of log is:

(a) SI (b) Smx1

(c) S_1+S_2/2x1 (d) S_1+S_2_______./S1S2/3x1

49. Calculate the basal area per ha if the number of tallied trees was 14 and the BAF of the wedge prism used is three:

(a) 42m^2 (b) 14 m^2

(c) 21 m^2 (d) None of the above

50. Normal form quotient is defined as:

(a) Mid diameter/d.b.h.

(b) Diameter at one half its height above breast height/d.b.h.

(c) Diameter at any point/d.b.h.

(d) Mid diameter x d.b.h.

51. Who coined the term form quotient?

(a) Schiffel (b) Tor-Jonson

(c) Schneider (d) Pressler

52. In Normal Form factor basal area is measured at:

(a) Breast height (b) Any Convenient height

(c) A constant proportion of the total height (d) None of the above

53. Who proposed the theory to explain variations in taper from tree to tree?

(a) Metzger (b) Huber

(c) Newton (d) Smalian

54. The instrument 'wedge prism' is mainly used in:

(a) Plot sampling (b) Horizontal point sampling

(c) Vertical point sampling (d) All of the above

55. One diopter is approximately equivalent to an angle of:

(a) 0.57° (b) 0.47°

(c) 0.75° (d) 0.74°

56. Linear growth model is preferred only for:

(a) Nursery stage (b) Sapling stage

(c) Pole stage (d) Tree stage

57. The standard length of a log expect the last section is:

(a) 5m (b) 4m

(c) 3m (d) 6m

58. If the canopy density is between 0.75 and 1.0, then the forest is referred as:

(a) Closed (b) Dense

(c) Thin (d) Open

59. Yield tables in which data are given for different grades of thinning:

(a) Volume yield table (b) Money yield table

(c) Single yield table (d) Multiple yield tables

60. Yield table is generally prepared for:

(a) Regular forest (b) Irregular forest

(c) Reserved forest (d) Mixed forest

61. The quarter girth formula to calculate volume of log is:

(a) $g^2 1/4$ (b) $4/g^2 \times 1$

(c) $g^2 1/16$ (d) $16/g^2 \times 1$

62. What are the variables require enough for general volume table preparation?

(a) Diameter only (b) Height only

(c) Diameter and height (d) Diameter, height and form

63. Sampling units are taken at two or more stages and sampling units gets smaller at each successive stage of sampling is called as:

(a) Unrestricted sampling (b) Stratified sampling

(c) Multistage sampling (d) Multiphase sampling

64. The height corresponding to the mean diameter of 250 biggest diameters per hectare as read from Diameter height curve is:

(a) Top height (b) Crop height

(c) Mean height (d) Mean basal area

65. The age diameter, age-height, age-volume relations throughout the life of the trees are analyzed by:

(a) Stump analysis (b) Stem analysis

(c) Pressler's increment borer (d) Mathematical formula

66. The instrument, which is used to measure diameter at any convenient height is:

(a) Tree caliper (b) Tailor's tape

(c) Wheeler's Penta Prism (d) Relaskop

67. Volume of stem from ground level down to 20 cm diameter over bark is:

(a) Commercial volume
(b) Standard stem small wood
(c) Standard stem timber
(d) Sawn out turn

68. In point sampling, P.P.S. is referred to as:

(a) Poly areal plot sampling
(b) Probability point sampling
(c) Probability plot sampling
(d) Probability proportional to size

69. Normally the figures of fractional quality running between the quality classes:

(a) 0.0 to 1.0
(b) 0.0 to 10
(c) 0.0 to 0.2
(d) 0.0 to 20

70. Which one is not considered as yield table:

(a) Volume yield table
(b) Volume table
(c) Money yield table
(d) Stand table

71. It is a measure of the relative productive capacity of a site for a particular species is called as:

(a) Site quality
(b) Site index
(c) Canopy density
(d) Crown density

72. An instrument, which is particularly used for vertical point sampling is:

(a) Wedge prism
(b) Spiegal relaskip
(c) Telerelskop
(d) Conimeter

73. Generally temporary sample plots are used for:

(a) Enumeration survey
(b) Preparation of yield table
(c) Measurement of crop volume and increment
(d) Comparison of growth in different qualities of locality

74. Schneider's formula is used for determining:

(a) Height of the tree
(b) Growth of the tree
(c) Volume of the tree
(d) Increment percentage

75. Which sampling method is random sampling method:

(a) Selective sampling
(b) Systematic sampling
(c) Stratified sampling
(d) Sequential sampling

76. When the tree is forked above the breast height, it is counted as,:

(a) One tree

(b) Each fork treated as separate tree

(c) Tree counted as one or two depend upon place measurement

(d) Cannot be determined

77. Site index is the relation between:

(a) Age and site
(b) Age and diameter
(c) Age and height
(d) Age and volume

78. The DBH was measured in United Kingdom is at:

(a) 1.37m
(b) 1.30m
(c) 1.39m
(d) 1.35m

79. On sloping ground the DBH is measured:

(a) Down hill side
(b) Uphill side
(c) Both up and down hill side
(d) None of the above

80. The figures obtained by Remote sensing visualizes facts by:

(a) Orographic view
(b) Orthographic view
(c) Perspective view
(d) Perpendicular projection

81. Normal form factor also called:

(a) True form factor
(b) Absolute form factor
(c) Artificial form factor
(d) Breast height form factor

82. In bamboo forest, the system of sampling adopted was:

(a) Multistage sampling
(b) Multiphase sampling
(c) Sequential sampling
(d) Random sampling

83. Solid volume of fire wood is measured by:

(a) Smalians formula
(b) Hubers formula
(c) Hunden hagen formula
(d) None of the above

84. Bole surface area calculated with the following parameters:

(a) Girth and length of log

(b) Length and Breadth of log

(c) Girth and length with form substitutes

(d) Both (a) and (b)

85. Horizontal projection on the ground of the tree crown is called:

(a) Crown cover
(b) Crown width
(c) Crown length
(d) Tree width

86. Vertical point sampling is attributable to:

(a) Bitterlich (b) Hirata

(c) Avery (d) Chaturvedi

87. Who introduced the horizontal point sampling technique?

(a) Hirata (b) Hirata

(c) Bitterlich (d) Smalin

88. The instrument 'wedge prism' mainly used in:

(a) Plot sampling (b) Horizontal point sampling

(c) Vertical point sampling (d) All of the above

89. If the canopy density is between 0.75 and 1.0, t hen the forest is referred as:

(a) Closed (b) Dense

(c) Thin (d) Open

90. The branch of forestry which deals with the determination of dimension, from and age and increment of single trees, stands or whole woods either standing or after felling is called:

(a) Dendrology (b) Forest Management

(c) Dendrometry (d) Forest Protection

91. What are the variables require enough for general volume table preparation?

(a) Diameter only (b) Height only

(c) Diameter and height (d) Diameter, height and form

92. Sampling units are taken at two or more stages and sampling units gets smaller at each successive stage of sampling is called as:

(a) Unrestricted sampling (b) Stratified sampling

(c) Multistage sampling (d) Multiphase sampling

93. If the focal length of the camera is 20 cm and the flying height above the ground level is 5000 m, calculate the photo scale:

(a) 1:10000 (b) 1:15000

(c) 1:20000 (d) 1:25000

94. The following are some clinomete5rs used in leveling. Pick the odd man out.

(a) Ghat tracer (b) Topographic Abney level

(c) De Lisle clinometer (d) Tangent clinometers

95. Hyposometric leveling is a method to find the:

(a) Height of mountains (b) Depth of water bodies

(c) Altitudes (d) Height of trees

96. Importance of Value Index (IVI) is used for:

(a) Estimation of biomass
(b) Mapping of forests
(c) Phyto-sociology of forests
(d) Survey of vegetation

97. Definition of 'Productivity' is:

(a) Biomass produced in one ha
(b) Biomass produced in one acre
(c) Biomass produced in one ha per year
(d) Biomass produced after one rotation

98. Volume of stem wood in round between 20 cm and 5 cm diameter over bark is called:

(a) Commercial volume
(b) Standard stem small wood
(c) Standard stem timber
(d) Sawn out turn

99. Volume of stem from ground level down to thin end of diameter over bark is:

(a) Commercial volume
(b) Standard stem small wood
(c) Standard stem timber
(d) Sawn out turn

100. The Latin name for Andaman Padauk tree is:

(a) *Veteria indica*
(b) *Pterocarpus dalbergioides*
(c) *Shorea robusta*
(d) *Dalbergia latifolia*

101. The butt end of a tree resembles frustum of:

(a) Trapezium
(b) Paraboloid
(c) Cone
(d) Neiloid

102. Top portion of tree resembles frustum of:

(a) Neiloid
(b) Paraboloid
(c) Cone
(d) Trapezium

103. Middle portion of tree resembles frustum of:

(a) Neiloid
(b) Paraboloid
(c) Cone
(d) Trapezium

104. Measure of the value of the angle expressed as its Sine is termed as:

(a) Diopter
(b) BAF
(c) Calibration distance factor
(d) Plot radius factor

105. The sample plot with minimum perimeter for a given area is:

(a) Square
(b) Rectangular
(c) Triangular
(d) Circular

106. Difficulty in applying yield tables to natural forests is:

(a) Lack of a representative age (b) Uniform site quality

(c) Over stocking (d) Over matured crop

107. Spectral signature of green vegetation in a satellite captured image is:

(a) White (b) Green

(c) Blue (d) Red

108. Vertical point sampling is used for estimating:

(a) BA/Ha (b) No of stems/Ha

(c) Mean stand height (d) All of the above

109. Panchromatic camera of IKNOS staellite has the highest ground resolution of:

(a) 1 m (b) 1.5m

(c) 5.8 m (d) 8.5 m

110. Stump analysis is used to assess the relation of:

(a) Age and growth (b) Diameter and volume

(c) Height and diameter (d) Height, diameter and volume

111. Tree caliper will give:

(a) Accurate measurement (b) Consistence measurement

(c) Absolute measurement (d) None of the above

112. Christens hypsometer is used to measure the tree:

(a) Volume (b) Taper

(c) Height (d) None of the above

113. Which of the following is correctly matched:

(a) Clinometers-density (b) Altimeter-atmospheric pressure

(c) Hypsometer-height (d) Relaskop-light intensity

114. Indian Institute of Remote Sensing is located at:

(a) Dehradun (b) Bangalore

(c) New Delhi (d) Chandigarh

115. The ideal time for taking Aerial photograph is:

(a) 9 to 11 am (b) 11 to 12 pm

(c) 2 to 3 pm (d) 3 to 4 pm

116. Metzgeris Theory or Girder theory discusses:

(a) Tree stem form (b) Tree volume
(c) Tree height (d) Species density

117. Horizontal Point is used for determination of:

(a) Basal area for hectare (b) Number of stems per hectare
(c) Volume per hectare (d) All of the above

118. Stem analysis is done to determine:

(a) Age of a tree (b) Height of a tree
(c) Volume of a tree (d) Past rate of growth of a tree

119. The rotation at which MAI is maximum:

(a) Rotation of height income (b) Financial rotation
(c) Rotation of maximum volume production (d) All of the above

120. In remote sensing, the scale for small scale photography ranges between:

(a) 1:40000 and 1:70000 (b) 1:20000 and 1:40000
(c) 1:5000 and 1:20000 (d) None of the above

121. The Forest Survey of India (FSI) conducts Forest Cover Survey once in every:

(a) 2 years (b) 5 years
(c) 10 years (d) 4 years

122. Small scale photograph has scale of:

(a) 1:40000 and 1:70000 (b) 1:20000 and 1:40000
(c) 1:5000 and 1:20000 (d) 1:50000 to 1:100000

123. The productive capacity of Indian Forest is:

(a) 2.1 m3/ha (b) 0.7 m3/ha
(c) 1 m3/ha (d) 0.5 m3/ha

124. The instrument not used in horizontal point sampling is:

(a) Simple angle gauge (b) Dumpy level
(c) Spiegel releskope (d) Wedge prism

125. Form factor is worked out using the formula:

(a) F=V/Sh (b) F=Sh/V
(c) F=V/S (d) F=V/h

126. The diameter corresponding to 250 biggest diameters per hectare is called:

(a) Crop diameter (b) Mean diameter
(c) Top diameter (d) Average diameter

127. The term normal form quotient is attributable to:

(a) Tor Jonson (b) Schiffel
(c) Hojer (d) Behre

128. Instrument used measuring the height of the tree based on trigonometical principles:

(a) Christen's hypsometer (b) Smithies hypsmeter
(c) Improvised calipers (d) Brandis hypsometer

129. Wooden scale is used to measure:

(a) Diameter of standing tree (b) Diameter of felled tree
(c) Girth of standing tree (d) Girth of felled tree

130. Which instrument is used to measure diameter at any height of a tree?

(a) Abney's level (b) Christen's hypsometer
(c) Tree caliper (d) Wheeler's pentaprism

131. Solid volume of the stack can be determined by:

(a) Timber calculators (b) Xylometric method
(c) Specific gravity method (d) None of the above

132. Local volume table are applicable to:

(a) Working circle (b) Block
(c) Compartment (d) Range

133. Girth is equal to:

(a) _______.d2 (b) _______d
(c) 2_______ (d) _______d/4

134. The error involved in using caliper is:

(a) Positive (b) Negative
(c) Both (a) and (b) (d) None of the above

135. Give the formula for basal area:

(a) Ah/2 (b) _______$d^2/_2$
(c) _______$d^2/_4$ (d) _______ab^2

136. Horizontal point is use for determination of:

(a) Volume/ha (b) Basal area/ha
(c) Height (d) None of the above

137. Random sampling (tick the odd man):

(a) Simple (b) Stratified
(c) Multistage (d) Systematic

138. Non random sampling:

(a) Systematic (b) Selective
(c) Sequential (d) List

139. A table showing for a given species average weight of trees or logs for one or more dimensions like girth, height or DBH is termed as:

(a) Stand table (b) Money table
(c) Biomass table (d) Yield table

140. The limit of standard small wood:

(a) Down to 20 cm diameter over bark
(b) Down to 20 inch diameter under bark
(c) Down to 5 cm diameter under bark
(d) Down to 5 cm diameter over bark

141. A table showing the distribution of stems by diameter classes is referred as:

(a) Stand table (b) Local volume table
(c) Yield table (d) None of the above

142. The abney level is used to measure _______.

(a) Tree height (b) Dbh
(c) Board feet (d) All of the above

143. Calipers are used to measure _______.

(a) Tree height (b) Dbh
(c) Board feet (d) All of the above

144. A tool used to help read aerial photographs is the _______:

(a) Haga altimeter (b) Clinometers
(c) Stereoscope (d) Abney level

145. Which tool is used for timber cruising?

(a) Wedge prism (b) Increment borer
(c) Cant hook (d) Chain saw

146. The haga altimeter will measure. _______.

(a) Tree height (b) Dbh
(c) Basal area (d) Board feet

147. A log of marketable size which is made unmarketable because of rot, defect or poor form is called a _______.

(a) Slash (b) Cull

(c) Scratch (d) Defects timber

148. A log rule is a table or formula used to determine tree:

(a) Age (b) Diameter

(c) Height (d) Volume

149. A bark gauge measures. _______:

(a) Diameter of the tree trunk

(b) Ratio of bark to wood

(c) The decibels generated when a felled hits the ground

(d) Tree bark thickness

150. A wedge prism is used in a type of inventory called:

(a) Cluster sampling (b) Double sampling

(c) Plot sampling (d) Point sampling

151. On a topographic map which of the following is a dark, heavy line with an elevation figure?

(a) Contour index (b) Contour interval

(c) Contour line (d) Horizontal control station

152. The Biltmore stick is used to measure. _______:

(a) Tree height (b) Dbh

(c) Board feet (d) Land area

153. The area of the end of a log is called the _______ area.

(a) Basal (b) End

(c) Stump (d) Butt

154. To perform a basal area sample, a forester would use which of the following tools:

(a) Diameter tape (b) Prism

(c) Realkascope (d) All of the above

155. The diameter of trees is usually measured at:

(a) Breast height

(b) 4 and ½ feet above the ground

(c) Dbh

(d) All of the above

156. Which of the following is/are an advantage of prism?

(a) Fast (b) Simple to use properly

(c) Reasonable accurate (d) All of the above

157. DBH can be measured by all of the following tools except _______.

(a) Diameter tape (b) Biltmore stick

(c) Merit hypsometer (d) Calipers

158. Which of the following will NOT measure the height of a tree?

(a) Merit hypsometer (b) Abney level

(c) Clinometers (d) Log scale stick

159. Diameter measurements for sawlogs are taken at:

(a) The large end of the log (b) The log center

(c) The small end of the log (d) The stump height mark

160. When using the prism, a tree to be counted should be:

(a) All trees in the plot regardless of offset (b) Offset completely

(c) Partially offset (d) All of the above

161. The general term used to describe instruments used to measure the diameter of a tree is:

(a) Altimeter (b) Clinometers

(c) Dendrometer (d) Hypsometer

162. A hand held instrument used for measuring tree heights, slopes and vertical angles is a/an _______.

(a) Abney level (b) Clinometers

(c) Stereoscope (d) Merritt hypsometer

163. A wedge prism is used to measure the _______ of trees in a stand.

(a) Basal area (b) Board foot volume

(c) Diameter (d) Cubic foot volume

164. When performing a forest inventory with a wedge prism you are selecting trees by:

(a) Species

(b) Their dbh

(c) Their distance form you

(d) Using a table of random numbers to identify the trees

165. An increment borer is NOT used to do which of the following?

(a) Estimate the age of a tree

(b) Determine the growth rate of a tree

(c) Eliminate knots in lumber

(d) An increment borer does all of the above

166. Tree diameter growth is related to _______.

(a) Age of the tree

(b) Availability of moisture and nutrients

(c) Stand density

(d) All of the above

167. A _______ is used to estimate basal area per acre.

(a) Clinometers (b) Hypsometer

(c) Planimeter (d) Wedge prism

168. A Jacob staff is used with a _______ to increase accuracy of tree measurement readings.

(a) Diameter tape (b) Planimeter

(c) Relaskop (d) Stereoscope

169. How many square feet are there in an acre?

(a) 40,000 (b) 43,522

(c) 43,560 (d) 10,000

170. Which of the following is the most limiting factor in the growth and development of an individual tree?

(a) Nutrients (b) Sunlight

(c) Rainfall (d) Temperature

171. Foresters recognize which of the following as the basic management unit in a forest?

(a) Stand (b) Lot

(c) Section (d) None of the above

172. DBH is:

(a) Diameter at breast height

(b) 4.5 feet above the ground

(c) A standard accepted in forestry Worldwide

(d) All of the above

173. General volume tab les which are applicable to large geographical area are based on:

(a) One variable
(b) Two variable
(c) Three variable
(d) Four variable

174. Which of the following tools would NOT be used to determine tree height?

(a) Clinometers
(b) Abney level
(c) Hypsometer
(d) None of the above

175. Which is the closest to a hectare?

(a) 1 acre
(b) 2.5 acres
(c) 5.2 acres
(d) 7 acres

176. The radius of a one-fifth acre plot is:

(a) 52.7 feet
(b) 37.2 feet
(c) 27.5 feet
(d) 73.2 feet

177. Of the tools for measuring DBH, which is the most precise?

(a) Caliper
(b) Biltmore stick
(c) Diameter tape
(d) Carpenter's rule

178. Th scaling diameter for saw logs is normally measured at:

(a) The big end inside bark
(b) The big end outside bark
(c) The small end inside bark
(d) The small end outside bark

179. The term stumpage refers to:

(a) Stumps left after harvesting
(b) The diameter of trees measured 1 foot off the ground
(c) The value of trees standing in a woodlot
(d) A measure of tree volume

180. Which of the following instruments measures angles as well as direction?

(a) Clinometers
(b) Abney level
(c) Realkascope
(d) None of the above

181. An abandoned field changes into a mature forest over an extended period of time through a process called:

(a) Succession
(b) Dominance
(c) Recycling
(d) All of the above

182. Management of forests for multiple uses involves which of the following uses:

(a) Timber
(b) Recreation
(c) Water
(d) All of the above

183. A log rule is a table or formula used to determine tree:

(a) Age (b) Diameter

(c) Height (d) Volume

184. Basal area of a woodlot is given in what units?

(a) Square feet/acre (b) Cubic feet/acre

(c) Board feet/acre (d) Number of trees/acre

185. Merchantable height is:

(a) Commonly measured with a log scale

(b) Measured at the first fork in the tree depending on the species

(c) Measured to the minimum diameter based on products to be taken

(d) Not normally considered in the volume table

(e) Both (b) and (c)

186. The number of stems falls off from one diameter class to the next in geometrical progression, which means that the percentage reduction in the stem number from one diameter class to the next is constant. This is referred to as:

(a) Brasnett's Law (b) Knuchel's Law

(c) De Liocourt's Law (d) Osmaston's Law

187. Tick the odd man out among the following technical terms connected with increfment:

(a) Mean Annual increment (b) Increment percent

(c) Current annual percent (d) Terminal annual increment

188. If girth over bark (g) and bark thickness (t) are respectively 30 cm and 2 cm find out the girth under bark (g). Assume? To be 3.0:

(a) 6cm (b) 12 cm

(c) 18 cm (d) 36 cm

189. One of the following instruments used for tree height measurement is not based on trigonometric principle. Identify the same.

(a) Brandis hypsometer (b) Smythie's hypsometer

(c) Blume Leiss hypsometer (d) Abney's level

190. The following are some methods developed for volume measurement in the sample plot. Identify the odd man out.

(a) Huber's method (b) Urich's method

(c) Gumple's method (d) Hartig's method

191. The product of form factor and total height of tree is defined as:

(a) Form quotient (b) Farm class

(c) Form height (d) Form point ratio

192. Which one of the following is not a formula for tree form?

(a) Behre's formula (b) Hojer's formula

(c) Metzger's formula (d) Smalian's formula

193. Volume table based on diameter, height and form quotient are called as:

(a) Form class volume table (b) General volume table

(c) Standard volume table (d) Commercial volume table

194. The following are some of the terms for point sampling. Identify the odd man out:

(a) Angle count cruising (b) Pointless cruising

(c) Variable plot cruising (d) Polyplot cruising

195. The following are some kinds of volume tables. Identify the odd man out:

(a) Multiple yield table (b) Variable yield table

(c) Single yield table (d) Money yield table

196. Crown length is vertical measurement of the crown of a tree from the tip of the crown of a tree to the:

(a) Lowest green branch on the bole

(b) Lowest green branch forming green crown all around

(c) Lowest branch green or dead

(d) Pont halfway between the lowest green branch forming green crown all around and the lowest green branch on the bole

197. Which height is used for determine the volume of a tree crop?

(a) Mean height (b) Mean height or crop height

(c) Top height (d) Mean height or top height

198. Thickness of bark can be estimated by:

(a) Increment borer (b) Swedish bark gauge

(c) Relascope (d) Callipers

199. The volume table gives the volume in round wood down to variable stated thin end diameters is:

(a) Commercial volume table (b) Standard volume table

(c) General volume table (d) None of the above

200. The instrument used for measuring diameter at unreadable height of standing tree is:

(a) Tape
(b) Calipers
(c) Dendrometer
(d) All of the above

201. The error caused by the instrument, the user and the weather condition is called:

(a) Instrumental error
(b) Personal error
(c) Error due to observation
(d) Error due to leaning of trees

202. Pick the odd one out from the following volume tables classified based on the scope of application:

(a) General volume table
(b) Regional volume table
(c) Local volume table
(d) Sawn out turn table

203. When CAI is equal to MAI:

(a) Maximum average volume per unit area per annum is obtained
(b) Average volume per unit area is lowest
(c) The CAI is at its peak
(d) Mai is at its minimum

204. Which of the following gives the correct volume of standing trees?

(a) DBH alone
(b) DBH, Height and form factor
(c) DBH and Height
(d) Height and form factor

205. Which of the following statement is false?

(a) Volume estimate of individual trees
(b) Estimation of volume of a group of trees
(c) Study the laws of growth
(d) Study stem form of trees

206. Land cover classification using three levels of the scale of 1:100000 was developed by which of the following agency:

(a) IIRS
(b) ISRO
(c) NRSA
(d) RRSC

207. Remote sensing system which provide their own source of energy for illumination are known as:

(a) Optical sensors
(b) Active sensors
(c) Passive sensors
(d) Thermal sensors

208. Land surface feature has specific response to radiation and it is termed as:

(a) Refraction (b) Reflection

(c) Spectral signature (d) Radiation

209. Scattering which occurs when the atmosphere particles are about the same size s the wave length of the radiation:

(a) Mie scattering (b) Non-selective

(c) Rayleigh scattering (d) None of the above

210. IIRS is located at:

(a) Hyderabad (b) New Delhi

(c) Dehradun (d) Bangalore

211. Ability of the sensors to differentiate different wavelength is known as:

(a) Spatial resolution (b) Spectral resolution

(c) Radiometric resolution (d) Temporal resolution

212. Scanning system which collects the data over a range of wavelengths is called as:

(a) TM (b) MSS

(c) WIFS (d) None of the above

213. Radiation which emitted from the Sun is in accordance with:

(a) Kirchoff's law (b) Stefan-Boltzman law

(c) Wien's displacement law (d) None of the above

214. Which is the first Indian satellite?

(a) IRS-IC (b) IRS-IA

(c) IRS-1 (d) IRS-p1

215. Which of the following wavelength band is used for vegetational studies?

(a) Visible only (b) Visible and IR

(c) UV and IR (d) IR, UV

216. Relative brightness of objects is an image is referred to:

(a) Texture (b) Tone

(c) Pattern (d) None of the above

217. Which of the following is the recently launched satellite with 1m spatial resolution?

(a) IRS-IE (b) LANDSAT-7

(c) IKNOS (d) SPOT-III

218. Which of the following is very useful in detecting and mapping forest fires:

(a) Visible sensors
(b) Thermal sensors
(c) Microwave sensors
(d) None of the above

219. Aerial photograph which has been taken with optical axis of approximately perpendicular to the horizontal plane and has been deviation upto <40 is:

(a) Oblique photograph
(b) High oblique photograph
(c) Vertical
(d) Low oblique photograph

220. RS and GIS has application in:

(a) Biodiversity assessment
(b) Vegetation classification
(c) Wild life management
(d) All of the above

221. Large scale photograph has scale of:

(a) 1:40000 to 1:70000
(b) 1:20000 to 1:40000
(c) 1:50000 o 1:20000
(d) 1:50000 to 1:100000

222. SOI Topo sheet is a:

(a) Perspective projection
(b) Orthogonal projection
(c) 3D- Projection
(d) None of the above

223. NNRMS head quarters in India is located at:

(a) Hyderabad
(b) Bangalore
(c) Dehradun
(d) Ahmedabad

224. Agency for aerial photography in India is:

(a) NRSA
(b) IIRS
(c) RRSC
(d) ISRO

225. High oblique photographs have tilt of:

(a) >4
(b) >60
(c) >30
(d) <30

226. Enhanced Thematic Mapper (ETM) has:

(a) 7-bands
(b) 4-bandsw
(c) 3-bands
(d) 8-bands

227. The oldest application of RS technology is in:

(a) Agriculture
(b) Military
(c) Forestry
(d) All of the above

Answer Sheet

1.	(b)	31.	(b)	61.	(c)	91.	(c)	121.	(a)
2.	(a)	32.	(c)	62.	(d)	92.	(c)	122.	(a)
3.	(c)	33.	(d)	63.	(c)	93.	(a)	123.	(b)
4.	(a)	34.	(a)	64.	(a)	94.	(b)	124.	(b)
5.	(c)	35.	(d)	65.	(b)	95.	(c)	125.	(a)
6.	(b)	36.	(b)	66.	(c)	96.	(d)	126.	(c)
7.	(c)	37.	(b)	67.	(c)	97.	(c)	127.	(b)
8.	(a)	38.	(c)	68.	(d)	98.	(b)	128.	(d)
9.	(a)	39.	(b)	69.	(c)	99.	(a)	129.	(b)
10.	(c)	40.	(d)	70.	(b)	100.	(b)	130.	(d)
11.	(c)	41.	(b)	71.	(a)	101.	(d)	131.	(b)
12.	(d)	42.	(b)	72.	(d)	102.	(c)	132.	(c)
13.	(c)	43.	(c)	73.	(a)	103.	(b)	133.	(b)
14.	(c)	44.	(d)	74.	(d)	104.	(a)	134.	(c)
15.	(d)	45.	(c)	75.	(c)	105.	(d)	135.	(c)
16.	(c)	46.	(b)	76.	(a)	106.	(a)	136.	(b)
17.	(a)	47.	(a)	77.	(c)	107.	(d)	137.	(d)
18.	(d)	48.	(c)	78.	(b)	108.	(c)	138.	(d)
19.	(c)	49.	(a)	79.	(b)	109.	(a)	139.	(d)
20.	(a)	50.	(a)	80.	(c)	110.	(a)	140.	(a)
21.	(a)	51.	(a)	81.	(a)	111.	(a)	141.	(b)
22.	(d)	52.	(c)	82.	(b)	112.	(c)	142.	(a)
23.	(d)	53.	(a)	83.	(d)	113.	(c)	143.	(a)
24.	(a)	54.	(b)	84.	(a)	114.	(a)	144.	(c)
25.	(b)	55.	(a)	85.	(a)	115.	(a)	145.	(a)
26.	(d)	56.	(a)	86.	(b)	116.	(a)	146.	(a)
27.	(b)	57.	(c)	87.	(c)	117.	(d)	147.	(b)
28.	(a)	58.	(b)	88.	(b)	118.	(d)	148.	(d)
29.	(a)	59.	(d)	89.	(b)	119.	(c)	149.	(d)
30.	(b)	60.	(a)	90.	(c)	120.	(a)	150.	(d)

151.	(a)	167.	(d)	183.	(d)	199.	(a)	215.	(b)
152.	(b)	168.	(b)	184.	(a)	200.	(c)	216.	(b)
153.	(a)	169.	(c)	185	E	201.	(a)	217.	(c)
154.	(b)	170.	(c)	186.	(c)	202.	(d)	218.	(b)
155.	(d)	171.	(a)	187.	(d)	203.	(a)	219.	(c)
156.	(d)	172.	(d)	188.	(c)	204.	(b)	220.	(d)
157.	(c)	173.	(c)	189.	(b)	205.	(a)	221.	(c)
158.	(d)	174.	(d)	190.	(c)	206.	(c)	222.	(d)
159.	(a)	175.	(b)	191.	(c)	207.	(b)	223.	(a)
160.	(c)	176.	(a)	192.	(d)	208.	(b)	224.	(a)
161.	(d)	177.	(c)	193.	(a)	209.	(a)	225.	(c)
162.	(b)	178.	(a)	194.	(d)	210.	(c)	226.	(a)
163.	(a)	179.	(c)	195.	(b)	211.	(b)	227.	(b)
164.	(b)	180.	(d)	196.	(a)	212.	(b)		
165.	(c)	181.	(a)	197.	(a)	213.	(b)		
166.	(d)	182.	(d)	198.	(b)	214.	(b)		

Chapter 4

Forest Management

1. The practical application of the scientific, technical and economical principles of forestry is called:

(a) Forest management (b) Forest mensuration

(c) Forest economics (d) None of the above

2. Primarily a person engaged in the profession of forestry is called _______.

(a) Ranger (b) Forester

(c) Guard (d) All of the above

3. The basis for establishing a management plan for a forest is:

(a) Money (b) Access

(c) Objectives (d) Time

4. Reserved forest are constituted under the:

(a) Indian Forest Act, 1927 (b) National Forest Policy, 1988

(c) Forest Policy, 1952 (d) None of the above

5. British rulers reserved the forest for:

(a) Industrial needs (b) To protect nature

(c) To meet the needs of local people (d) None of the above

6. Forest Conservation Act was enacted during:

(a) 1880 (b) 1980

(c) 1988 (d) 1998

7. **The smallest permanent working plan unit of management:**
   (a) Block (b) Compartment
   (c) Annual coupe (d) Periodic block

8. **_______ is the head of the forest administration in India:**
   (a) PCCF (b) DFO
   (c) Inspector General of Forest (IGF) (d) Ranger

9. **_______ is a very import ant unit in the management and administration of forest:**
   (a) Range (b) Beat
   (c) Circle (d) Division

10. **_______ is the smallest functional territorial unit and is the foundation of Indian forest administration:**
    (a) Beat (b) Range
    (c) Circle (d) Blocks

11. **_______ is the main territorial division of forest:**
    (a) Compartment (b) Block
    (c) Periodic block (d) Coupe

12. **Based on territorial classification which forest is temporary nature?**
    (a) Sub-compartment (b) Periodic block
    (c) Coupe (d) None of the above

13. **Who is the kingpin of forest administration in India?**
    (a) District Forest Officer (DFO) (b) PCCF
    (c) Ranger (d) ACF

14. **_______ is the backbone of forest management:**
    (a) Even aged forest (b) Maximum sustained yield
    (c) Regular forest (d) All of the above

15. **What is a tool of forest management?**
    (a) Working plan (b) Working circle
    (c) Felling series (d) Coupe

16. **Unit of a working plan in India is:**
    (a) Forest division (b) Felling series
    (c) Cutting section (d) Coupe

17. **A written scheme of management aiming at continuity of policy, controlling the treatment of a forest is called:**
   (a) Working circle (b) Working plan
   (c) Felling series (d) Coupe

18. **Working circles divided into _______:**
   (a) Felling series (b) Division
   (c) Cutting section (d) Coupe

19. **Working plans are invariably based on _______ principle:**
   (a) Sustained yield (b) Progressive yield
   (c) All of the above (d) None of the above

20. **Felling series is divided into _______:**
   (a) Cutting section (b) Division
   (c) Periodic block (d) None of the above

21. **Felling series is divided into number of _______:**
   (a) Cutting section (b) Periodic block
   (c) Working circle (d) Annual coupes

22. **The material that a forest can yield annually or periodically in perpetuity is called:**
   (a) Sustained yield (b) Progressive yield
   (c) All of the above (d) None of the above

23. **The concept of progressive yield originally advocated by:**
   (a) Hartig (b) Smith
   (c) Von montle (d) Howards

24. **The concept of progressive yield envisages raising:**
   (a) Productivity of soil and crop (b) Only productivity of the soil
   (c) Only productivity of the crop (d) None of the above

25. **An age class with one year as the interval is called:**
   (a) Age class (b) Age gradation
   (c) All of the above (d) None of the above

26. **A forest area forming a part or whole of a working plan area is called:**
   (a) Division (b) Periodic block
   (c) None (d) Working circle

27. The period which a forest crop takes between its formation and final felling is known as:

(a) Rotation or production period
(b) Increment
(c) Conversion period
(d) None of the above

28. The term rotation strictly speaking is correctly applicable to:

(a) Regular crops only
(b) Irregular crop
(c) Both
(d) None of the above

29. Physical rotation is applicable to:

(a) Protection
(b) Amenity forests
(c) Road side avenues
(d) All of the above

30. Silvicultural rotation may be useful in forests managed primarily for:

(a) Aesthetic
(b) Recreational purpose
(c) Both (a) and (b)
(d) None of the above

31. Production of maximum material of specific dimensions or quality is the aim of:

(a) Technical rotation
(b) Financial rotation
(c) Silvicultural rotation
(d) None of the above

32. Pulp and paper industries favours adaptation for which type of rotation:

(a) Technical rotation
(b) Financial rotation
(c) Rotation of maximum volume production
(d) Silviculural

33. Which type of rotation yields largest volume per unit are per annum:

(a) Technical rotation
(b) Financial rotation
(c) Silviculural
(d) Rotation of maximum volume production

34. For production of firewood best rotation is:

(a) Rotation of Maximum volume
(b) Technical rotation production
(c) Physical rotation
(d) Financial rotation

35. Common practice in forestry is to adopt a combination of:

(a) Financial rotation and technical rotation
(b) Technical and rotation of highest income
(c) Maximum volume production and financial rotation
(d) None of the above

36. **From the overall national point of view which type of rotation is important:**
   (a) Technical rotation
   (b) Rotation of maximum
   (c) Silvicultural
   (d) Rotation of highest Income/ Revenue

37. **The rotation which is most profitable:**
   (a) Physical rotation
   (b) Technical rotation
   (c) Rotation of maximum volume production
   (d) Financial rotation

38. **Which type of rotation is useful for aesthetic and recreational purpose?**
   (a) Silvicultural rotation
   (b) Physical rotation
   (c) Technical rotation
   (d) Rotation of maximum volume production

39. **Normal forest under a given set of condition does not depend of this factor:**
   (a) A normal increment
   (b) A normal distribution of age classes
   (c) A normal distribution of tree species
   (d) A normal growing stock

40. **Maximum average volume per unit area per annum in Eucalyptus can be obtained at:**
   (a) MAI=CAI
   (b) CAI>MAI
   (c) CAI=PAI
   (d) MAI=PAI

41. **Who should draw up the working plan?**
   (a) Local DFO
   (b) Working plan officer
   (c) Conservator of forests
   (d) PCCF

42. **Faustamann/s formula is used for calculating:**
   (a) Soil expectation value
   (b) Sustained yield
   (c) Progressive yield
   (d) None of the above

43. **Length of rotation is affected by:**
   (a) Rate of growth
   (b) Silvicultural characteristics of species
   (c) Economic and social conditions
   (d) All of the above

44. **Which rotation controlling the output of material/forest products in form or quality?**
   (a) Silvicultural
   (b) Physical rotation
   (c) Both (a) and (b)
   (d) None of the above

45. Soil expectation value (Se) is ______.

(a) X/0.0p
(b) Faustmann's formula
(c) Both (a) and (b)
(d) None of the above

46. Which rotations controlling the financial returns 47:

(a) Silvicultural and physical rotation
(b) Rotation of maximum gross or net income and financial rotation
(c) Technical and maximum volume rotation
(d) All of the above

47. The concept of normal forest was first introduced by:

(a) German forester's
(b) American forester
(c) French forester
(d) None of the above

48. Which is the forest management is territorial division:

(a) Block
(b) Beat
(c) Compartment
(d) Coupe

49. The CVP index developed by Paterson indicates:

(a) Stand productivity
(b) Site productivity
(c) Commercial productivity
(d) None of the above

50. What is the percentage of forest owned by Indian government?

(a) 60
(b) 75
(c) 85
(d) 95

51. A favorable balance between timber growth and harvest (growth equals or exceeds harvest) is referred to as:

(a) Sustained yield timber growth
(b) Increment
(c) Growing stock
(d) None of the above

52. The volume or number of stems that can be removed annually or periodically is called:

(a) Yield
(b) Rotation
(c) Growing stock
(d) Increment

53. All material from thinning is called:

(a) Intermediate yield
(b) Final yield
(c) Yield capacity
(d) None of the above

54. The yield from a Normal forest is referred called:

(a) Normal forest
(b) Growing stock
(c) Abnormal forest
(d) None of the above

55. In Howard's modification crop is measured down to diameter corresponding to the rotation of:

(a) 3/8 (b) 2/3

(c) 1/3 (d) ½

56. Which of the forms basis for yield regulation in French method of 1883:

(a) Volume (b) Diameter

(c) Increment (d) All of the above

57. Von Montal's formula is best applicable to:

(a) Regular evenaged normal forest (b) Irregular unevenaged forest

(c) Abnormal forest (d) None of the above

58. Masson's formula is also known as:

(a) Exploitation percent (b) Masson's ratio

(c) Both (a) and (b) (d) None of the above

59. In which method of yield regulation, enumeration is carried out down to the diameter corresponding to half the rotation:

(a) Howard's formula (b) Masson's formula

(c) Von mantle formula (d) All of the above

60. French method of 183 is also called as:

(a) Quartier bleu method (b) Exploitation per cent

(c) U.P. safeguarding formula (d) All of the above

61. Which is taken as unit of W.P.M. in India?

(a) Working circle (b) Coupe

(c) Beat (d) Felling series

62. The scale of the stock maps is:

(a) 1:15000 (b) 1:10000

(c) 1:1500 (d) None of the above

63. The regeneration survey maps have scale of:

(a) 1:5000 (b) 1:4000

(c) Both (a) and (b) (d) None of the above

64. The scale of working plan map is:

(a) 1:50000 (b) 1:40000

(c) 1:10000 (d) None of the above

65. The enumeration maps have scale of:

(a) 1:50000 (b) 1:20000

(c) 1:10000 (d) 1:5000

66. The scale of forest type maps is:

(a) 1:5000 (b) 1:20000
(c) 1:10000 (d) 1:50000

67. The term capital in forest management usually refers to:

(a) Growing stock (b) Sustained yield
(c) Progressive yield (d) All of the above

68. Normal growing stock in clear felling systems is:

(a) (I X r)/2 (b) [(I X r)/2] + (I/2)
(c) [(I X r)/2] - (I/2) (d) All of the above

69. Basic pre-requisites to achieve the objective of sustained yield is:

(a) Normal increment (b) Age gradations
(c) Growing stock (d) All of the above

70. Commonly used reducing factor in India is:

(a) Volume (b) Diameter
(c) Area (d) Area and volume

71. Term rotation is strictly applicable to:

(a) Regular crop only (b) Irregular crop
(c) Both (d) None of the above

72. The main objective of yield regulation is to achieve _______ from the forests:

(a) Sustained yield (b) Growing stock
(c) Increment (d) None of the above

73. Basis of yield regulation in Von Mantels formula is:

(a) Area (b) Increment
(c) Volume only (d) Area and volume

74. Von Mantels formula of yield regulation is:

(a) 2GS/r (b) 200/r
(c) 8V/3r (d) 9V/4r

75. Stock maps are prepared by working plan rangers with scale of:

(a) 1:15000 (b) 1:10000
(c) 1:20000 (d) 1:5000

76. The production period for any species is slightly longer on:

(a) Richer sites (b) Marginal sites
(c) Poorer sites (d) Both (b) and (c)

77. Normally felling cycle in selection forests is:

(a) 5 years (b) 10 years

(c) Both (a) and (b) (d) 5-10 years

78. Technological development have greater implications on:

(a) Sustained yield (b) Progressive yield

(c) Normal yield (d) None of the above

79. Standard length of a log except the last section is:

(a) 4.5 m (b) 3 m

(c) 1.5 m (d) 1 m

80. One cubic meter of timber in metric system is equal to:

(a) 39.000 cft in FPS system (b) 30.000 cft in FPS system

(c) 35.3147 cft in FPS system (d) None of the above

81. Recently formulated national forest policy of 1988 laid emphasis on:

(a) Maximum timber production (b) Sustainable management

(c) Joint forest management (d) None of the above

82. Crop diameter is:

(a) Mean diameter of trees in a stand

(b) Diameter corresponding to the mean basal area of a uniform crop

(c) Mean diameter of the dominant trees

(d) None of the above

83. JFM first implemented at:

(a) Jabalpur (b) Ayyalur

(c) Arabari (d) Jhansi

84. Lowest administrative unit:

(a) Beat (b) Coupe

(c) Compartment (d) None of the above

85. Which type of rotation is most favourable for natural reproduction:

(a) Silvicultural rotation (b) Physical rotation

(c) Technical rotation (d) Financial rotation

86. A term generally applied to the determination of the yield and the prescribed means realizing it is called:

(a) Financial yield (b) Yield regulation

(c) Yield maximization (d) None of the above

87. A material that a forest can yield annually in perpetuity:
(a) Total yield (b) Sustained yield
(c) Periodic yield (d) Intermittent yield

88. The concept of progressive yield envisage:
(a) Raising the productivity of soil (b) Judicious tending
(c) Appropriate silvicultual treatments (d) All of the above

89. Which social forestry programmes is purely managed by private efforts:
(a) Farm forestry (b) Community forestry
(c) Rehabilitation of degraded forest area (d) Urban forestry

90. Which forestry practice is not included in the scope of social forestry:
(a) Farm forestry (b) Extension forestry
(c) Recreational forestry (d) Protection forestry

91. The rotation followed for blue gum in the Nilgiris forest division was:
(a) 8 (b) 10
(c) 12 (d) 4

92. The smallest permanent working plan unit in India is:
(a) Coupe (b) Felling series
(c) Compartment (d) Circle

93. Type of rotation (pick the odd man out):
(a) Regular (b) Silvicultural
(c) Technical (d) Financial

94. The rotation followed for sandal wood in South India is:
(a) Physical rotation (b) Technical rotation
(c) Financial rotation (d) Silvicultural rotation

95. The smallest functional territorial unit in forest management is:
(a) Circle (b) Division
(c) Section (d) Beat

96. The period required to regenerate the whole of Periodic block is called as:
(a) Regeneration interval (b) Regeneration period
(c) Rotation period (d) Periodic block

97. TAP programme in Tamilnadu was funded by:
(a) Sweden (b) Canada
(c) Japan (d) USA

98. Enumeration maps are prepared in the scale of:

(a) 1:10000 (b) 1:15000

(c) 1:20000 (d) 1:50000

99. Howeard's formula for yield regulation is:

(a) 8v/3r (b) 9v/4r

(c) 6v/3r (d) 2gs/r

100. The rotation in which a species retains satisfactory vigour of growth and reproduction on a given site is:

(a) Silvicultural rotation (b) Financial rotation

(c) Rotation for maximum volume production (d) Technical rotation

101. The concept of progressive yield was attributable to:

(a) Osmaston (b) Hartig

(c) Nautiyal (d) None of the above

102. An uneven aged crop containing theoretically all age classes is said to be:

(a) Selection forest (b) Shelter wood forest

(c) Even aged forest (d) None of the above

103. The oldest forest act of Tamilnadu state is enacted during:

(a) 1894 (b) 1864

(c) 1882 (d) 1927

104. The total increment up to a given age divided by that age is called:

(a) Current annual increment (b) Periodic annual increment

(c) Mean annual increment (d) Final mean annual increment

105. Which of the following is correct?

(a) To begin with the CAI keeps below MAI

(b) The CAI attains the maximum before the MAI

(c) The MAI is falling when the CAI is still raising

(d) While the MAI is more than CAI, MAI is rising

106. The MAI at rotation age is called:

(a) Current annual increment (b) Mean annual increment

(c) Final mean annual increment (d) Periodic annual increment

107. Presslers formula to find out increment per cent is:

(a) V-v/V+vX200/n (b) V+v/V-vX200/n

(c) V-v/V+vXn/200 (d) V+v/V-vX n/200

108. Schneiders formula to find out increment per cent is:

(a) P= 200/nD (b) P=400/nD

(c) P=D/200n (d) P=400/D

109. The total volume of trees in a fully stocked forest with normal distribution of age classes for a given rotation is called:

(a) Normal growing stock (b) Growing stock

(c) Regular growing stock (d) Even growing stock

110. The yield from normal forest is called:

(a) Normal yield (b) Sustained yield

(c) Yield capacity (d) Final yield

111. The material that a forest can yield annually in perpetuity is called:

(a) Normal yield (b) Sustained yield

(c) Progressive yield (d) Fixed yield

112. Which among the following statements is incorrect regarding sustained yield concept?

(a) Local labour is fully employed

(b) Market can be developed by sustained supplies

(c) There is no wide fluctuation in the yield of wood

(d) Wood used industries have assured supply of raw material

113. How many types of rotation are recognized in forestry?

(a) 2 (b) 4

(c) 6 (d) 8

114. Howard's modification is based on:

(a) Volume only (b) Area only

(c) Area and volume (d) Increment

115. Increment method of yield regulation is based on:

(a) Area only (b) Volume only

(c) Area and volume only (d) Increment only

116. Brandis diameter class method of yield regulation also called as:

(a) Indian method (b) US method

(c) United kingdom method (d) Swiss method

117. GS=Ixrxc. Here c refers to:

(a) Adjustment factors (b) Flurry constant

(c) Swiss constant (d) Bavarian constant

118. Working plan map is generally prepared on the scale:

(a) 1:50000 (b) 1:60000

(c) 1:40000 (d) 1:25000

119. The first Inspector general of forest was:

(a) Sir Dietrich Brandis (b) Howard

(c) Chaturvedi (d) Alfred moller

120. The National Commission on Agriculture submitted its report during:

(a) 1970 (b) 1976

(c) 1967 (d) 1980

121. The crop composition 50-75 per cent of teak may be classified as:

(a) Pure teak (b) Mainly teak

(c) Mixed teak (d) Misc. with some teak

122. Pick the odd man out:

(a) Normal series of age gradations (b) Normal growing stock

(c) Normal distribution of quality classes (d) Normal increment

123. The time that elapses between the successive main fellings of the same area is called:

(a) Felling cycle (b) Felling series

(c) Felling interval (d) Regeneration fellings

124. The Office Incharge of Administrative Unit Section:

(a) Ranger (b) Forester

(c) Guard (d) Watcher

125. Legally the forest in the country are classified as:

(a) National forest and village forest

(b) Reserve forest and protected forest

(c) Farm forest and social forest

(d) Moisture deciduous and dry deciduous

126. The productivity of Indian Forest is:

(a) 2.1 m^3/ha (b) 0.5 m^3/ha

(c) 0.7 m^3/ha (d) 1.5 m^3/ha

127. It is a measure of the relative productive capacity of a site for a particular species is called as:

(a) Site quality (b) Site index

(c) Canopy density (d) Crown density

128. Site index is the relation between:

(a) Age and site
(b) Age and diameter
(c) Age and height
(d) Age and volume

129. Joint Forest Management was first introduced in:

(a) West Bengal
(b) Tamil Nadu
(c) Odisha
(d) Uttarakhand

130. The Working Plan Officer is:

(a) Deputy conservator of forests
(b) Conservator of forests
(c) Assistant conservator of forests
(d) Ranger

131. In working plan, tropical wet evergreen are coloured with:

(a) Violet
(b) Purple
(c) Yellow
(d) Green

132. In working plan, tropical thorn are coloured with:

(a) Violet
(b) Red
(c) Yellow
(d) Green

133. The first Independent National Forest Policy was enunciated during:

(a) 1950
(b) 1952
(c) 1954
(d) 1956

134. Forest and wildlife were brought under concurrent list as per:

(a) 40th amendment
(b) 42nd amendment
(c) 44th amendment
(d) 48th amendment

135. The Head Quarters of Inspector General of Forest:

(a) Dehradun
(b) Missouri
(c) Delhi
(d) Mumbai

136. The sum of all the trees growing in a forest is called:

(a) Selection forest
(b) Growing stock
(c) Normal forest
(d) None of the above

137. The MAI at rotation age is called:

(a) Current annual increment
(b) Mean annual increment
(c) Final mean annual increment
(d) Periodic annual increment

138. When CAI is equal to MAI:

(a) Maximum average volume per unit area per annum is obtained
(b) Minimum average volume per unit
(c) All of the above
(d) None of the above

139. CAI/MAI relationship (Pick the odd out):
(a) CAI is greater than MAI, the MAI is raising
(b) To begin MAI keeps below CAI
(c) CAI attains its maximum before MAI
(d) MAI is fully under CAI

140. Average annual increment for any short period q:
(a) CAI (b) PAI
(c) MAI (d) IP

141. Fast growing species is one which yields an increment of:
(a) $7m^3$/ha/yr (b) 10 m^3/ha/yr
(c) 15 m^3/ha/yr (d) 20 m^3/ha/yr

142. The rotation followed for teak in moist deciduous forest:
(a) 40 to 60 years (b) 20 to 40 years
(c) 60 to 80 years (d) 80 to 100 years

143. Simmons formulae is a based on the following method of yield regulation:
(a) Area (b) Increment
(c) Volume (d) All of the above

144. Schneider formula is to determined:
(a) Volume increment per cent (b) Height increment per cent
(c) Diameter increment per cent (d) Age increment

145. The first Forest Policy was enunciated during:
(a) 1864 (b) 1874
(c) 1884 (d) 1894

146. Functionally forests are classified into:
(a) Two types (b) Three types
(c) Four types (d) None of the above

147. Working circle is a classification based on:
(a) Administrative (b) Territorial
(c) Management (d) Functional

148. Third World Forestry Congress was held in:
(a) Dehradun (b) Helsinki
(c) Delhi (d) Kyoto

149. Masson's formula is based on:

(a) Volume (b) Increment
(c) Volume and increment (d) Diameter

150. Symthies safe guarding formula is based on:

(a) Volume (b) No of trees
(c) Volume and increment (d) Diameter

151. Schneider's formula is used for determining:

(a) Increment percentage (b) Diameter
(c) Volume (d) None of the above

152. Which one of the following is correctly matched?

(a) CAI- It is the increment that a tree puts on from origin upto that age at which the tree is cut
(b) PAI- It is the average annual increment for any short period
(c) MAI- It is the increment which a tree puts on a single year
(d) TAI- It is the total increment upto a given age divided by that age

153. Legally the forest in the country are classified as:

(a) Reserve and protected forest (b) Protection and social forest
(c) Production and village forest (d) None of the above

154. Joint forest management was first introduced in:

(a) West Bengal, 1988 (b) Tamil Nadu, 1988
(c) Karnataka, 1988 (d) Gujarat, 1988

155. The working plan officer is:

(a) Deputy conservator of forest (b) Conservator for forest
(c) District forest officer (d) Chief conservator of forest

156. In working plan, tropical wet evergreen are coloured with:

(a) Violet (b) Red
(c) Blue (d) Black

157. In working plan, tropical thorn forest are coloured with:

(a) Purple (b) Yellow
(c) Blue (d) Red

158. In working plan, Himalayan dry temperate forest are coloured with:

(a) Purple (b) Yellow
(c) Blue (d) Red

159. Average annual increment for any short period:

(a) MAI (b) PAI
(c) MAI and PAI (d) None of the above

160. The rotation at which MAI is maximum:

(a) Rotation of highest income/revenue
(b) Rotation of maximum volume production
(c) Financial rotation
(d) All of the above

161. A change of one Silvicultural system or one species to another is called _______:

(a) Conversion (b) Conversion period
(c) Rotation (d) All of the above

162. The period during which a change from one silvicultural system to another is called_______.

(a) Conversion period (b) Rotation
(c) Both (a) and (b) (d) None of the above

163. Who is called the Trinity of Norms for normal forest?

(a) Osmaston (b) Sir. Dietrich Brandis
(c) Hartig (d) Nautiyal

164. What is the Trinity of Norms in forestry?

(a) Normal series of age gradations (b) Normal growing stock
(c) Normal increment (d) All of the above

165. A forest which corresponds in every way to the objects of management is called _______:

(a) Normal forest (b) Increment
(c) Abnormal forest (d) None of the above

166. The crop composition >75 per cent of teak may be classified as:

(a) Pure teak (b) Mainly teak
(c) Mixed teak (d) Misc. with some teak

167. The sum of all trees growing in the forest or a specified part is called:

(a) Increment (b) Rotation
(c) Growing stock (d) None of the above

168. Tangible benefits are _______.

(a) Wood (b) Industrial raw material
(c) Both (a) and (b) (d) None of the above

169. Intangible Benefits are:

(a) Protective (b) Regulative

(c) Socio cultural (d) All of the above

170. Forests are classified into various categories for purposes of _______.

(a) Description (b) Administration

(c) Management and record (d) All of the above

171. The NCA (1976) has suggested a modified functional classification _______.

(a) Production forest (b) Protection forest

(c) Social forest (d) All of the above

172. Based on legal forests are classified _______.

(a) Reserved (b) Protected

(c) Un-classed forest (d) All of the above

173. Based on Territorial forests are classified _______.

(a) Block (b) Compartment

(c) Sub-compartment (d) All of the above

174. India formulated its first National Forest Policy in the year:

(a) 1949 (b) 1952

(c) 1964 (d) 1971

175. The percent ratio of normal yield to the normal growing stock is called. _______:

(a) Utilization percent (b) Exploitation percent

(c) Masson's ratio (d) All of the above

176. National Forest Policy (NFP), 1988 envisages that _______. Per cent of the total geographical area of India should be under forest.

(a) 33 (b) 40

(c) 42 (d) 66

177. CAI and MAI curves are used to determine _______:

(a) Rotation age (b) Regeneration period

(c) Increment (d) Growing stock

178. The rotation followed for sandal wood in south India is:

(a) Physical rotation (b) Financial rotation

(c) Technical rotation (d) Rotation for maximum volume production

179. Rotation of roadside and avenue trees is:

(a) Technical rotation
(b) Financial rotation
(c) Rotation for maximum volume production
(d) Physical rotation

180. The rotation followed for blue gum in the Nilgiris forest division was:

(a) 10
(b) 2
(c) 5
(d) 8

181. The forest survey of India (FSI) conducts forest cover survey once in every:

(a) Two years
(b) Ten years
(c) Five years
(d) None of the above

182. Schneider's formula is used for determining:

(a) Increment Percentage
(b) Quality increment
(c) Price increment
(d) All of the above

183. Who was the first person in India to be named a "Conservator of Forests":

(a) Brandis
(b) Cleghorn
(c) Watson
(d) None of the above

184. Social forestry programme was launched in the:

(a) Fifth five year plan
(b) Fourth five year plan
(c) Sixth five year plan
(d) Seventh five year plan

185. Quality increment per cent is _______.

(a) P=V-v/V+vx200/n
(b) V+v/V-vx200/n
(c) V-v/V+vxn/200
(d) V+v/V-vxn/200

186. Masson's formula for yield is _______.

(a) 200/r
(b) 2/r
(c) Both (a) and (b)
(d) None of the above

187. Smithies Modification formula for yield is _______.

(a) 2V/(r-x)
(b) 2GS/r
(c) 200/r
(d) None of the above

188. Burma formula for yield is _______.

(a) 9v/4r
(b) 2v(r-x)
(c) 8v/3r
(d) None of the above

189. Floating periodic block is also known as:

(a) Single allotment method
(b) Revocable allotment method
(c) Quartier Bleu Method
(d) All of the above

190. FPB was coloured blue on the map, the rest of the area was uncoloured called as:

(a) Quartier Bleu
(b) Quartier Blanc
(c) Uncoloured quartier bleu
(d) All of the above

191. Which yield regulation method is widely adopted in India:

(a) Annual coupes by gross area
(b) Annual coupes by reduced area
(c) Both (a) and (b)
(d) None of the above

192. Brandis diameter class method is otherwise called as:

(a) Indian method
(b) French method
(c) German method
(d) None of the above

193. A favourable balance between timber growth and harvest (growth equals or exceeds harvest) is referred to as:

(a) Sustained yield timber growth
(b) Silviculture
(c) Timber stand improvement
(d) All of the above

194. The discount rate at which NPV is zero is called:

(a) BC ratio
(b) IRR
(c) GNP
(d) None of the above

195. In forest, the standard colour given for the regeneration status Excellent (81-100 per cent) is:

(a) Green
(b) Blue
(c) Yellow
(d) Red

196. In forest, the standard colour given for the regeneration status Fair (21-40 per cent) is:

(a) Green
(b) Blue
(c) Yellow
(d) Red

197. In forest, the standard colour given for the regeneration status Deficient (0-20 per cent) is:

(a) Green
(b) Blue
(c) Black
(d) Red

198. Joint Forest Management (JFM) is a concept of developing partnerships between fringe forest user groups:

(a) Forest department (f(d)
(b) Forest based industry
(c) Local panchayat
(d) None of the above

199. Social Forestry Include:

(a) Form forestry
(b) Extension forestry
(c) Recreation forestry
(d) All of the above

200. Joint forest management evolved in early:

(a) 1950
(b) 1960
(c) 1970
(d) 1980

Answer Sheet

1.	(a)	31.	(a)	61.	(a)	91.	(b)	121.	(b)
2.	(b)	32.	(d)	62.	(a)	92.	(c)	122.	(c)
3.	(c)	33.	(c)	63.	(c)	93.	(a)	123.	(a)
4.	(a)	34.	(a)	64.	(a)	94.	(a)	124.	(b)
5.	(a)	35.	(c)	65.	(a)	95.	(d)	125.	(b)
6.	(b)	36.	(d)	66.	(a)	96.	(b)	126.	(c)
7.	(b)	37.	(d)	67.	(a)	97.	(c)	127.	(a)
8.	(c)	38.	(a)	68.	(d)	98.	(d)	128.	(c)
9.	(a)	39.	(c)	69.	(d)	99.	(a)	129.	(a)
10.	(a)	40.	(a)	70.	(c)	100.	(a)	130.	(a)
11.	(b)	41.	(b)	71.	(a)	101.	(b)	131.	(a)
12.	(a)	42.	(a)	72.	(a)	102.	(a)	132.	(b)
13.	(a)	43.	(d)	73.	(c)	103.	(c)	133.	(b)
14.	(b)	44.	(c)	74.	(a)	104.	(c)	134.	(b)
15.	(a)	45.	(c)	75.	(a)	105.	(b)	135.	(c)
16.	(a)	46.	(b)	76.	(d)	106.	(c)	136.	(b)
17.	(b)	47.	(a)	77.	(d)	107.	(a)	137.	(c)
18.	(a)	48.	(a)	78.	(b)	108.	(d)	138.	(a)
19.	(a)	49.	(b)	79.	(b)	109.	(a)	139.	(d)
20.	(a)	50.	(d)	80.	(c)	110.	(a)	140.	(b)
21.	(d)	51.	(a)	81.	(b)	111.	(b)	141.	(b)
22.	(a)	52.	(a)	82.	(b)	112.	(c)	142.	(a)
23.	(a)	53.	(a)	83.	(c)	113.	(c)	143.	(c)
24.	(a)	54.	(a)	84.	(a)	114.	(a)	144.	(c)
25.	(b)	55.	(d)	85.	(a)	115.	(d)	145.	(d)
26.	(d)	56.	(a)	86.	(b)	116.	(a)	146.	(c)
27.	(a)	57.	(a)	87.	(b)	117.	(b)	147.	(c)
28.	(a)	58.	(c)	88.	(d)	118.	(a)	148.	(b)
29.	(d)	59.	(a)	89.	(a)	119.	(a)	149.	(a)
30.	(c)	60.	(a)	90.	(d)	120.	(b)	150.	(b)

151.	**(a)**	**161.**	**(a)**	**171.**	**(d)**	**181.**	**(a)**	**191.**	**(a)**
152.	**(b)**	**162.**	**(a)**	**172.**	**(d)**	**182.**	**(a)**	**192.**	**(a)**
153.	**(a)**	**163.**	**(a)**	**173.**	**(d)**	**183.**	**(c)**	**193.**	**(a)**
154.	**(a)**	**164.**	**(d)**	**174.**	**(b)**	**184.**	**(a)**	**194.**	**(b)**
155.	**(a)**	**165.**	**(a)**	**175.**	**(a)**	**185.**	**(a)**	**195.**	**(a)**
156.	**(a)**	**166.**	**(a)**	**176.**	**(a)**	**186.**	**(c)**	**196.**	**(d)**
157.	**(d)**	**167.**	**(c)**	**177.**	**(a)**	**187.**	**(a)**	**197.**	**(c)**
158.	**(b)**	**168.**	**(c)**	**178.**	**(a)**	**188.**	**(a)**	**198.**	**(a)**
159.	**(b)**	**169.**	**(d)**	**179.**	**(d)**	**189.**	**(c)**	**199.**	**(d)**
160.	**(b)**	**170.**	**(d)**	**180.**	**(a)**	**190.**	**(b)**	**200.**	**(c)**

Chapter 5

Forest Ecology and Environment

1. **The Word 'Ecology' is derived from:**
   (a) Greek
   (b) Latin
   (c) French
   (d) None of the above
2. **The word *'oikos'* in greek means:**
   (a) House
   (b) Land
   (c) Monkey
   (d) Area
3. **The term Ecology was fist coined by:**
   (a) Charles Darwin
   (b) Ernst Haeckel
   (c) Thomas Edison
   (d) Tansely
4. **He branch of biology that deals with environment:**
   (a) Ecology
   (b) Ecosystem
   (c) Ethnology
   (d) Ecobotany
5. **Basic unit of ecological hierarchy is:**
   (a) Organisms
   (b) Biomass
   (c) Microbes
   (d) Ecology
6. **A distinct unit of life in nature:**
   (a) Organisms
   (b) Biomes
   (c) Microbes
   (d) Ecology

7. The term Niche was first coined by:

(a) Charles Darwin (b) Ernst Haeckel

(c) Grinnel (d) Element

8. Forests and wildlife are:

(a) Renewable resources (b) Non renewable resources

(c) Both (a) and (b) (d) None of the above

9. Biological species are:

(a) Renewable resources (b) Non renewable resources

(c) Both (a) and (b) (d) None of the above

10. The residence time of water in t he air is:

(a) 10 days (b) 15 days

(c) 100 days (d) 150 days

11. The term 'Ecosystem' was introduced by:

(a) Ernst Haeckel (b) Arthur Tansley

(c) Element (d) Whittakar

12. The structural and functional unit of biosphere is:

(a) Mass (b) Environment

(c) Ecosystem (d) All of the above

13. The relationship between biotic and abiotic environment:

(a) Biogenesis (b) Symbiosis

(c) Holocoenosis (d) None of the above

14. Key industry animals are:

(a) Primary consumers (b) Tertiary consumers

(c) Secondary consumers (d) Primary producers

15. The term 'key industry animals' was coined by:

(a) Elton (b) Priestley

(c) Tansley (d) Element

16. The rate of biomass production is called:

(a) Productivity (b) Production

(c) Biodiversity (d) All of the above

17. The rate of total production of organic matter by the producer per unit area and time is:

(a) Primary productivity (b) Secondary productivity

(c) Net primary productivity (d) Gross primary productivity

18. **The rate at which organic matter is stored after the basic metabolic processes is:**
    (a) Primary productivity (b) Secondary productivity
    (c) Net primary productivity (d) Gross primary productivity

19. **The productivity that is left after meeting the cost of respiration and maintenance of producers is:**
    (a) Primary productivity (b) Secondary productivity
    (c) Net primary productivity (d) Gross primary productivity

20. **The rate of increase in the biomass of consumers per unit area and time is:**
    (a) Primary productivity (b) Secondary productivity
    (c) Net primary productivity (d) Gross primary productivity

21. **Ecological pyramids were developed by:**
    (a) Charles Elton (b) Priestley
    (c) Tansley (d) Element

22. **The graphical representation of ecological parameters is called:**
    (a) Ecological pyramid (b) Ecology
    (c) Ecological hill (d) Ecological mountain

23. **Ecological pyramids are also called as:**
    (a) Eltonian pyramids (b) Charles pyramids
    (c) Hierarchical pyramids (d) All of the above

24. **_______ Ecological pyramid is always straight:**
    (a) Pyramid of biomass (b) Pyramid of numbers
    (c) Pyramid of energy (d) All of the above

25. **Name of the pioneer plants os xerarch:**
    (a) Lichens (b) Grasses
    (c) Toads (d) Frogs

26. **A population genetically adapted to its particular environment is known as:**
    (a) Ecotone (b) Ecotype
    (c) Ecosystems (d) Ecology

27. **The environment which has been modified by human activities is called:**
    (a) Anthropogenic environment (b) Natural environment
    (c) Manmade environment (d) Both (a) and (c)

28. Groups of species exploiting a common resources are called:

(a) Builds (b) Guilds

(c) Community (d) Population

29. In our country the Van Mahotsav Day is observed on:

(a) Second of October (b) First week of June

(c) Twenty first of March (d) First of July

30. The process of conversion of gas into a liquid is called:

(a) Condensation (b) Sublimation

(c) Vaporization (d) Solidification

31. The drying up of wet cloths represents the process of:

(a) Evaporation (b) Boiling

(c) Condensation (d) Solidification

32. During change of state in a matter the temperature remains same due to:

(a) Loss of heat (b) Less supply of heat

(c) Lattice energy (d) Latent heat

33. Which biome is characterized by very large, mostly evergreen trees and ultisols?

(a) Equatorial (b) Tropical forests

(c) Both (a) and (b) (d) None of the above

34. Which Biome is characterized by rapid nutrient cycling and high biomass?

(a) Deserts (b) Deciduous

(c) Tropical rain forests (d) All of the above

35. Biotic environment includes:

(a) Producers (b) Consumers

(c) Decomposers (d) All of the above

36. Decomposers include:

(a) Bacteria (b) Fungi

(c) Both (a) and (b) (d) None of the above

37. Abiotic environment does not include:

(a) air (b) Water

(c) Soil (d) Plants

38. Vermicomposting is done by:

(a) Fungus (b) Bacteria

(c) Worms (d) Animals

39. The group of organisms which convert light into food are called:

(a) Autotrophs (b) Heterotrophs
(c) Decomposers (d) Omnivores

40. Example of non-biodegradable?

(a) Vegetables (b) Fruits
(c) Earthworm (d) Aluminium foil

41. Air pollution is caused by:

(a) Insecticides (b) Sewage
(c) Smoke (d) Loud speakers

42. An antibacterial chemical present in wood smoke is:

(a) Formaldehyde (b) Carbon
(c) Chlorine (d) Carbohydrate

43. About half of earth's photosynthesis is carried out by:

(a) Cyanobacteria (b) Rain forest
(c) Protists (d) Marine phytoplankton

44. Ramsar convention is associated with:

(a) Forest conservation (b) Soil conservation
(c) Wetland conservation (d) Wildlife conservation

45. Plants are green because of the presence of a pigment called:

(a) Glucose (b) Nitrogen
(c) Chlorophyll (d) Oxygen

46. Air is composed of gases, water vapours and:

(a) Dust particles (b) Rain fall
(c) Snowfall (d) Light

47. Medicine of quinine is provided by:

(a) Eucalyptus (b) Aconite
(c) Cinchona (d) Pine

48. Chief source of energy in environment is:

(a) Fire (b) Moon
(c) Sun (d) Stars

49. When trees are cut, amount of oxygen:

(a) Decreases (b) Increases
(c) Both (a) and (b) (d) None of the above

50. Process through which plants reproduce:

(a) Eating (b) Evaporation

(c) Pollination (d) Condensation

51. 71 per cent of earth surface is covered with:

(a) Land (b) Air

(c) Nitrogen (d) Water

52. CNG stands for:

(a) Compressed Natural gas (b) Composed Natural gas

(c) Compressed Nitrogen gas (d) Composed Nitrogen gas

53. Green Petrol is:

(a) Leaded petrol (b) Unleaded petrol

(c) Gasohol (d) Low sulphur diesel

54. Bio-indicator of underground water is:

(a) *Impatiens blasamina* (b) *Prosopis cineraria*

(c) Astragalus (d) Haplopappus

55. Air pollution is highest in:

(a) Kolkata (b) Lucknow

(c) Panchmarhi (d) Bhopal

56. Which is a major pollutant causing acid rain?

(a) SO_2 (b) NO_2

(c) CO (d) O_3

57. The phenomenon of corrosion of marble due to acid rain:

(a) Marble fever (b) Marble rain

(c) Marble pain (d) Marble cancer

58. An alternative source of energy to fuel energy:

(a) Tidal energy (b) Petroleum

(c) Nuclear energy (d) Compressed natural gas

59. The best way to dispose plant waste:

(a) Burning (b) Composting

(c) Dumping in a water body (d) Incineration

60. An ecofriendly slogan:

(a) Do or die (b) Birds of same feather fly together

(c) Time and tide wait for none (d) Say no to crackers

61. The term 'dead' with respect to a water body refers to:
(a) The inability of a water body to sustain aquatic life
(b) The ability of a water body to sustain aquatic life
(c) The inability of a water body to flow
(d) The heating or cooling of a water body

62. Example of waster borne disease:
(a) Small pox (b) Meningitis
(c) Diarrhea (d) Cholera

63. The water suitable for drinking:
(a) Pure water (b) Portable water
(c) Potable water (d) Pungent water

64. The best household method to obtain potable water:
(a) Filtration (b) Boiling
(c) Chemical treatment (d) Decantation

65. The branch of biology which is concerned with the interrelationship between plants and animals is called:
(a) Ecology (b) Botany
(c) Physiology (d) Ethnobotany

66. The largest unit of living organisms on earth is:
(a) Biosphere (b) Ecotone
(c) Ecosystem (d) Biome

67. What is the components of an ecosystem are:
(a) Biotic (b) Abiotic
(c) Both (a) and (b) (d) None of the above

68. The green plants are called:
(a) Producers (b) Consumers
(c) Decomposers (d) All of the above

69. For domestic drinking water supply the total dissolved should not exceed:
(a) 200ppm (b) 500ppm
(c) 700ppm (d) 900ppm

70. Which of the chemicals is responsible for the reduction of ozone content of the atmosphere?
(a) Carbon dioxide (b) Sulphur dioxide
(c) Carbon monoxide (d) Chlorofluoro carbon

71. Acid rains occur when atmosphere is heavily polluted with:
(a) CH_4 and O_3 (b) O_2 and NO_2
(c) SO_2 and NO_2 (d) CO_2 and SO_2

72. Spraying of DDT on crops causes pollution of:
(a) Soil (b) Water
(c) Both (a) and (b) (d) None of the above

73. Soil erosion can be prevented by:
(a) Overgrazing (b) Afforestation
(c) Removal of vegetation (d) Deforestation

74. The main source of water pollution is:
(a) Smog (b) Sewage
(c) Industries (d) Detergents

75. Which gas is mainly responsible for green house effect ?
(a) CO_2 (b) SO_2
(c) O_2 (d) N_2

76. Cutting of trees in a forest is called:
(a) Deforestation (b) Afforestation
(c) Selection cutting (d) All of the above

77. The sound of which intensity is regarded as the safe noise level for a city?
(a) 45dB (b) 75dB
(c) 80dB (d) 100 dB

78. Eco-labelling is governed by:
(a) ISO 14001 (b) ISO 14010
(c) ISO 14020 (d) ISO 14002

79. Earth day is observed on:
(a) April 21 (b) April 22
(c) June 5 (d) September 16

80. State with the highest percentage of forests is:
(a) Mizoram (b) Tamil Nadu
(c) Kerala (d) Arunachal Pradesh

81. When was Project Tiger launched in India?
(a) 1973 (b) 1975
(c) 1981 (d) 1983

82. World Wildlife Fund was founded in:

(a) 1947 (b) 1961
(c) 1952 (d) 1988

83. World Environment Day is observed on:

(a) April 21 (b) April 22
(c) June 5 (d) September 16

84. The branch of biology dealing with interactions among organisms and between organisms and their environment is called:

(a) Economy (b) Modeling
(c) Recycling (d) Ecology

85. All of the members of a particular species that live in one area are called a (an):

(a) Biome (b) Population
(c) Community (d) Ecosystem

86. Eco-Auditing focuses on:

(a) Assessment of actual effects of exist activities (b) Pollution effects
(c) Climatic effects (d) Water management

87. What is the original source of almost all the energy in most ecosystems?

(a) Carbohydrates (b) Water
(c) Carbon (d) Sunlight

88. International ozone day is celebrated on:

(a) 5th June (b) 16th September
(c) 8th March (d) 21st March

89. An organism that uses energy to produce its own food supply from inorganic compounds is called a:

(a) Autotroph (b) Heterotroph
(c) Chemotroph (d) Monotroph

90. Which organisms does NOT require sunlight to live?

(a) Chemosynthetic bacteria (b) Algae
(c) Trees (d) Photosynthetic bacteria

91. An organism that cannot make its own food is called a (an):

(a) Heterotroph (b) Chemotroph
(c) Autotroph (d) Producer

92. Organisms that obtain nutrients by breaking down dead and decaying plants and animals are called:

(a) Decomposers (b) Omnivores
(c) Autotrophs (d) Producers

93. What is an organism that feed only on plants called?

(a) Carnivore (b) Herbivore
(c) Omnivore (d) Detritivore

94. What is an ecological model of the relationships that form a network of complex interactions among organisms in a community from producers to decomposers?

(a) Food web (b) An ecosystem
(c) Food chain (d) A population

95. What animals eat both producers and consumers?

(a) Omnivores (b) Decomposers
(c) Both (a) and (b) (d) None of the above

96. Only 10 per cent of the energy stored in an organism can be passed on to the next trophic level. Of the remaining energy, some is used for the organism's life processes, and the rest is:

(a) Used in reproduction (b) Stored as body tissue
(c) Stored as fat (d) Eliminated as heat

97. The repeated movement of water between Earth's surface and the atmosphere is called:

(a) Water cycle (b) Nitrogen cycle
(c) Carbon cycle (d) Phosphorus

98. Five R polices for waste minimization:

(a) Reduction, Replacement, Recovery, Recycling and Reutilization
(b) Regain, Reuse, Resend, Re oxidation, Reduce
(c) Regain, Reuse, Resend, Reimbursement, Reduce
(d) Reduction, Reuse, Resend, Re oxidation, Reduce

99. The Calorific value of natural gas varies from:

(a) 10,000 – 18,000 k Cal/m^3 (b) 12,000 – 14,000 k Cal/m^3
(c) 13,000 – 18,000 k Cal/m^3 (d) 16,000 – 18,000 k Cal/m^3

100. Organisms need nutrients in order to:

(a) Utilize hydrogen and oxygen (b) Recycle chemical compound
(c) Carry out essential life functions (d) Carry out nitrogen fixation

101. The average year after year conditions of temperature and precipitation in a particular region is the region's:

(a) Weather
(b) Climate
(c) Both (a) and (b)
(d) None of the above

102. BOD of the effluents discharged on land for irrigation should not exceed:

(a) 20 mg/l
(b) 25 mg/l
(c) 30 mg/l
(d) 35 mg/l

103. The greenhouse effect is:

(a) The result of an excess of carbon dioxide in the atmosphere
(b) A natural phenomenon that maintains earth's temperature range
(c) The result of the differences in the angle of the sun's rays
(d) An unnatural phenomenon that causes heat energy to be radiated back into the atmosphere

104. Ecotone is:

(a) A zone between two ecosystem
(b) A zone of an ecosystem
(c) An ecological study
(d) Ecosystem

105. Which is a biotic factor that affects the size of a population in a specific ecosystem?

(a) Average temperature of the ecosystem
(b) Type of soil in the ecosystem
(c) Number and kinds of predators in the ecosystem
(d) Concentration of oxygen in the ecosystem

106. Biogas is composed of:

(a) Ethane
(b) Methane
(c) CO_2
(d) All of the above

107. An interaction in which one organism captures and feeds on another organism is called:

(a) Predation
(b) Parasites
(c) Symbionts
(d) Commensale

108. No two species can occupy the same niche in the same habitat at the same time:

(a) Because of the interactions that shape the ecosystem
(b) Unless the species require different abiotic factors
(c) Because of the competitive exclusion principle
(d) Unless the species require different biotic factors

109. A symbiosis in which both species benefit is:

(a) Commensalism
(b) Mutualism
(c) Predation
(d) Parasitism

110. The series of predictable changes that occurs in a community over time is called:

(a) Population growth
(b) Ecological succession
(c) Climax community
(d) Climate change

111. Which two biomes have the least amount of precipitation?

(a) Tropical rain forest and temperate grassland
(b) Tropical savanna and tropical dry forest
(c) Tundra and desert
(d) Boreal forest and temperate woodland and shrubland

112. A biome is identified by its particular set of abiotic factors and its:

(a) Average precipitation and temperature
(b) Characteristic ecological community
(c) Distance from the equator
(d) Specific geographical location

113. One of the main characteristics of a population is its:

(a) Change over time
(b) Geographic distribution
(c) Dynamics
(d) Habitat

114. Which are two ways a population can decrease in size?

(a) Immigration and emigration
(b) Increased death rat and immigration
(c) Decreased birthrate and emigration
(d) Emigration and increased birthrate

115. When individuals in a population reproduce at a constant rate, it is called:

(a) Logistic growth
(b) Growth density
(c) Exponential growth
(d) Multiple growth

116. As resources in a population become less available, population growth:

(a) Declines rapidly
(b) Increases slowly
(c) Reaches carrying capacity
(d) Enters a phase of exponential growth

117. In a logistic growth curve, exponential growth is the phase in which the population:

(a) Reaches carrying capacity
(b) Grows quickly and few animals are dying
(c) Growth begins to slow down
(d) Growth stops

118. A biotic or an abiotic resource in the environment that causes population size to decrease is a:

(a) Carrying capacity
(b) Limiting nutrient
(c) Limiting factor
(d) Growth factor

119. Example of are limiting factors:

(a) Competition
(b) Predation
(c) Human disturbances
(d) All of the above

120. If a population grows larger than the carrying capacity of the environment, the:

(a) Death rate may rise
(b) Birth rate may rise
(c) Death rate must fall
(d) Birth rate must fall

121. Which of the following is a density independent limiting factor?

(a) Earthquake
(b) Disease
(c) Emigration
(d) Parasitism

122. Example of density dependent limiting factor:

(a) Competition
(b) Crowding
(c) Disease
(d) All of the above

123. Methanization is carried out by the process of:

(a) Anaerobic fermentation
(b) Aerobic fermentation
(c) Composting
(d) All of the above

124. A disease resulting in the deaths of one-third of a dense population of bats in a cave would be a:

(a) Density – dependent limiting factor
(b) Result of exponential growth
(c) Density independent limiting factor
(d) Nutrient limiting factor

125. What is the harm from the depletion of Earth's ozone layer:

(a) The average temperature of earth's surface will increase gradually
(b) The oxygen content of the atmosphere will decrease
(c) Increased amount of Ultra violet radiation will reach earth's surface
(d) Sea levels will rise as the polar ice caps will gradually melt

126. Rapid EIA is called:
(a) A study of three months (b) A study of four months
(c) A study of six months (d) A study of one year

127. Liquid ammonia can be used in refrigeration because of its:
(a) High basicity (b) High dipole moment
(c) High heat of vaporization (d) Non-toxic nature

128. Which of the following pollutants causes ozone holes in the ozone layer?
(a) CO_2 (b) SOv
(c) CO (d) CFC

129. The most serious environmental effect posed by hazardous wastes is:
(a) Contamination of groundwater (b) Air pollution
(c) Increased use of land for landfills (d) Destruction of habitat

130. The concentration of which gas is highest in our environment?
(a) Nitrogen (b) Oxygen
(c) Carbon (d) Hydrogen

131. The use of anthropoid hazard chemicals like DDT causes reduction in fish population because the chemical mimics the effect of:
(a) Luteinizing hormone (b) Testosterone
(c) Estrogen (d) All of the above

132. Biogas production is the outcome of:
(a) Methanogensis (b) Gasification
(c) Both (a) and (b) (d) None of the above

133. The increase in the concentration of CO_2 in our environment in last fifty years; since 1960 is about:
(a) 20 per cent (b) 10 per cent
(c) 14 per cent (d) 6 per cent

134. The depletion in the Ozone layer is caused by:
(a) Nitrous oxide (b) Carbon dioxide
(c) Chlorofluorocarbons (d) Methane

135. A major in stream use of water is for:
(a) Producing hydroelectric power (b) Dissolving industrial wastes
(c) Agricultural irrigation (d) Domestic use

136. The example of Municipal and industrial discharge pipes:

(a) Non-point sources of pollution (b) Point sources of pollution

(c) Violations of the clean water act (d) Irrigation

137. The presence of high coliform counts in water indicate:

(a) Contamination by human wastes (b) Phosphorus contamination

(c) Decreased biological oxygen demand (d) Hydrocarbon contamination

138. How the biological oxygen demand gets affected with the increased presence of organic matter in water?

(a) The oxygen demand increases

(b) The oxygen demand decreases

(c) The oxygen demand remains unchanged

(d) None of the above

139. Maximum density of water is at:

(a) 4°C (b) 0°C

(c) 40°C (d) 100°C

140. The aggregation of all ecosystems on the earth is referred to as:

(a) Atmosphere (b) Ecosphere

(c) Stratosphere (d) Ionosphere

141. The stage in which the biological processes is used to purify water in a wastewater treatment plants is called:

(a) Primary sewage treatment (b) Secondary sewage treatment

(c) Tertiary sewage treatment (d) Tetra sewage treatment

142. Groundwater mining in coastal areas can result into:

(a) Increase in the salinity of groundwater

(b) Decrease in the toxicity of groundwater

(c) Decrease in the salinity of groundwater

(d) Increase in the water table

143. Thiobacillus and Beggiatoa play an important role in the:

(a) Water cycle on earth (b) Phosphorus cycle

(c) Sulphur cycle (d) Breakdown of Sewage

144. The primary soil macronutrients are:

(a) Potassium (b) Phosphorus

(c) Nitrogen (d) All of the above

145. Bulking of sewage sludge is frequently associated with:

(a) High C: P ratio (b) Medium C:P ratio

(c) Low C; P ratio (d) None of the above

146. The process of mixing two or more gases is called:

(a) Condensation (b) Diffusion

(c) Sublimation (d) Solidification

147. A matter is defined as one which has:

(a) No mass and occupies no space (b) Mass but occupies no space

(c) Mass and occupies space (d) No mass but occupies space

148. When water freezes into solid, its density:

(a) Decreases (b) Increases

(c) Neutral (d) None of the above

149. The water when converted into gaseous form is called as:

(a) Water gas (b) Fog

(c) Snow (d) Water vapour

150. When salt is added into water, its boiling point:

(a) Decreases (b) Increases

(c) Neutral (d) None of the above

151. The miscibility of two liquids is determined by:

(a) Molecular size: molecules of similar sizes tend to mix together

(b) Density: liquids of similar densities usually mix well together

(c) Polarity: polar and nonpolar liquids are not miscible while polar liquids can usually be mixed with each other

(d) All of the above

152. The particles vibration in gases takes place:

(a) About their mean position (b) About a vertical axis

(c) About a horizontal axis (d) In any direction

153. The noise index Lgo refers to:

(a) Background noise level (b) Peak noise level

(c) Average noise level (d) 90 per cent

154. Which of the following property doesn't represent the fluid characteristics:

(a) It can flow (b) It has mass

(c) Can be perceived by our senses (d) It has a definite shape

155. The optimum range of wind speeds for wind power generation is:

(a) 2-4m/sec (b) 4-12 m/sec

(c) 12-14 m/sec (d) 2-20 m/sec

156. Asbestos use is banned in many countries.

(a) Cardiac disease (b) Cataract

(c) Lung cancer (d) Urinary disease

157. Sublimation is defined as:

(a) The process in which a solid transitions directly into the gas phase

(b) The process in which a gas transitions directly into the soil phase

(c) The process in which a gas transitions into the liquid phase

(d) The process in which a liquid transitions into the gas phase

158. Mobile phone frequencies are in the range of:

(a) 1kHz-100MHz (b) 100kHz-100MHz

(c) 100kHZ-1000MHz (d) 1000kHz-10000MHz

159. Ozone "hole" in the stratosphere was discovered over the continent of:

(a) North America (b) Australia

(c) Green land (d) Antarctica

160. Population growth is:

(a) Exponential (b) Linear

(c) Logistic (d) All of the above

161. Tertiary consumers are called:

(a) Carnivores (b) Herbivores

(c) Omnivores (d) All of the above

162. Which one of the following is the cleanest source of energy?

(a) Hydropower (b) Fossil fuel

(c) Nuclear power (d) Wind energy

163. Which waves carry most energy?

(a) UV light (b) Infra-red light

(c) Microwaves (d) Millimeter waves

164. Who developed the word hot spots?

(a) Odum (b) Norman Myes

(c) Clement (d) Brandis

165. The reason for signing 1987 Montreal Protocol was:

(a) To stop global trade of products made from endangered animals

(b) To do away with the use of CFC's, which were found to be responsible for depletion of the ozone layer

(c) To prohibit and ban nuclear testing in tropical deserts and oceans

(d) To start using renewable sources of energy instead of fossil fuels to reduce the anthropogenic greenhouse effect

166. Which of the countries are the largest source of carbon dioxide emissions from fossil fuel burning?

(a) Countries of the former U.S.S.R. (b) Developing nations

(c) Japan and china (d) North America

167. Residence time of water molecule in the ocean is:

(a) 3.5 years (b) 35 years

(c) 3.5 million years (d) 350000 years

168. Group of individuals of the same species that share common attributes are called:

(a) Community (b) Population

(c) Ecotype (d) Society

169. Which is the cleanest source of energy?

(a) Hydropower (b) Fossil fuel

(c) Nuclear power (d) Wind energy

170. The atmospheric layer of ozone normally absorbs what proportion of UV radiation from sun?

(a) More than 99 per cent (b) Around a quarter

(c) Less than 10 per cent (d) About half

171. In order to address the problem of ozone depletion, the United Nations convened a meeting in 1987 in Canada. In that meeting, all the member nations agreed to reduce down the CFC production and use. What was the name of this agreement?

(a) Kyoto Agreement (b) Montreal protocol

(c) Chemical stewardship program (d) Vancouver convention

172. Acid rain is caused by:

(a) CO and CO_2 (b) SO_2 and O_2

(c) SO_2 and NO_2 (d) NO_2 and O_2

173. What will be the outcome of eutrophication of surface waters?

(a) Overproduction of biomass

(b) Decrease in nitrogen concentration

(c) Decrease in phosphorus concentration

(d) Decrease in both nitrogen and phosphorus concentration

174. Example of non-biodegradable product:

(a) Animal bones (b) Pine leaves

(c) Wool (d) Nylon

175. The chemical that is used to ripen mangoes is:

(a) Calcium sulphide (b) Calcium carbide

(c) Calcium carbonate (d) Calcium chloride

176. Alpha diversity represents:

(a) Species richness (b) Species evenness

(c) Species richness and evenness (d) Species richness and dominance

177. The combination of primary and secondary treatment reduces the original sewage BOD by:

(a) 30-40 per cent (b) 50-60 per cent

(c) 60-70 per cent (d) 80-90 per cent

178. A natural phenomenon that becomes harmful due to pollution is:

(a) Global warming (b) Ecological balance

(c) Green house effect (d) Desertification

179. The pollutant responsible for ozone holes is:

(a) CO_2 (b) SO_2

(c) CO (d) CFC

180. One of the best solutions to get rid of non-biodegradable wastes is:

(a) During (b) Dumping

(c) Burying (d) Recycling

181. Animal dung is _______. Waste.

(a) Biodegradable (b) Non-biodegradable

(c) Hazardous (d) Toxic

182. Petroleum consists mainly of:

(a) Hydrocarbons (b) Sulphur

(c) Nitrogen (d) All of the above

183. Largest producer of Natural Gas in the World:

(a) Russia (b) Ameria

(c) India (d) Australia

184. Name one non-biodegradable waste which may pollute the earth to dangerous levels of toxicity, if not handled properly.

(a) DDT (b) CFC

(c) Radioactive substances (d) PAN

185. In a lake polluted with pesticides, which will contain t he maximum amount of pesticides?

(a) Small fish (b) Microscopic animals

(c) Big fish (d) Water birds

186. Name the substance whose accumulation in pelicans of Lake Michigan led to the formation of thin shells of their eggs.

(a) CFC (b) PAN

(c) DDT (d) PAC

187. Name the process in which a harmful chemical enters the food chain and gets concentrated at each level in the food chain.

(a) Biomagnifications (b) Biological magnification

(c) Bioillumination (d) Both (a) and (b)

188. Which is one is non-renewable resources:

(a) Coal and minerals (b) Plants and coal

(c) Water and natural gas (d) Energy and water

189. Example of renewable resource?

(a) Soil (b) Water

(c) Flora (d) All of the above

190. _______ of stratosphere provides protection to our life.

(a) Ozone (b) Nitrogen

(c) Carbon (d) Hydrogen

191. The life supporting gases such as O_2, CO_2 and N_2 are chiefly concentrated is the _______.

(a) Troposphere (b) Stratosphere

(c) Mesosphere (d) Ionosphere

192. Which soil is the best for plant growth?

(a) Loamy soil (b) Sandy soil

(c) Sand soil (d) Clay soil

193. Both power and manure are provided by _______.

(a) Biogas plants
(b) Compost
(c) Both (a) and (b)
(d) None of the above

194. In the atmosphere, the layer above the troposphere is _______.

(a) Stratosphere
(b) Mesosphere
(c) Ionosphere
(d) Thermosphere

195. _______ is the major raw material for biogas.

(a) Cow dung
(b) Plants
(c) Earthworms
(d) All of the above

196. A biosphere reserve conserves and preserves _______:

(a) Wild animals
(b) Wild land
(c) Natural vegetation
(d) All of the above

197. Atomic energy is obtained by using ores of _______.

(a) Copper
(b) Uranium
(c) Neither a nor b
(d) Both (a) and (b)

198. The fission material for atomic explosion is:

(a) Uranium 235
(b) Plutonium 239
(c) Radio Iodine
(d) Both (a) and (b)

199. The area reserved for the welfare of wildlife is called:

(a) National park
(b) Botanical garden
(c) Sanctuary
(d)

200. The death of the last individual of a species is called:

(a) Extinction
(b) Endangered
(c) Vulnerable
(d) Disappearance

201. Which is a fossil fuel?

(a) Natural gas
(b) Petrol
(c) Coal
(d) All of the above

202. Biogas generation is mainly based on the principle of _______.

(a) Fermentation
(b) Composting
(c) Gasification
(d) Landfills

203. Red Data Book provides a list of _______.

(a) Rare
(b) Endangered
(c) Endemic species
(d) All of the above

204. Floods can be prevented by _______.

(a) Afforestation (b) Erosion

(c) Overgrazing (d) Deforestation

205. Who coined the term 'Chronobiology'?

(a) Franz Halberg (b) MK Chandrasekhar

(c) DS Farner (d) E Bunning

206. What is meaning of Chronobiology?

(a) The study of the effects of time on living systems (b) Study of living systems

(c) Study of rocks (d) Study of time

207. Energy flow in an ecosystem is:

(a) Cyclic (b) Unidirectional

(c) Multidirectional (d) Sequential

208. The major types of flora and fauna found in aquatic ecosystems are determined primarily by:

(a) Light (b) Temperature

(c) Soil (d) All of the above

209. The most abundant alkaline component of ht atmosphere is:

(a) Carbon (b) Nitrogen

(c) Hydrogen (d) All of the above

210. Oceans cover about _______. of the earth's surface.

(a) 70 per cent (b) 30 per cent

(c) 10 per cent (d) 50 per cent

211. The ocean zone that covers the continental shelf is the:

(a) Estuary (b) Coastal zone

(c) Littoral zone (d) Benthic zone

212. The ocean sea contains _______ per cent of the surface area of the oceans and _______ per cent of its plant and animal life.

(a) 90-10 (b) 10-90

(c) 80-20 (d) 20-80

213. The ecosystems with the world's highest net primary productivities per unit area are found in the:

(a) Estuary (b) Coastal zone

(c) Littoral zone (d) Benthic zone

214. Which species is characteristics of tropical coastal wetlands?

(a) Cypress
(b) Coconut
(c) Mangrove
(d) Palm

215. The main pollutant in vehicle emission is:

(a) CO_2
(b) SO_2
(c) NO
(d) Ozone

216. Organisms of the intertidal zone:

(a) Are constantly being swept away
(b) Are exposed to constant salinity levels
(c) Are exposed to both air and underwater conditions
(d) Must avoid drowning at high tide

217. Most of the wetlands that are lost are used for:

(a) Mining
(b) Urban development
(c) Agriculture
(d) Forestry

218. "Mitigation banking" refers to:

(a) The federal policy of zero net loss of wetlands
(b) Creating or restoring as much wetlands as are destroyed
(c) The world bank's support of wet lands protection policies
(d) Debt-for-nature swaps

219. What is CITES?

(a) The treaty banning barrier island development
(b) A treaty banning trade in endangered species
(c) A treaty banning dumping plastic in the ocean
(d) A treaty banning overfishing

220. What is the largest wetlands restoration project?

(a) The okeefenokee
(b) The everglades
(c) Dismal swamp
(d) Chesapeake bay

221. Desalination may be accomplished by:

(a) Distillation
(b) Osmosis
(c) Salt–eating bacteria
(d) Macrofiltration

222. The first Environmental law in India was enacted in:

(a) 1950
(b) 1960
(c) 1970
(d) 1980

223. Nitrates and phosphates are examples of:

(a) Disease – causing agents (b) Oxygen-demanding wastes

(c) Organic chemicals (d) Inorganic plant nutrients

224. One class of pollutants that can cause excessive growth of algae is:

(a) Radioactive substances (b) Oxygen-demanding wastes

(c) Inorganic plant nutrients (d) Organic chemicals

225. Thermal pollution is more prevalent near:

(a) Hot water springs (b) Coal based power plants

(c) Temperate zones (d) Tropical zones

226. Which is a point source of water pollution?

(a) Offshore oil wells (b) Livestock feedlots

(c) Urban lands (d) Croplands

227. Which is a nonpoint source of water pollution?

(a) Sewage treatment plant (b) Electric power plant

(c) Active and inactive coal mines (d) Logged forest

228. The permissible Arsenic levels in drinking water is:

(a) 20./l (b) 30./l

(c) 40./l (d) 50./l

229. Over half of the total mass of pollutants entering streams and lakes comes from:

(a) Point pollution from sewage treatment plants

(b) Point pollution from industrial plants

(c) Non point pollution from roadside runoff

(d) Non point pollution from agriculture

230. Due to the increasing energy use the ozone layer over the Arctic is:

(a) Decreasing (b) Increasing

(c) Both (a) and (b) (d) None of the above

231. In cultural eutrophication, game fish die from:

(a) Suffocation from lack of oxygen (b) Acid deposition

(c) Toxic substances in the water (d) Salt

232. To establish wind mill for commercial production of electricity, the optimum range of wind speeds required is:

(a) 100 m/sec (b) <100 m/sec

(c) 200 m/sec (d) <100 m/sec

233. Sea water is _______ nature:

(a) Acidic (b) Alkaline

(c) Saline (d) All of the above

234. In an aquatic ecosystem, ht e depth to which light penetrates is called:

(a) Aphotic (b) Photic

(c) Euphotic (d) Nonphotic

235. High BOD values in a water sample reflects a:

(a) High biodegradable organic content (b) Low organic content

(c) High phosphorous content (d) High nitrogen content

236. Zeitgeber means:

(a) Time giving (b) Photoperiod information

(c) Temperature information (d) Rainfall information

237. Environmental Impact Assessment (EIA) seeks to achieve goals pertaining to:

(a) Sustainable development (b) Prevention of pollution only

(c) Conservation of resources only (d) Social benefits

238. The criteria indicators of water pollution are:

(a) pH (b) Coliform

(c) COD (d) All of the above

239. True xerophytes have:

(a) Lower rate of transpiration than mesophytes

(b) Transpiration rate equal to mesophytes

(c) Transpiration rate higher than mesophytes

(d) No transpiration

240. Plants growing in extremely cold soils are:

(a) Psychrophytes (b) Halophytes

(c) Lithophytes (d) Oxylophytes

241. Xerophytes have long roots:

(a) To draw water from deep water beds (b) Due to light

(c) To give mechanical support (d) None of the above

242. The orchid plants, which are found growing on the trees, are:

(a) Parasites (b) Saprophytes

(c) Lithophytes (d) Epiphytes

243. In submerged hydrophytes functional stomata are found:

(a) On lower surface of leaf
(b) On both the surface of leaf
(c) No where on the plant
(d) On upper surface of leaf

244. What is the relationship between water fern Azoll and nitrogen-fixing cyanobacteria Anabaena called?

(a) Symbiotic
(b) Predation
(c) Parasitic
(d) Epiphytic

245. When animals feed on other dead animals, which had died naturally or had been killed by another animals, the relationship is termed as:

(a) Predation
(b) Competition
(c) Symbiosis
(d) Scavenging

246. The first plants to reappear in a badly burned forest area will most probably be:

(a) Liverworts
(b) Grasses
(c) Mosses
(d) Ferns

247. The transfer of energy from organisms to organism in a natural community establishes:

(a) Food habitat
(b) Food chains
(c) Food nexus
(d) All of the above

248. Which one is a primary consumer?

(a) Carnivore
(b) Scavenger
(c) Saprophyte
(d) Herbivore

249. Energy and nutrients enter a community by way of the:

(a) Producers
(b) Consumers
(c) Detrivores
(d) Both (a) and (b)

250. The food chain that starts from plants and goes from smaller to larger animals is:

(a) Predator food chain
(b) Saprophytic food chain
(c) Detritus food chain
(d) Parasitic food chain

251. Driving force of ecosystem is:

(a) Solar energy
(b) Water
(c) Wind
(d) All of the above

252. The leguminous plants are important in agriculture because:

(a) They need very little water to grow

(b) They are disease resistant

(c) They help in nitrogen economy

(d) Six crops of leguminous of the plants are harvested in a year

253. National Biodiversity Strategy and action plant (NBSAP) was established in:

(a) 1986 (b) 1999

(c) 2002 (d) 2012

254. MIC and Chernobyl tragedies occurred at:

(a) Bhopal 1984, Ukraine 1986 (b) Bhopal 1986, Ukraine 1984

(c) Bhopal 1988, Ukraine 1986 (d) Bhopal 1984, Ukraine 1988

255. Ecofriendly detergents do not contain:

(a) PO4 and bleaches (b) Alcohols

(c) Water softeners (d) Fluoresers

256. Nitrogen oxide can be absorbed by plants like:

(a) Pinus (b) Quercus

(c) Junipers (d) All of the above

257. Which satellite recorded the presence of an ozone hole?

(a) TIROS-N (b) GOES

(c) NIMBUS-7 (d) LANDSAT-3

258. Natural sink of stratospheric ozone layer is:

(a) SO_2 (b) HCF

(c) Sulphur flux of oceans (d) Freons

259. Which plant brought from America for controlling water pollution has become a problem in India:

(a) Opuntia (b) Lantana

(c) Kalmath weed (d) Eichhornia

260. The intensity of noise pollution is measured in:

(a) Decibels (b) Mach

(c) Hertz (d) Noy

261. What is the intensity of sound in normal conversation?

(a) 10x20dB (b) 70x90dB

(c) 0x60dB (d) 30x60 dB

262. Range of human hearing is:

(a) 20x1000Hz (b) 20x1500Hz

(c) 20x10000Hz (d) 20x15000Hz

263. The nutrient enrichment is known as:

(a) Symbiosis (b) Eutrophication

(c) Nitrification (d) Both (a) and (b)

264. Eutrophication (mineral pollution) is maximum in:

(a) Bottom layer of deep lakes (b) Middle layer of deep lakes

(c) Upper layer of deep lakes (d) All layers

265. A sewage is labeled as strong if its biological oxygen demand is around:

(a) 300 mg/litre (b) 500 mg/litre

(c) 600 mg/litre (d) 700 mg/litre

266. Sound limit in industrial area is:

(a) 30dB (b) 40dB

(c) 50dB (d) 75dB

267. Ambient noise standard in India for silence zone in day time is:

(a) 30dB (b) 40dB

(c) 50dB (d) 75dB

268. Ambient noise standard in India for residential area at night time is:

(a) 30dB (b) 45dB

(c) 50dB (d) 75dB

269. The standard permissible noise limits for refrigerator is:

(a) 60dB (b) 46dB

(c) 68dB (d) 75dB

270. BOD is:

(a) Amount of O_3 altered by organisms in water

(b) Amount of O_2 present in water at 370C

(c) Amount of O_2 altered by microbes for decomposition of organic matter

(d) All of the above

271. Safe limit of water fluoride level is:

(a) 0.1x0.4 mg/litre (b) 0.2x0.8 mg/litre

(c) 0.5x0.8 mg/litre (d) 0.5x8.0 mg/litre

272. Mass killing of fishes occur in:
(a) Eutrophic lakes
(b) Benthos lakes
(c) Both (a) and (b)
(d) None of the above

273. Species ecology is:
(a) Autecology
(b) Synecology
(c) Community ecology
(d) Palaeo ecology

274. Biosphere consists of:
(a) Lithosphere
(b) Atmosphere
(c) Hydrosphere
(d) All of the above

275. Biogeochemical cycles are also known as:
(a) Sedimentary cycle
(b) Gaseous cycle
(c) Material cycling
(d) Cycle of water

276. Biosphere consists of interacting functional units called:
(a) Ecosystems
(b) Populations
(c) Communities
(d) Species

277. Maximum oxygen comes from:
(a) Phytoplanktons
(b) Grassland
(c) Forests
(d) Herbs and shrubs

278. The reservoir pool of carbon is:
(a) Air and water
(b) Fossil fuels
(c) Rocks
(d) All of the above

279. How many green house gases exist?
(a) Two
(b) Three
(c) Four
(d) Six

280. Annual world precipitation is:
(a) $7x10^{10}g$
(b) $7x10^{20}g$
(c) $4.46x10^{20}g$
(d) $4.46x10^{20}$

281. Degradation of proteins plays a part in:
(a) Calvin cycle
(b) Water cycle
(c) Sulphur cycle
(d) Nitrogen cycle

282. Biome is a:
(a) Community of plants
(b) Community of animals
(c) Small sized ecosystems
(d) Large sized ecosystem

283. Ecological niche of an organism denotes:

(a) Habitat of the organisms
(b) On whom the organisms feeds
(c) The status of an organism within a particular community
(d) All of the above

284. The transition zone where two different types of communities meet is called:

(a) Ecotone (b) Ecocline
(c) Ecotype (d) Ecosystem

285. Total organic matter present in an ecosystem is called:

(a) Biomass (b) Biome
(c) Litter (d) Food

286. The most important organisms for an ecosystem are:

(a) Green plants (b) Herbivores
(c) Carnivores (d) Protozoa

287. A food chain is:

(a) Group of organisms which eat the same type of food
(b) Animals eating the plants
(c) Series of plants/animals which are interrelated in the form of organism being eaten as food by the other
(d) None of the above

288. A food chain consists of:

(a) Producers (b) Consumers
(c) Decomposers (d) Both (a) and (b)

289. A food chain always starts with:

(a) Photosynthesis (b) Respiration
(c) Nitrogen fixation (d) Decay

290. Relationship between different organisms can best studied by:

(a) Eltonian pyramids (b) Charles pyramids
(c) Hierarchical pyramids (d) All of the above

291. Controlling factor in ecosystem is:

(a) Predation (b) Light
(c) Precipitation (d) Nutrients

292. Micro consumers are called:

(a) Omnivores
(b) Saprophyte
(c) Both (a) and (b)
(d) None of the above

293. In pond ecosystem the shape of pyramid of number is:

(a) Upright
(b) Inverted
(c) Spindle like
(d) Variable

294. In parasitic food chain the pyramid of number is:

(a) Upright
(b) Inverted
(c) Spindle like
(d) Variable

295. In grassland ecosystem the pyramid of biomass is:

(a) Upright
(b) Inverted
(c) Spindle like
(d) Variable

296. Pyramid of energy in forest ecosystem is:

(a) Always inverted
(b) Always upright
(c) Spindle shaped
(d) Both (a) and (b)

297. In pyramid of food, the producers occupy:

(a) The base
(b) Top
(c) Middle
(d) All of the above

298. Rate of storage of organic matter not used by heterotrophs is termed as:

(a) Net primary productivity
(b) Gross primary productivity
(c) Secondary productivity
(d) None of the above

299. Graphic representation of biomass relationship between the producers and consumers in an ecosystem is called:

(a) Ecological pyramid
(b) Ecological hill
(c) Ecological mountain
(d) All of the above

300. Organisms living at the bottom of a lake are:

(a) Benthos
(b) Biocoen
(c) Biocide
(d) All of the above

301. A recently discovered ecosystem is:

(a) Tundra
(b) Iceberg
(c) Vent
(d) Crater

302. Factor related to structure of earth's surface is:

(a) Topographic
(b) Edaphic
(c) Physiographic
(d) Both (a) and (c)

303. In upright pyramid of biomass, herbivores occupy the position:

(a) One (b) Two

(c) Three (d) Four

304. Which is amphibious?

(a) Monogonum (b) Digonum

(c) Trigonum (d) Polygonum

305. Which forests have maximum diversity?

(a) Tropical forests (b) Temperate forests

(c) Alpine forest (d) Thorn forest

306. A treeless biome is:

(a) Tundra (b) Grassland

(c) Desert (d) All of the above

307. Lentic ecosystem occurs in:

(a) Standing water (b) Running water

(c) River water (d) All of the above

308. Ecosystem of a pond is:

(a) Lentic (b) Lotic

(c) Both (a) and (b) (d) None of the above

309. Alpine tundra is found in:

(a) Siberia (b) Greenland

(c) Himalyas (d) Both (a) and (b)

310. Type of plants having adaptations to check transpiration is:

(a) Xerophytes (b) Lithophytes

(c) Halophytes (d) Epiphytes

311. Viviparity and Pneumatophores are features of:

(a) Hydrophytes (b) Mesophytes

(c) Xerophytes (d) Haloaphytes

312. Which part of the world has high density of organisms?

(a) Savannahs (b) Grassland

(c) Deciduous forest (d) Tropical rain forest

313. The most favoured habitat of a plant is called:

(a) Ecological habitat (b) Microhabitat

(c) Ecotype (d) Ecoline

314. Alpine plants show:

(a) Halophytism (b) Mesophytism

(c) Luxurian growth (d) Xerophytism

315. Temperate forest occur in India in:

(a) Indo-Gangetic plants (b) Himalyas

(c) Eastern India (d) Southern peninsula

316. The forest colours in autumn is:

(a) Temperate evergreen forest (b) Temperate deciduous forest

(c) Tropical evergreen forest (d) Tropical deciduous forest

317. Ephemerals are a type of xerophytes:

(a) Drought escaping (b) Drought resistant

(c) Drought enduring (d) None of the above

318. Succulents occur in:

(a) Deserts (b) Tundra

(c) Temperate deciduous forest (d) Tropical rain forest

319. The proximate causes of behavior are interactions with the environment, but behavior is ultimately shaped by:

(a) Evolution (b) Mutation

(c) Natural selection (d) Genetic drift

320. Which of these ecosystems has the highest primary productivity per square meter?

(a) Thorn forest (b) Temperate forest

(c) Deciduous forest (d) Tropical rain forest

321. The total biomass of photosynthetic autotrophs present in an ecosystem is known as the:

(a) Standing state (b) Standing crop

(c) Standing vegetation (d) All of the above

322. Starting on bare rock, what is the usual ecological succession of organisms?

(a) Lichens → grasses → shrubs → trees

(b) Trees → grasses → shrubs → Lichens

(c) Grasses → shrubs → trees → Lichens

(d) None of the above

323. The stable stage that is established in an area as a result of the process of ecological succession is known as the:

(a) Population growth
(b) Ecological succession
(c) Climate change
(d) Climax community

324. The place an organism lives is called its _______.

(a) Habit
(b) Habitat
(c) Niche
(d) Community

325. This is a symbiotic relationship between organisms where both are helped (+,+) This relationship is called _______:

(a) Mutualism
(b) Parasitism
(c) Commensalism
(d) Parasite

326. This is a symbiotic relationship between organisms where one organisms is helped, while the other is harmed. (+,-) This relationship is called _______.

(a) Mutualism
(b) Parasitism
(c) Commensalism
(d) Parasite

327. This is a symbiotic relationship between organism where one organism is helped, while the other it is associated with is basically unaffected. (+,0) This relationship is called:

(a) Mutualism
(b) Parasitism
(c) Commensalism
(d) Parasite

328. The feeding role of an organism is called a (n) _______:

(a) Biche
(b) Niche
(c) Crèche
(d) None of the above

329. As a population reaches its carrying capacity, there is an increase in competition for:

(a) Food
(b) Shelter
(c) Mates
(d) All of the above

330. An ecosystem consists of:

(a) A community of organisms
(b) Energy
(c) The soil, water and weather
(d) All of the above

331. Organisms that manufacture organic nutrients for an ecosystem are called:

(a) Primary producers
(b) Primary consumers
(c) Secondary producers
(d) Secondary consumers

332. In a food web, which type of organism receives energy from every other type?

(a) Herbivore
(b) Producers
(c) Carnivore
(d) None of the above

333. Animals that feed on plants are at least in the:

(a) Second trophic level
(b) Third trophic level
(c) Fourth trophic level
(d) Fifth trophic level

334. Because energy diminishes at each successive trophic level, few ecosystems can contain more than:

(a) Two trophic levels
(b) Three trophic levels
(c) Four trophic levels
(d) Five trophic levels

335. Precipitation and evaporation are important components of the:

(a) Nitrogen cycle
(b) Carbon cycle
(c) Potassium cycle
(d) Water cycle

336. The paths of water, carbon, nitrogen, and phosphorus pass from the non-living environment to living organisms and back to the non-living environment in closed circles called:

(a) Biogeochemical cycles
(b) Chemical cycle
(c) Atmosphere cycle
(d) None of the above

337. The process by which species evolve in response to other living members of their ecosystem is called:

(a) Competition
(b) Coevolution
(c) Coincidence
(d) Predation

338. The diversity that enables life on earth to adapt and survive environmental changes is called:

(a) Functional diversity
(b) Species diversity
(c) Ecosystem diversity
(d) Genetic diversity

339. The variety of processes, including matter cycling and energy flow within ecosystems, that result from species interacting with one another in food webs is called:

(a) Functional diversity
(b) Species diversity
(c) Ecosystem diversity
(d) Genetic diversity

340. A change in the genetic characteristics of a population from one generation to another is called:

(a) Emigration
(b) Mutation
(c) Genetic drift
(d) Evolution

341. Some individuals in a population have genetic traits that enhance their ability to survive and produce offspring, a process called:

(a) Natural selection (b) Emigration

(c) Mutation (d) Genetic drift

342. Biodiversity is believed to be the result of the interaction between:

(a) Geographic isolation and reproductive isolation

(b) Speciation and extinction

(c) Mass extinction and background extinction

(d) Speciation and genetic engineering

343. The number of different species an ecosystem contains is its:

(a) Speciation (b) Species evenness

(c) Species niche (d) Species richness

344. A species in an ecosystem that plays a central role in the health of that ecosystem and whose removal may cause the collapse of the ecosystem, is called a(n):

(a) Indicator species (b) Keystone species

(c) Specialist species (d) Foundation species

345. A species in an ecosystem that shapes communities by creating and enhancing habitats in ways that benefit other species is called:

(a) Indicator species (b) Keystone species

(c) Specialist species (d) Foundation species

346. In which eco systems, producers are of large size:

(a) Grassland ecosystem (b) Tropical deciduous

(c) Temperate (d) Desert

347. The test has self purification capacity of water body:

(a) BOD test (b) DOD test

(c) COD test (d) All of the above

348. Sources of soil pollutants are:

(a) Herbicides (b) Pesticides

(c) Weedicides (d) All of the above

349. Most poisonous pollutant in water is:

(a) Arsenic (b) Nickel

(c) Cobalt (d) All of the above

350. Pollutant free alternative to petrol for automobiles:

(a) Propane (b) Methane

(c) Ethane (d) All of the above

351. Upright pyramid of biomass in forest ecosystem is:

(a) Producer – herbivores –carnivores

(b) Carnivores – herbivores – producer

(c) Both (a) and (b)

(d) None of the above

352. Thickness of ozone layer is measured in:

(a) Pascal (b) Dabson units

(c) Cm/sec (d) Both (a) and (c)

353. The gas responsible for Bhopal tragedy:

(a) Methyl isocyanide (b) Methyl isocyanate

(c) BHC (d) All of the above

354. The gas does not pollute air:

(a) Nitrogen Dioxide (b) SO_2

(c) CO (d) NO

355. Term ecology was proposed by:

(a) Odum (b) Reiter

(c) Grinnel (d) Lack

356. Who coined the term symbiosis?

(a) De Bary (b) Clements

(c) Tansley (d) McDougall

357. Plants which are attached to the stones:

(a) Halophytes (b) Psychrophytes

(c) Lithophytes (d) Oxylophytes

358. Plants which grow in clod areas are:

(a) Halophytes (b) Psychrophytes

(c) Lithophytes (d) Oxylophytes

359. Plants growing in soils having sand and gravel a re:

(a) Psammophytes (b) Halophytes

(c) Psychrophytes (d) Oxylophytes

360. Plant growing well in full sunlight are called:

(a) Halophytes (b) Psychrophytes
(c) Lithophytes (d) Oxylophytes

361. Plants growing in saline soil water are called:

(a) Halophytes (b) Psychrophytes
(c) Lithophytes (d) Oxylophytes

362. The ministry of Environment was set up in:

(a) 1952 (b) 1980
(c) 1986 (d) 1988

363. The National Institute of Oceanography is presently situated at:

(a) Kerala (b) Calicut
(c) Cochin (d) Goa

364. The process used for removal of water hardness is:

(a) Zeolite process (b) Heber's process
(c) Ostwald process (d) None of the above

365. The first biosphere reserve in India was:

(a) Nilgiri (b) Gulf of Manna
(c) Sundarbans (d) Nanda Devi

366. Permanent hardness of water is caused by:

(a) Carbonates and chlorides (b) Bicarbonates and chlorides
(c) Carbonates and sulphates (d) Chlorides and sulphates

367. In which year Biodiversity act was proposed:

(a) 2000 (b) 2002
(c) 2012 (d) 1998

368. Kyoto protocol came into force on:

(a) 16 February 2004 (b) 14 February 2006
(c) 14 February 2004 (d) 16 February 2006

369. The compound used for artificial rain making/cloud seeding:

(a) AgI (b) Ni
(c) Fe (d) Hg

370. Number species in single community:

(a) Alpha diversity (b) Beta diversity
(c) Gamma diversity (d) All of the above

371. The degree to which species composition changes along an environmental gradient:

(a) Alpha diversity
(b) Beta diversity
(c) Gamma diversity
(d) All of the above

372. Which diversity applied to larger geographical scales?

(a) Alpha diversity
(b) Beta diversity
(c) Gamma diversity
(d) All of the above

373. What is frequency?

(a) Total number of quadrats in which species has occurred/total number of quadrats studied
(b) Total number of individual of the species/total number of quadrats studied
(c) Total number of individual of the species/total number of quadrats in which species has occurred
(d) None of the above

374. What is density?

(a) Total number of quaqrats in which species has occurred/total number of quadrats studied
(b) Total number of individual of the species/total number of quadrats studied
(c) Total number of individual of the species/total number of quadrats in which species has occurred
(d) None of the above

375. What is abundance?

(a) Total number of quaqrats in which species has occurred/total number of quadrats studied
(b) Total number of individual of the species/total number of quadrats studied
(c) Total number of individual of the species/total number of quadrats in which species has occurred
(d) None of the above

376. Study of energy transformation in living systems called:

(a) Bioenergetics
(b) Biotic potential
(c) Bioconcentration
(d) Energy flow

377. Symbiosis in which one organisms in inhibited by other but the latter remains unaffected is called:

(a) Commensalism (b) Amensalism

(c) Neutralism (d) Mutualism

378. The science of relationship of life to weather is called:

(a) Biometeorology (b) Biotic potential

(c) Biogeosphere (d) Biological amplification

379. A plant which grows on ice or snow is called:

(a) Cryophyte (b) Hydrophyte

(c) Halophyte (d) Hygrophyte

380. Science of communication and control mechanisms in systems is called:

(a) Cybernetics (b) Sciophyte

(c) Toxicology (d) Genecology

381. The ration between number of species and number of individuals is called:

(a) Diversity (b) Frequency

(c) Diversity index (d) Abundance

382. The process of successful establishment of a species in a new area:

(a) Exesis (b) Ecad

(c) Both (a) and (b) (d) None of the above

383. Maintenance of equilibrium between living components of an ecosystem, so that it remains stable system is called:

(a) Ecological balance (b) Ecological succession

(c) Ecological equivalents (d) Ecotone

384. One or more ecotype in a single coenospecies is called:

(a) Ecospecies (b) Ecospecies

(c) Endemic species (d) Indicator species

385. Science of improving the human race by improving environment:

(a) Euthenics (b) Ecology

(c) Ecospecies (d) Eutrophic

386. Science which deals with the genetic of population in relation the environment is called:

(a) Genecology (b) Toxicology

(c) Autecology (d) Synecology

387. Study of changes in population densities in an area:

(a) Population dynamics | (b) Population
(c) Both (a) and (b) | (d) None of the above

388. Species diversity is lowest in ecosystems:

(a) Tundra | (b) Temperate grasslands
(c) Desert | (d) Tropical forest

389. Best soil for growth of plants in summer is:

(a) Clay | (b) Sand
(c) Red | (d) Loamy

390. Ecological succession on sand is:

(a) Halosere | (b) Xerosere
(c) Hydrosere | (d) Psammomere

391. Succession showing changes in communities at a place is called:

(a) Biotic succession | (b) Climatic succession
(c) Geographical succession | (d) Edaphic succession

392. Secondary succession is not completed if the barren area is invaded by:

(a) Sphagnum | (b) Lantana
(c) Eichhornia | (d) All of the above

393. Xerosere is:

(a) Xerophytes | (b) Halophyte
(c) Ecological succession on rock | (d) None of the above

394. Maximum diversity is found in:

(a) Pioneer community | (b) Seral community
(c) Climax community | (d) Moss stage

395. Minimum diversity is found in:

(a) Pioneer community | (b) Seral community
(c) Climax community | (d) Moss stage

396. Long term stability of a biotic community is governed by its:

(a) More diversity | (b) Population
(c) Pioneer community | (d) Environment

397. Which seral stage in a succession takes maximum time?

(a) Pioneer stage | (b) Climax stage
(c) Seral stage | (d) None of the above

398. Regeneration of community is due to:

(a) Primary succession
(b) Secondary succession
(c) Climax community
(d) Both (a) and (b)

399. Last succession in a sere is called:

(a) Autogenic succession
(b) Allogenic succession
(c) Biotic climax
(d) Climatic climax

400. Most wide spread succession is:

(a) Autotrophic succession
(b) Autogenic succession
(c) Allogenic succession
(d) None of the above

401. Largest vegetation group is known as:

(a) Association
(b) Consociation
(c) Formation
(d) Ecosystems

Answer Sheet

1.	(a)	31.	(a)	61.	(a)	91.	(a)	121.	(b)
2.	(a)	32.	(d)	62.	(d)	92.	(a)	122.	(d)
3.	(b)	33.	(c)	63.	(c)	93.	(b)	123.	(a)
4.	(a)	34.	(c)	64.	(b)	94.	(a)	124.	(a)
5.	(a)	35.	(d)	65.	(a)	95.	(a)	125.	(c)
6.	(a)	36.	(c)	66.	(a)	96.	(d)	126.	(a)
7.	(c)	37.	(d)	67.	(c)	97.	(a)	127.	(b)
8.	(a)	38.	(c)	68.	(a)	98.	(a)	128.	(d)
9.	(b)	39.	(a)	69.	(b)	99.	(b)	129.	(a)
10.	(a)	40.	(d)	70.	(d)	100.	(c)	130.	(a)
11.	(b)	41.	(c)	71.	(c)	101.	(a)	131.	(a)
12.	(c)	42.	(a)	72.	(c)	102.	(c)	132.	(c)
13.	(c)	43.	(d)	73.	(b)	103.	(b)	133.	(c)
14.	(a)	44.	(c)	74.	(b)	104.	(a)	134.	(a)
15.	(a)	45.	(c)	75.	(a)	105.	(c)	135.	(c)
16.	(a)	46.	(a)	76.	(a)	106.	(d)	136.	(a)
17.	(d)	47.	(c)	77.	(a)	107.	(a)	137.	(a)
18.	(c)	48.	(c)	78.	(a)	108.	(c)	138.	(a)
19.	(c)	49.	(a)	79.	(b)	109.	(b)	139.	(a)
20.	(b)	50.	(c)	80.	(a)	110.	(b)	140.	(b)
21.	(a)	51.	(d)	81.	(a)	111.	(c)	141.	(b)
22.	(a)	52.	(a)	82.	(b)	112.	(b)	142.	(a)
23.	(a)	53.	(b)	83.	(c)	113.	(b)	143.	(d)
24.	(b)	54.	(b)	84.	(d)	114.	(c)	144.	(d)
25.	(a)	55.	(a)	85.	(b)	115.	(c)	145.	(a)
26.	(b)	56.	(a)	86.	(a)	116.	(c)	146.	(b)
27.	(d)	57.	(d)	87.	(d)	117.	(b)	147.	(c)
28.	(b)	58.	(a)	88.	(b)	118.	(c)	148.	(a)
29.	(d)	59.	(b)	89.	(a)	119.	(d)	149.	(d)
30.	(a)	60.	(d)	90.	(a)	120.	(a)	150.	(b)

151.	(c)	202.	(a)	253.	(b)	304.	(d)	355.	(b)
152.	(d)	203.	(d)	254.	(a)	305.	(a)	356.	(a)
153.	(b)	204.	(a)	255.	(a)	306.	(d)	357.	(c)
154.	(d)	205.	(a)	256.	(d)	307.	(a)	358.	(b)
155.	(b)	206.	(a)	257.	(c)	308.	(a)	359.	(a)
156.	(c)	207.	(b)	258.	(d)	309.	(c)	360.	(d)
157.	(a)	208.	(a)	259.	(d)	310.	(a)	361.	(a)
158.	(b)	209.	(b)	260.	(a)	311.	(d)	362.	(b)
159.	(d)	210.	(a)	261.	(d)	312.	(a)	363.	(d)
160.	(a)	211.	(b)	262.	(d)	313.	(b)	364.	(a)
161.	(a)	212.	(a)	263.	(b)	314.	(d)	365.	(a)
162.	(d)	213.	(b)	264.	(a)	315.	(b)	366.	(d)
163.	(b)	214.	(c)	265.	(a)	316.	(b)	367.	(a)
164.	(b)	215.	(a)	266.	(d)	317.	(a)	368.	(a)
165.	(b)	216.	(c)	267.	(c)	318.	(a)	369.	(a)
166.	(d)	217.	(c)	268.	(b)	319.	(a)	370.	(a)
167.	(b)	218.	(b)	269.	(b)	320.	(d)	371.	(b)
168.	(c)	219.	(b)	270.	(c)	321.	(b)	372.	(c)
169.	(d)	220.	(b)	271.	(c)	322.	(a)	373.	(a)
170.	(a)	221.	(a)	272.	(a)	323.	(d)	374.	(b)
171.	(b)	222.	(a)	273.	(a)	324.	(b)	375.	(c)
172.	(c)	223.	(d)	274.	(d)	325.	(a)	376.	(a)
173.	(a)	224.	(c)	275.	(c)	326.	(b)	377.	(b)
174.	(d)	225.	(b)	276.	(a)	327.	(c)	378.	(a)
175.	(b)	226.	(a)	277.	(a)	328.	(b)	379.	(a)
176.	(a)	227.	(d)	278.	(d)	329.	(d)	380.	(a)
177.	(c)	228.	(a)	279.	(d)	330.	(d)	381.	(c)
178.	(c)	229.	(d)	280.	(d)	331.	(a)	382.	(a)
179.	(d)	230.	(a)	281.	(d)	332.	(c)	383.	(a)
180.	(d)	231.	(a)	282.	(d)	333.	(a)	384.	(b)
181.	(a)	232.	(a)	283.	(c)	334.	(d)	385.	(a)
182.	(a)	233.	(a)	284.	(a)	335.	(d)	386.	(a)
183.	(a)	234.	(b)	285.	(a)	336.	(a)	387.	(a)
184.	(c)	235.	(a)	286.	(a)	337.	(b)	388.	(a)

185.	(d)	236.	(b)	287.	(c)	338.	(d)	389.	(a)
186.	(c)	237.	(a)	288.	(d)	339.	(a)	390.	(d)
187.	(d)	238.	(d)	289.	(a)	340.	(d)	391.	(a)
188.	(a)	239.	(a)	290.	(a)	341.	(a)	392.	(d)
189.	(d)	240.	(a)	291.	(a)	342.	(b)	393.	(c)
190.	(a)	241.	(a)	292.	(b)	343.	(d)	394.	(c)
191.	(a)	242.	(d)	293.	(a)	344.	(b)	395.	(a)
192.	(a)	243.	(c)	294.	(b)	345.	(d)	396.	(a)
193.	(a)	244.	(a)	295.	(a)	346.	(a)	397.	(a)
194.	(a)	245.	(d)	296.	(b)	347.	(a)	398.	(c)
195.	(a)	246.	(d)	297.	(a)	348.	(d)	399.	(d)
196.	(d)	247.	(b)	298.	(a)	349.	(a)	400.	(c)
197.	(b)	248.	(d)	299.	(a)	350.	(a)	401.	(b)
198.	(d)	249.	(a)	300.	(a)	351.	(a)		
199.	(a)	250.	(a)	301.	(d)	352.	(b)		
200.	(a)	251.	(a)	302.	(a)	353.	(b)		
201.	(d)	252.	(c)	303.	(b)	354.	(a)		

Chapter 6

Forest Utilization and NTFPs

1. **The process of harvesting, conversion and disposal of standing forest produce is called:**
   (a) Forest mensuration
   (b) Forest management
   (c) Forest utilization
   (d) Forest economics

2. **Aim of the forest utilization is:**
   (a) Complete and efficient utilization of standing timber
   (b) Extraction of timbers
   (c) Storage of timbers
   (d) Marketing of timbers

3. **The tree price generally represents _______ value of the saleable products:**
   (a) 1-2 per cent
   (b) 10-22 per cent
   (c) 25-31 per cent
   (d) 25-50 per cent

4. **Total cost of felling, conversion and transport is approximately _______ per cent of the total cost of timber in the market:**
   (a) 20 per cent
   (b) 30 per cent
   (c) 40 per cent
   (d) 50 per cent

5. **The wastages in felling and conversion will be more in _______ region:**
   (a) Hilly
   (b) Waste land
   (c) Forest areas
   (d) Terrain

6. **The total volume of usable rough timber from the average tree is only:**
   (a) 17 per cent to 22 per cent
   (b) 27 per cent to 45 per cent
   (c) 37 per cent to 52 per cent
   (d) 47 per cent to 65 per cent

7. **Logging branch at Forest Research Institute, Dehradun came into existence during:**
   (a) April, 1947
   (b) June, 1947
   (c) April, 1957
   (d) June, 1957

8. **Logging branch at Forest Research Institute, Dehradun established due to the recommendations of:**
   (a) Dr. A. Huber
   (b) Dr. A. Koroleff
   (c) Both (a) and (b)
   (d) None of the above

9. **Logging Training Centre was established at:**
   (a) Batote
   (b) Dehradun
   (c) Delhi
   (d) Mumbai

10. **Logging training centre at Batote was established by:**
   (a) Dr. A. Huber
   (b) Dr. A. Koroleff
   (c) Dr. H.G. Winklemann
   (d) All of the above

11. **Logging Training Centre at Batote was created during:**
   (a) 1948-49
   (b) 1958-59
   (c) 1968-69
   (d) 1978-79

12. **Logging Training Centre Project implemented throughout India mainly for:**
   (a) Improvement in logging methods
   (b) Timber harvesting
   (c) Both (a) and (b)
   (d) None of the above

13. **The extent of utilization of any species closely related with its:**
   (a) Market price
   (b) Demand
   (c) Supply price
   (d) All of the above

14. **The felling tools and implements were characterized by:**
   (a) Strength
   (b) Durability
   (c) Efficiency
   (d) All of the above

15. **Axes are used mainly for:**
   (a) Felling
   (b) Trimming
   (c) Splitting and grubbing
   (d) All of the above

16. The portion of axe head in front of the eye is known as:

(a) Blade (b) Back

(c) Face (d) Thickness

17. The portion of axe head behind the eye is known as:

(a) Blade (b) Back

(c) Face (d) Thickness

18. The sides of the blade should be:

(a) Back (b) Convex

(c) Curves (d) Thickness

19. The edge of the blade should be:

(a) Back (b) Convex

(c) Curves (d) Thickness

20. Generally the weight of the axe will be:

(a) To 1.5 kg (b) 1.5 to 2.0 kg

(c) 2.5 to 3.0 kg (d) To 2.5 kg

21. In India, the axe handles will be in _______ shape:

(a) Round (b) Square

(c) Rectangular (d) Oval

22. European and American axes have _______ shaped handle:

(a) Round (b) Square

(c) Rectangular (d) Oval

23. Bill hook is often used or felling:

(a) Teak (b) Bamboos

(c) *Terminalia* spp. (d) All of the above

24. Saws are widely used for:

(a) Felling (b) Cross cutting

(c) Ripping of logs (d) All of the above

25. The edge of the tooth which faces the cutting direction is called:

(a) Blade (b) Back

(c) Face (d) Thickness

26. The opposite edge of the tooth is called:

(a) Blade (b) Back

(c) Face (d) Thickness

27. The distance between the two adjacent teeth is known as:

(a) Blade (b) Back

(c) Face (d) Space

28. The entire opening between two adjacent teeth is called:

(a) Space (b) Gullet

(c) Pitch (d) Set

29. The extent to which the teeth are bent is termed as:

(a) Space (b) Gullet

(c) Pitch (d) Set

30. The width of the saw cut is:

(a) Kerf (b) Gauge

(c) Set (d) Gullet

31. The thickness of the blade is called:

(a) Kerf (b) Gauge

(c) Set (d) Gullet

32. The suitable stroke length of for cross cut saws is:

(a) 50 to 80 cm (b) 60 to 80 cm

(c) 80 to 100 cm (d) 100 to 120 cm

33. _______ different tasks of the tooth are:

(a) Cut (b) Break

(c) Carry away the fibres (d) All of the above

34. Per toothed saw consists of all teeth in the form of:

(a) Isosceles triangles (b) Oval

(c) Square (d) Circular

35. The raker toothed saw consisting of teeth in the form of:

(a) Rakers (b) Cutters

(c) Both (a) and (b) (d) None of the above

36. Bow man saw is _______. Man saw:

(a) One man (b) Two man

(c) Both (a) and (b) (d) None of the above

37. The pruning saw will have tooth pattern of _______teeth cutting:

(a) Reverse (b) Straight

(c) Both (a) and (b) (d) None of the above

38. Power saws are mainly used for:

(a) Felling (b) Cross cutting
(c) Both (a) and (b) (d) None of the above

39. Wedges are used while felling to prevent:

(a) Jamming (b) Cutting
(c) Splitting (d) All of the above

40. Cant hooks are used as a:

(a) Lever in rolling (b) Stopping
(c) Turning (d) All of the above

41. Pickaroons are used for:

(a) Drawing (b) Pulling of small logs
(c) Both (a) and (b) (d) None of the above

42. Debarking spades are used for:

(a) Cutting of logs (b) Splitting the logs
(c) Felling the logs (d) Debarking the logs

43. Log hooks are used for:

(a) Dragging (b) Lifting
(c) Rolling of logs (d) All of the above

44. The use of Stem Tightener is to prevent:

(a) Splitting of stem (b) Cutting of stem
(c) Felling of stem (d) Debarking of stem

45. The best felling season is:

(a) Winter (b) Summer
(c) Rainy (d) All of the above

46. Felling should begin with _______ of the slope:

(a) Top (b) Bottom
(c) Centre (d) Any part of tree

47. The waste attributed due to stump is:

(a) 1 per cent to 2 per cent (b) 2 per cent to 5 per cent
(c) 5 per cent to 7 per cent (d) 7 per cent to 10 per cent

48. Notch or undercut is made to decide the:

(a) Direction of fall (b) Opposite direction of fall
(c) Both (a) and (b) (d) None of the above

49. **The undercut should be _______ diameter of the tree:**
(a) 1/3 (b) 2/3
(c) (d) 2/4

50. **Felling by roots is practiced in:**
(a) *Santalum album* (b) *Acacia catechu*
(c) *Juglens regia* (d) All of the above

51. **The length of the log depends on:**
(a) Market requirement (b) Transport
(c) Both (a) and (b) (d) None of the above

52. **The rounded logs split with axe is called:**
(a) Squares (b) Hakries
(c) Poles (d) Waney baulks

53. **Pieces of short lengths with square surfaces is called:**
(a) Beams (b) Scantlings
(c) Planks (d) Blocks

54. **Quarter sawing otherwise termed as:**
(a) Radially sawing (b) Hand sawing
(c) Machine sawing (d) None of the above

55. **In quarter sawn timber _______ will be exposed:**
(a) Medullary rays (b) Soft fibrous tissues
(c) Haof theserd fibrous tissues (d) None of the above

56. **In tangentially sawn timber _______ will be exposed:**
(a) Medullary rays (b) Soft fibrous tissues
(c) Hard fibroust issues (d) None of the above

57. **The transport constitutes _______ per cent of the total cost of delivery:**
(a) 20 to 30 (b) 30 to 40
(c) 40 to 60 (d) 60 to 70

58. **Minor transportation is also called:**
(a) Off road Transportation (b) Transport by land
(c) Water transportation (d) Overhead transportation

59. **Floating method of water transportation is applicable mostly to:**
(a) Softwoods (b) Hardwoods
(c) Both (a) and (b) (d) None of the above

60. Telescopic floating attributed to:
(a) Upper steams with plenty of water
(b) Free floating
(c) Both (a) and (b)
(d) None of the above

61. Nullas can be effectively utilized for:
(a) Telescopic floating
(b) River floating
(c) Free floating
(d) All of the above

62. Construction over and around obstruction in telescopic floating is called:
(a) Pathru
(b) River floating
(c) Free floating
(d) All of the above

63. River floating is otherwise called as:
(a) Free floating
(b) Sweeping
(c) Both (a) and (b)
(d) None of the above

64. Major hazard in the river floating is:
(a) Jamming
(b) Splitting
(c) Debarking
(d) All of the above

65. The average distance covered per day in river floating is:
(a) 1 to 2 km
(b) 2 to 3 km
(c) 3 to 4 km
(d) 4 to 5 km

66. A quantity of wood firmly bounded together is called:
(a) Raft section
(b) Floating
(c) Booms
(d) Wet slides

67. Collection of raft section is called:
(a) Raft section
(b) Floating
(c) Booms
(d) Wet slides

68. The average number of bamboo culms per raft is around:
(a) 100 culms
(b) 1000 culms
(c) 10000 culms
(d) 100000 culms

69. Construction of floating obstruction across the river is termed as:
(a) Raft
(b) Floating
(c) Booms
(d) Wet slides

70. Permanent booms are otherwise called:
(a) Fixed booms
(b) Two way boom
(c) V-shaped boom
(d) All of the above

71. Two way boom is otherwise called:

(a) Single arm boom
(b) V-shaped boom
(c) Both (a) and (b)
(d) None of the above

72. The arms of two way boom should form _______ degree of angle at the apex on the down stream side:

(a) 200
(b) 300
(c) 400
(d) 500

73. To place the boom proper in right position _______ is used:

(a) Guy ropes
(b) Log line
(c) Line dori
(d) All of the above

74. To extract timber from nullas where quantity of water is less_______ used:

(a) Floating
(b) Rafting
(c) Booms
(d) Wet slides

75. _______ transportation method can be deployed where surface transportation is not possible:

(a) Transport by land
(b) Water transportation
(c) Over had transportation
(d) Minor transportation

76. A place where wood and other forest produce is stored for disposal is:

(a) Forest
(b) Biomass
(c) Saw mill
(d) Depot

77. A place which is mainly intended for collection and checking of forest produce is called:

(a) Forest depots
(b) Transit depot
(c) Sale depot
(d) None of the above

78. An intermediate depots which may inside or outside the forest is called:

(a) Forest depots
(b) Transit depot
(c) Sale depot
(d) None of the above

79. A place where the timber and other forest produce stored for disposal by sale is called:

(a) Forest depots
(b) Transit depot
(c) Sale depot
(d) None of the above

80. Situation where continuous supply of forest produce is necessary _______. Felling system can be deployed:

(a) Felling by government agency
(b) Felling by private agency
(c) Government and purchasers combined
(d) Purchasers alone

81. Situation which demands care and control over felling _______. Felling system is highly suited:

(a) Felling and extraction by government agency

(b) Felling and extraction by government and purchasers combined

(c) Felling and extraction by purchasers alone

(d) All of the above

82. Most satisfactory system of felling and extraction:

(a) Working by government agency

(b) Working by purchasers alone

(c) Working by government and purchasers combined

(d) All of the above

83. _______ is the system of sale where the exact quantity of produce is not known:

(a) Lump-sum sale (b) Payment on out turn

(c) Both (a) and (b) (d) None of the above

84. _______ is the system of sale where the amount fixed with the actual produce extracted or sold:

(a) Lump-sum sale (b) Payment on out turn

(c) Both (a) and (b) (d) None of the above

85. A price fixed which prevents unreasonable low price during the sale of produce is:

(a) Upper price (b) Reserve price

(c) Both (a) and (b) (d) None of the above

86. Sale of forest produce based on individual consultation is called:

(a) Sale by private bargain (b) Sale to highest bidder

(c) Sale by royalty (d) None of the above

87. _______ is the method of sale of forest produce where keen competition exists:

(a) Sale by private bargain (b) Sale to highest bidder

(c) Sale by royalty (d) None of the above

88. _______ is the auction method which commences with prohibitively high price:

(a) Upset price (b) Reverse price

(c) Dutch auction (d) None of the above

89. Earnest Money Deposit is usually _______.

(a) 2 per cent
(b) 5 per cent
(c) 10 per cent
(d) 14 per cent

90. Security Deposit is _______ per cent of the contract value:

(a) 2 per cent
(b) 5 per cent
(c) 10 per cent
(d) 14 per cent

91. Intelligent choice of timber for specific purpose is called:

(a) Logging
(b) Grading
(c) Both (a) and (b)
(d) None of the above

92. In India, _______ is the organization responsible for grading of timbers:

(a) Indian standards institution
(b) Grading
(c) Quality
(d) Storage

93. Separation of fibres along the grain is called:

(a) Check
(b) Shake
(c) Flutes
(d) Spiral grain

94. Patches of bark either partially or wholly embedded inside the wood is called:

(a) Shake
(b) Flutes
(c) Spiral grain
(d) Inbark

95. Difference in girth at the butt and top is called:

(a) Taper
(b) Flutes
(c) Knot
(d) Surface cracks

96. Deviation from straightness of log is called:

(a) Crookedness
(b) Curvature bend
(c) Both (a) and (b)
(d) None of the above

97. Tree that shows departure from its cylindrical shape is called:

(a) Knor
(b) Flute
(c) Spiral grain
(d) Spilt

98. A portion of branch that embedded in the wood by natural growth is called:

(a) Knor
(b) Flute
(c) Spiral grain
(d) Spilt

99. A knot in which fibres are completely inter grown with surrounding wood is called:

(a) Live knot
(b) Dead knot
(c) Both (a) and (b)
(d) None of the above

100. A knot in which fibres are not completely inter grown with surrounding wood is called:

(a) Live knot (b) Dead knot
(c) Both (a) and (b) (d) None of the above

101. A type of growth in which the fibres take a spiral course about the bole is called:

(a) Spiral grain (b) Crack
(c) Spilt (d) Rot

102. Separation of fibres along the longitudinal directions is called:

(a) Spiral grain (b) Crack
(c) Spilt (d) Rot

103. Separation of fibres along the grain which extends from one surface to another is called:

(a) Spiral grain (b) Crack
(c) Spilt (d) Rot

104. A small soft core occurring in the structural centre is called:

(a) Centre-heart (b) Hollow centre
(c) Rot (d) Shakes

105. The partial or complete or complete removal of pith with or without removal of heartwood is called:

(a) Centre-heart (b) Hollow centre
(c) Rot (d) Shakes

106. A partial or complete separation between adjoining layers of fibres is called:

(a) Centre-heart (b) Hollow centre
(c) Rot (d) Shakes

107. A crack origination from centre and extending radially in one or opposite direction is called.

(a) Heart shake (b) Star shake
(c) Radial shake (d) Ring shake

108. Several splits or cracks originate from the centre of the timber is called:

(a) Heart shake (b) Star shake
(c) Radial shake (d) Ring shake

109. All cracks appearing along the radii is called:

(a) Heart shake (b) Star shake
(c) Radial shake (d) Ring shake

110. Shakes that are formed rupture of tissues in a circular direction is called:

(a) Heart shake (b) Star shake
(c) Radial shake (d) Cup and Ring shake

111. Complex knot formed at points where dormant buds show abnormal vigour is called:

(a) Burr (b) Twisted fibre
(c) Waviness (d) Checks

112. A defect caused in timber due to inclusion with the wood of local wounds that have healed up is termed as:

(a) Checks (b) Split
(c) Rindgalls (d) Radial shakes

113. Wood boring marine animals are belonging to the family:

(a) Crustaceans (b) Molluses
(c) Both (a) and (b) (d) None of the above

114. Important Molluces genera that attacks timber in sea water are:

(a) Teredo (b) Brankia
(c) Martesia (d) All of the above

115. Crustaceans and Molluces that attacks timber in sea water are collectively called as:

(a) Shipworms (b) Earthworms
(c) Both (a) and (b) (d) None of the above

116. Important Crustaceans genera that attacks timber in sea water are:

(a) Limnoria (b) Chof theseelura
(c) Sphaeroma (d) All of the above

117. In which rot wood is subjected to _______ and _______ which imparts brown colour to wood:

(a) Oxidation (b) Hydrolysis
(c) Both (a) and (b) (d) None of the above

118. World' best aircraft timber is:

(a) *Ochrama pyramidalis* (b) *Abies pindrow*
(c) *Artocarpus hirsute* (d) *Acer* spp.

119. Wooden battery separators are commonly made from:

(a) Conifers (b) Teak
(c) Casuarinas (d) Sal

120. The best wood for making the wooden bearing is:
(a) *Guaiacum officinale* (b) Guaiacum sanctum
(c) Both (a) and (b) (d) None of the above

121. The wood that is widely used for making dugouts is:
(a) *Duabanga sonneratioides* (b) *Trewia nudiflora*
(c) Both (a) and (b) (d) None of the above

122. The best suited wood for tight cooperage is:
(a) *Quercus* spp. (b) Teak
(c) *Casuarinas* (d) All of the above

123. Harmonium is generally made out of:
(a) *Tectona grandis* (b) *Pinus roxhurghii*
(c) *Canarium eupjyllum* (d) *Articarpus heterophyllus*

124. _______ is the most suited wood for pencil industry:
(a) *Juniperus virginiana* (b) J. Procera
(c) Both (a) and (b) (d) None of the above

125. Widely used species for making pattern are:
(a) *Tectona grandis* (b) *Cedrus deodara*
(c) Both (a) and (b) (d) None of the above

126. Species suitable for timber purposes (Pick the odd man out):
(a) *Pterocarpus santalinus* (b) *Tectona grandis*
(c) *Dalbergia sissoo* (d) *Ailanthus excelsa*

127. Pick out the tree which yields soft fibre:
(a) *Sterculia villosa* (b) *Albizia lebbeck*
(c) *Azadirachta indica* (d) *Acacia nilotica*

128. Wavy-grain wood is attributable to:
(a) *Dalbergia latifolia* (b) *Tectona grandis*
(c) *Pterocarpus santalinus* (d) *Santalum album*

129. Timbers which are capable of withstanding rapid seasoning are termed as:
(a) Most refractory (b) Non refractory
(c) Medium refractory (d) Refractory

130. The tree suitable for timber as well as fodder purpose, is:
(a) *Tectona grandis* (b) *Pterocarpus marsupium*
(c) *Acacia nilotica* (d) *Ceiba pentendra*

131. The Latin name for golden bamboo is:

(a) *Bambusa bamboos*
(b) *Bambusa vulgaris*
(c) *Bambusa tulda*
(d) *Dendrocalamus strictus*

132. Indian Kino tree is:

(a) *Pterocarpus santalinus*
(b) *Pterocarpus dalbergioides*
(c) *Pterocarpus marsupium*
(d) None of the above

133. Name the proprietary preservative containing high boiling creosote as base:

(a) Curpinol
(b) Hardiproof
(c) Solignum
(d) Celcure

134. Henna dye is obtained from:

(a) *Lawsoniai nermis*
(b) *Punica granatum*
(c) *Bixaora nalla*
(d) *Morinda tinctona*

135. Highly refractory wood (Pick out the odd man out):

(a) *Xylia xylocarpa*
(b) *Madhuca latifolia*
(c) *Shorea robusta*
(d) *Vateria indica*

136. The Latin name for satin wood:

(a) *Chloroxylon swietenia*
(b) *Lagerstroemia lanceolata*
(c) *Simarouba glauca*
(d) *Pterocarpus dalbergioides*

137. Oleo resins are obtained from:

(a) *Pinus roxburghii*
(b) *Carcinia Morella*
(c) *Vateria indica*
(d) *Canarium strictum*

138. Cutch is obtained from:

(a) *Acacia nilotica*
(b) *Acacia catechu*
(c) *Pterocarpus marsupium*
(d) *Acacia coincinna*

139. An example for wood oil is:

(a) *Cedrus deodara*
(b) *Calophyllum inophyllum*
(c) *Eucalyptus globules*
(d) *Lannea coramendellica*

140. Indian butter tree is:

(a) *Calophyllum inophyllum*
(b) *Madhuca latifolia*
(c) *Mimusops elengi*
(d) *Pongami apinnata*

141. Fruit-tan is obtained form:

(a) *Anogeissus latifolia*
(b) *Terminalia chebula*
(c) *Cassia auriculata*
(d) *Acacia mollissima*

142. 'Chilgoza' is obtained from:

(a) *Pinus roxburghii*
(b) *Pinuskesiya*
(c) *Pinusgerardiana*
(d) *Pinuselliotі*

143. Which one of the following is correctly matched?

(a) *Dysoxylum malabaricum* – Elm
(b) *Ulmus Wallichiana* – White cedar
(c) *Tectona grandis* – Ven teak
(d) *Picea smithiana* – Spruce

144. Antiaris toxicaria (upas) yielding a strong fibre belongs to the family:

(a) Urticaceae
(b) Moraceae
(c) Ulmaceae
(d) Asclepiadaceae

145. The Kino of Indian pharmacopoeia often termed as 'Malabar Kino' is obtained from:

(a) *Bombax malabaricum*
(b) *Dysoxylum malabaricum*
(c) *Pterocarpus marsupium*
(d) *Ailanthus grandis*

146. The common host plant of lac insect is:

(a) *Azadarichta indica*
(b) *Dalbergia sisso*
(c) *Butea monosperma*
(d) *Acacia auriculiformis*

147. The scientific name of Divi-Divi is:

(a) *Mimusops littooalis*
(b) *Caesalpinia coriaria*
(c) *Berberis nepalensis*
(d) *Morinda tinctoria*

148. Trade/Common name of *Dalbergia latifolia* is:

(a) Satin wood
(b) Rose wood
(c) Axle wood
(d) Amaltas

149. Deodar (*Cedrus Deodara*) oil, which is used to cure ulcer and skin diseases obtained from its:

(a) Wood
(b) Seed
(c) Leaves
(d) Roots

150. Indigoferatinctoria is used in:

(a) Medicine
(b) Pulp industry
(c) Ship building industry
(d) Dye industry

151. In the manufacture of Braille paper which of the following species is used?

(a) *Eucalyptus grandis*
(b) *Eucalyptus tereticornis*
(c) *Eucalyptus globules*
(d) *Eucalyptus citriodora*

152. Veneer is utilized for the manufacture of:

(a) Playwood (b) Black board

(c) Compregnated wood (d) All of the above

153. Palmrosa oil is:

(a) Flower oil (b) Grass oil

(c) Root oil (d) None of the above

154. Commercial cultivation of *Saussuria lappa* is found in the state of:

(a) Himachal Pradesh (b) Tamil Nadu

(c) Rajasthan (d) Odisha

155. Which of the following is correctly matched?

(a) *Cinnamomum tamela* – Cardamom (b) *Butea monosperma* – Silk

(c) *Diospyrus melanoxylum* – Bidi leaf (d) *Rauwolfia* – Climber

156. The trees species suitable for charcoal production is:

(a) *Tectona grandis* (b) *Terminalis arjuna*

(c) *Proosopi sjuliflora* (d) *Mangifera indica*

157. Which of the following is correctly matched?

(a) Black babool – *Acacia leucopholea*

(b) White babool – *Acacia nilotica*

(c) Green wattle – *Acacia dalbata*

(d) Black wattle – *Acacia decurrens*

158. The part of Chinchona plant which yields quinine is:

(a) Root (b) Leaf

(c) Flower (d) Bark

159. The raw materials for tanning industries are from:

(a) *Acacia nilotica* (b) *Acacia leucopholea*

(c) *Acacia mearnsii* (d) None of the above

160. Costus oil is obtained from:

(a) *Veleriana wallichi* (b) *Saussuria lappa*

(c) *Canarium strictum* (d) *Pandanus tectorius*

161. The average tannin content from myrobalans is:

(a) 28 per cent (b) 30 per cent

(c) 32 per cent (d) 36 per cent

Answer Sheet

1.	(c)	31.	(b)	61.	(a)	91.	(b)	121.	(c)
2.	(a)	32.	(c)	62.	(a)	92.	(a)	122.	(a)
3.	(b)	33.	(d)	63.	(c)	93.	(a)	123.	(a)
4.	(d)	34.	(a)	64.	(a)	94.	(d)	124.	(c)
5.	(a)	35.	(c)	65.	(b)	95.	(a)	125.	(c)
6.	(c)	36.	(a)	66.	(a)	96.	(c)	126.	(d)
7.	(d)	37.	(a)	67.	(a)	97.	(b)	127.	(a)
8.	(c)	38.	(c)	68.	(c)	98.	(a)	128.	(c)
9.	(a)	39.	(a)	69.	(c)	99.	(a)	129.	(b)
10.	(c)	40.	(d)	70.	(a)	100.	(b)	130.	(b)
11.	(b)	41.	(c)	71.	(b)	101.	(a)	131.	(b)
12.	(c)	42.	(d)	72.	(c)	102.	(b)	132.	(c)
13.	(a)	43.	(d)	73.	(a)	103.	(c)	133.	(c)
14.	(d)	44.	(a)	74.	(d)	104.	(a)	134.	(a)
15.	(d)	45.	(a)	75.	(c)	105.	(b)	135.	(d)
16.	(a)	46.	(a)	76.	(d)	106.	(d)	136.	(a)
17.	(b)	47.	(b)	77.	(a)	107.	(a)	137.	(a)
18.	(b)	48.	(a)	78.	(b)	108.	(b)	138.	(b)
19.	(c)	49.	(b)	79.	(c)	109.	(c)	139.	(a)
20.	(b)	50.	(d)	80.	(a)	110.	(d)	140.	(b)
21.	(a)	51.	(c)	81.	(b)	111.	(a)	141.	(b)
22.	(d)	52.	(b)	82.	(b)	112.	(c)	142.	(c)
23.	(b)	53.	(d)	83.	(a)	113.	(c)	143.	(d)
24.	(d)	54.	(a)	84.	(b)	114.	(d)	144.	(b)
25.	(c)	55.	(a)	85.	(c)	115.	(a)	145.	(c)
26.	(b)	56.	(b)	86.	(a)	116.	(d)	146.	(b)
27.	(d)	57.	(d)	87.	(b)	117.	(c)	147.	(b)
28.	(b)	58.	(a)	88.	(c)	118.	(a)	148.	(a)
29.	(d)	59.	(a)	89.	(c)	119.	(a)	149.	(d)
30.	(a)	60.	(a)	90.	(c)	120.	(c)	150.	(a)

151.	**(c)**	**154.**	**(a)**	**156.**	**(c)**	**158.**	**(d)**	**160.**	**(b)**
152.	**(a)**	**155.**	**(c)**	**157.**	**(c)**	**159.**	**(a)**	**161.**	**(a)**
153.	**(b)**								

Chapter 7

Wood Science and Technology

1. **Identify the hardwood timber species:**
   (a) Sal (b) Conifer
   (c) Poplar (d) Semal
2. **The type of cells present in soft wood is:**
   (a) Tracheids (b) Vessels
   (c) Fibre (d) None of the above
3. **An example for ring porous wood is:**
   (a) *Dipterocarpus* (b) *Gmelina*
   (c) *Hardwickia* (d) *Quercus*
4. **Pick out the tree which yields soft fibre:**
   (a) *Sterculia villosa* (b) *Albizia lebbeck*
   (c) *Azadirachta indica* (d) *Acacia nilotica*
5. **The tree that serves as raw material for the manufacture of pencil is:**
   (a) *Adina* (b) *Juniperus*
   (c) *Terminalia* (d) *Mesua*
6. **An example for wood oil is:**
   (a) *Cedrus deodara* (b) *Calophyllum inophyllum*
   (c) *Eucalyptus globules* (d) *Lannea coramendellica*

7. **Indian butter tree is:**
   (a) *Calophyllum inophyllum* (b) *Madhuca latifolia*
   (c) *Mimusops elengi* (d) *Pongamia pinnata*

8. **Fruit – tan is obtained form:**
   (a) *Anogeissus latifolia* (b) *Terminalia chebula*
   (c) *Cassia auriculata* (d) *Acacia mollissima*

9. **'Chilgoza' is obtained from:**
   (a) *Pinus roxburghii* (b) *Pinus kesiya*
   (c) *Pinus gerardiana* (d) *Pinus ellioti*

10. **Pick the odd man out: Defects due to rupture of tissues:**
   (a) Twist (b) Split
   (c) Check (d) Shake

11. **Except one, the following refer to natural defects in a log. Identify the exception:**
   (a) Shake (b) Flute
   (c) Spring (d) Check

12. **Wavy-grain wood is attributable to:**
   (a) *Dalbergia latifolia* (b) *Tectona grandis*
   (c) *Pterocarpus santalinus* (d) *Santalum album*

13. **Timbers which are capable of withstanding rapid seasoning are termed as:**
   (a) More refractory (b) Non refractory
   (c) Medium refractory (d) Refractory

14. **One of the following species comes under highly refractory wood for seasoning:**
   (a) *Tectona grandis* (b) *Shorea robusta*
   (c) *Dalbergia sissoo* (d) *Bombax ceiba*

15. **In softwood trees the conduction and mechanical functions are performed by:**
   (a) Vessels (b) Fibres
   (c) Tracheids (d) Pores

16. **The Latin name for Andaman Padauk tree is:**
   (a) *Veteria indica* (b) *Pterocarpus dalbergioides*
   (c) *Shorea robusta* (d) *Dalbergia latifolia*

17. The Latin name for golden bamboo is:

(a) *Bambusa bamboos* (b) *Bambusa vulgaris*

(c) *Bambusa tulde* (d) *Dendrocalamus strictus*

18. Indian Kino tree is:

(a) *Pterocarpus santalinus* (b) *Pterocarpus dalbergioides*

(c) *Pterocarpus marsupium* (d) None of the above

19. Name the proprietary preservative containing high boiling creosote as base:

(a) Curpinol (b) Hardiproof

(c) Solignum (d) Celcure

20. The chief characteristics of an ideal wood preservative are:

(a) High permanency and stability (b) High toxicity

(c) High penetrability (d) All of the above

21. Henna dye is obtained from:

(a) *Lawsonia inermis* (b) *Punica gransatum*

(c) *Bixa oranalla* (d) *Morinda tinctona*

22. Oleo resins are obtained from:

(a) *Pinus roxburghii* (b) *Carcini amorella*

(c) *Vateria indica* (d) *Canarium strictum*

23. Cutch is obtained from:

(a) *Acacia nilotica* (b) *Acacia catechu*

(c) *Pterocarpus marsupium* (d) *Acacia coincinna*

24. The tree t hat serves as raw material for the manufacture of pencil is:

(a) Adina (b) Juniperus

(c) Terminalia (d) Mesua

25. Density of trees in a forest stand is normally expressed as:

(a) Cubic feet per acre (b) Square feet per acre

(c) Tons per acre (d) Basal area of the stand in square feet

26. What is a choker?

(a) A wire cable or chain used to move logs

(b) An instrument for measuring tree diameters

(c) An obstruction in a stream

(d) A tree that does not fall after being cut

27. Nutrients and water absorbed from the soil move up through a tree in the:

(a) Xylem
(b) Root hairs
(c) Phloem
(d) All of the above

28. The practice of measuring the volume of logs or bolts after they are harvested is called:

(a) Cruising
(b) Scaling
(c) Tallying
(d) All of the above

29. Kerf refers to:

(a) A type of log scale
(b) Cutting width of a saw blade
(c) The distance between the chain and the bar
(d) Log volume

30. Treating wood posts and lumber with penta, creosote or CCA improves its:

(a) Resistance to decay and insects
(b) Attractiveness
(c) Susceptibility to decay and insects
(d) Strength

31. A table showing the estimated or calculated amount of lumber which may be sawed from logs of given length and diameter is called _______:

(a) Biltmore stick
(b) Log rule
(c) Merritt Hypsomete
(d) Scale stick

32. Frilling involves _______.

(a) Cutting the tree down
(b) Encircling a tree stem with downward, overlapping axe cuts
(c) Spraying herbicide on the tree leaves
(d) Using tubular injection

33. Which tree part listed below is considered non-living?

(a) Cambium
(b) Heartwood
(c) Inner bark
(d) Sapwood

34. An articulated log skidder is _______:

(a) A modified farm tractor
(b) Less maneuverable than-a non-articulated skidder
(c) One with a hinge in the midsection
(d) Suitable only for gentle terrain

35. A source of revenue for a logging operation is _______:

(a) Depreciation
(b) Income tax
(c) Log sales
(d) Vehicle insurance

36. A false ring is _______.

(a) An extremely narrow ring
(b) An extremely wide ring
(c) A second ring for a given year
(d) A black ring

37. A Pulaski is used in firefighting for _______.

(a) Felling and clearing
(b) Grubbing and chopping
(c) Raking and sweeping
(d) Scraping and raking

38. Reducing a felled tree into marketable size logs is called _______.

(a) Bucking
(b) Felling
(c) Limbing
(d) Skidding

39. The function of the cambium of the tree is _______.

(a) Development of leaves
(b) Growth in height of the tree
(c) Growth of wood and bark cells
(d) Production of carbohydrates

40. Chromate zinc chloride is superior to zinc chloride as a preservative because of its _______.

(a) Lowest cost
(b) Lower fire hazard
(c) Resistance to flaking
(d) Resistance to leaching

41. Chromate zinc chloride is superior to zinc chloride due to its _______.

(a) Lowest cost
(b) Lower fire hazard
(c) Resistance to flaking
(d) Resistance to leaching

42. The part of the logging operation in which logs are dragged from where they were cut to a central location for loading is called_______.

(a) Bucking
(b) Felling
(c) Snagging
(d) Skidding

43. Water contained within the cell wall of wood is called _______.

(a) Bound water
(b) Fiber water
(c) Free water
(d) Suppressed water

44. Which tree part carries food made in the leaves down to branches, trunk and roots:

(a) Phloem
(b) Sapwood
(c) Heartwood
(d) Cambium

45. In order to "girdle a tree" which of the following must be completely severed?

(a) Phloem
(b) Xylem
(c) Cambium
(d) Phloem and cambium

46. Which of the following statements best describes "heartwood"?

(a) Lighter colored wood just inside the bark

(b) Part o the tree stem which is always live wood

(c) Carries nutrients and water from the roots to the crown

(d) None of the above

47. Both the inner bark and sapwood are formed by cell divisions of the _______. layer.

(a) Phloem (b) Xylem

(c) Cambium (d) All of the above

48. Which of the following is a direct product of photosynthesis?

(a) Oxygen (b) Carbohydrates

(c) Fats and protein (d) All of the above

49. Annual rings on a tree are composed of:

(a) Summer wood (b) Spring wood

(c) Both (a) and (b) (d) None of the above

50. The only living part of the tree trunk is the _______.

(a) Heartwood (b) Sapwood

(c) Cambium layer (d) Outer bark

51. _______. carries the sap from the roots to the leaves.

(a) Heartwood (b) Xylem

(c) Phloem (d) Cambium layer

52. Closely spaced annual rings indicate _______:

(a) Fast growth (b) Slow growth

(c) Very tall trees (d) Very short trees

53. _______ carries plant food manufactured in the leaves down to the branches trunk and roots.

(a) Heartwood (b) Sapwood

(c) Phloem (d) Xylem

54. The _______. builds wood tissue in the turn.

(a) Cambium layer (b) Heartwood

(c) The leaf (d) Phloem

55. _______ gives support and strength to the tree.

(a) Outer bark (b) Sapwood

(c) Heartwood (d) Cambium layer

56. _______ helps to protect the tree trunk from injury.

(a) Cambium layer (b) Sapwood

(c) Heartwood (d) Outer bark

57. The manufacture of plant food in the leaves of a plant is called _______.

(a) Respiration (b) Transpiration

(c) Photosynthesis (d) Respiration

58. The center of a twig is called the _______.

(a) Cambium layer (b) Sapwood

(c) Pith (d) Bole

59. The word "buck" is often used in the discussion of forest harvesting activities. What does this word mean?

(a) To putlogs in a pile

(b) To cut trees into logs

(c) To push a log truck out of the mud

(d) To move logs from the stump to the landing

60. The three "E's" associated with the forest stewardship program are:

(a) Environmental, esthetics and economic

(b) Energy, esthetics and economics

(c) Esthetics, environmental and energy

(d) Everlasting, esthetics and energy

61. The profession dealing with maintenance of ornamental trees is called:

(a) Silviculture (b) Dendrology

(c) Arboriculture (d) Hydrology

62. In forestry the letters TSI stand for the words _______.

(a) Total saleable investment (b) Timber stand improvement

(c) Total site income (d) Taxable site index

63. The first step in forest harvesting is _______.

(a) Skidding (b) Felling and bucking

(c) Loading (d) Hauling

64. Bucking is _______.

(a) Cutting a tree down (b) Cutting a down tree into log lengths

(c) Cutting the limbs off a standing tree (d) Cutting the limbs off a down tree

65. Debris left after a logging operation is called_______.

(a) Duff (b) Waste
(c) Shalsh (d) Stubble

66. Which among these preservations is the most eco-friendly?

(a) Copper-chrome arsenic (ASCU) (b) Acid cupric chrome (ACC)
(c) Chromate zinc chloride (CZC) (d) Copper chrome boric (CCB)

67. Kraft paper is made entirely from unbleached pulp produced from:

(a) Sulphite pulp (b) Mechanical pulp
(c) Soda pulp (d) Sulphate pulp

68. Soft woods are found to be most suitable for the production of:

(a) Chemical process (b) Mechanical process
(c) Soda process (d) Chemical process

69. Paper, which is more or less moisture content of the wood should be about:

(a) Filler (b) Sizing
(c) Beating (d) Loading

70. The addition of non-fibrous mineral matter to stock prior to sheet formation is known as:

(a) Ageing (b) Loading
(c) Opacity (d) Beating

71. For the manufacture of rayon, the pulp should be contain more:

(a) Ligning (b) Cellulose
(c) Hemi cellulose (d) Acetic acid

72. The cross banded construction of wood is called as:

(a) Laminated wood (b) Laminae wood
(c) Soft wood (d) Plywood

73. The caustic soda treated cellulose is called as:

(a) Cellulose (b) Hemi cellulose
(c) Alkali cellulose (d) Cellulose xanthation

74. Method of assembling of wood components in ply wood manufacture is called as:

(a) Cross banded construction (b) Balanced construction
(c) Lamination (d) None of the above

75. For the construction of arched or curved assemblies, which of the following is used:

(a) Plywood (b) Laminated wood

(c) Core board (d) Particle board

76. Most common commercial chemical agents used for bleaching are:

(a) Sodium compounds (b) Chlorine compounds

(c) Acids compounds (d) None of the above

77. For manufacture of news print paper, which of the following species is used:

(a) *Bambusa arundinacea* (b) *Pinus roxburghii*

(c) *Boswellia serrate* (d) *Eulaliopsis binata*

78. Commonly used chemical pulping process in india is:

(a) Sulphate process (b) Sulphite process

(c) Acidic process (d) Semi chemical process

79. Excellent material for Aero place propeller blades is:

(a) Impreg (b) Compreg

(c) Staypak (d) Staybord

80. ASCU is which of the following type of preservative:

(a) Oil type (b) Organic solvent type

(c) Tar type (d) fixed water soluble

81. The wood preservative Arsenic –Copper Chromate in the ratio of 1:3:4 comes under the category of:

(a) Oil type (b) Water soluble leachable type

(c) Water soluble fixed type (d) Organic solvent type

82. The caustic soda treated cellulose is called as:

(a) Hemi-cellulose (b) Cellulose

(c) Alkali cellulose (d) Cellulose xanthation

83. Heating of wood of these in absence of oxygen to produce methanol is called as:

(a) Wood saccharification (b) Wood distillation

(c) Wood hydrolysis (d) None of the above

84. Paper, which is more or less impervious to the penetration of inks, is said to be:

(a) Filler (b) Sizing

(c) Beating (d) Loading

85. Which of the following glue is commonly used in plywood industries:

(a) Phenolic glue (b) Animal glue

(c) Starch glue (d) Casein glue

86. Yield in sulphate pulping posses is:

(a) 90-95 per cent (b) 45-50 per cent

(c) 70 per cent (d) 50-60 per cent

87. Mechanical treatment given to fibres to increase tensile strength, bursting strength *etc.* is:

(a) Loading (b) Beating

(c) Bleaching (d) Sizing

88. Wood that is specially treated to increase its strength is called as:

(a) Improved (b) Composite wood

(c) Soft wood (d) Plywood

89. Glues once cured are not softened by heat are called as:

(a) Thermosetting glues (b) Thermoplastic glue

(c) Synthetic glues (d) None of the above

90. For treatment of Bamboo poles commonly used wood preservation process is:

(a) Bethal process (b) Boucherie process

(c) Lowry process (d) Boulton process

91. Estimating the amount of board feet, cords, or pieces in a timber stand is called:

(a) Scaling (b) Guessing

(c) Browsing (d) Cruising

92. Which is the most durable part of a wooden fence post?

(a) Sapwood (b) Heartwood

(c) Both are equally durable (d) None of the above

93. Decay or wood rot is caused by which of which of the following organisms?

(a) Bacteria (b) Fungi

(c) Soil insects (d) All of the above

94. A butt log is:

(a) A log with excessive swelling at one end

(b) A log with many defects

(c) The first log cut above the stump

(d) The smallest log (in diameter) cut from the tree

95. Logging costs typically represent what per cent of the total cost of saw timber delivered to a mill?

(a) 33 per cent (b) 50 per cent

(c) 75 per cent (d) 80 per cent

96. When timber buyers openly bid against one another at a timber sale it is called a (an):

(a) Auction sale (b) Mill scale sale

(c) Sealed bid sale (d) Stumpage sale

97. The length of time needed to air dry or season wood depends on:

(a) Climate (b) Method of stacking

(c) Wood size (d) All of the above

98. A cant hook is used to _______:

(a) Roll logs (b) Load logs

(c) Fell logs (d) Skid logs

99. The darker colored interior wood near the tree's center, often no longer living, is called:

(a) Core wood (b) Heartwood

(c) Sapwood (d) Terminal wood

100. A Pulaski is used in forest fire fighting for _______:

(a) Felling and clearing (b) Grubbing and chopping

(c) Raking and sweeping (d) Scraping and raking

101. The width of the cut made by a saw is called the _______.

(a) Cant (b) Gap

(c) Kerf (d) Notch

102. Paper was first made from ground mulberry bark by the _______.

(a) American (b) Chinese

(c) Egyptians (d) Indians

103. The tree which has more than one leader stem is called:

(a) Forked (b) Buttressed

(c) Epicormic branches (d) Fluted

104. Species suitable for structural timber purpose (Pick the odd man out):

(a) *Pterocarpus santalinus* (b) *Tectona grandis*

(c) *Dalbergia latifolia* (d) *Ailanthus excelsa*

105. An example for ring porous wood is:

(a) *Dipterocarpus* (b) *Gmelina*

(c) *Hardwickia* (d) *Quercus*

106. Pick out the tree which yields soft fibre:

(a) *Sterculia villosa* (b) *Albizia lebbek*

(c) *Azadirachta indica* (d) *Acacia nilotica*

107. Root rot in sal is caused by:

(a) *Pseudomonas* (b) *Polyporus*

(c) *Ganoderma* (d) *Pythium*

108. The species, which express rapid growth:

(a) Late wood (b) Early wood

(c) Spring wood (d) Summer wood

109. In broad-leaved species the mechanical function is performed by:

(a) Vessels (b) Fibres

(c) Tracheids (d) Tyloses

110. In softwoods conduction and mechanical functions are performed by:

(a) Vessels (b) Fibres

(c) Tracheids (d) Tyloses

111. In broad leaved species the conduction function is performed by:

(a) Vessels (b) Fibres

(c) Tracheids (d) Tyloses

112. In ring porous wood the vessels are:

(a) Larger (b) Smaller

(c) Medium (d) Intermediate

113. Pick the wavy grained wood:

(a) *Pterocarpus santalinus* (b) *Santalum album*

(c) *Pterocarpus marsupium* (d) *Terminalia chebula*

114. Thecapacity of wood tobend out the shape without rupture is called:

(a) Elasticity (b) Flexibility

(c) Fissibility (d) None of the above

115. The capacity of wood to split along with fibres is termed as:

(a) Elasticity (b) Flexibility

(c) Fissibility (d) None of the above

116. The fibres saturation point for tropical timbers ranged between:

(a) 0-15 per cent (b) 15-30 per cent

(c) 30-45 per cent (d) 45-60 per cent

117. Pick out the soft wood:

(a) *Tectona grandis* (b) *Albizzia lebbeck*

(c) *Chukrasia tabularis* (d) *Toona ciliata*

118. Good fuel woods (Pick the odd man out):

(a) *Acacia nilotica* (b) *Casuarinas equisetifolia*

(c) *Chloroxylon swietenia* (d) *Ailanthus excels*

119. Species suitable for cricket bat:

(a) *Acacia nilotica* (b) *Ailanthus excels*

(c) *Diospyros melanoxylon* (d) *Salix alba*

120. Tyloses found only in the heart wood of:

(a) Soft wood (b) Hard wood

(c) Reaction wood (d) Ring-porous wood

121. Ringporous wood species:

(a) *Acacia nilotica* (b) *Hardwickia binnata*

(c) *Azadiracta indica* (d) *Tectona grandis*

122. Majority of Indian dicotyledonous timbers are:

(a) Ring porous wood (b) Semi ring porous wood

(c) Diffuse porous wood (d) Reaction wood

123. Horizontally arranged cells, running radially from bark towards the centre of tree:

(a) Vessels (b) Fibres

(c) Tracheids (d) Medullary rays

124. The layer that produced between xylem and phloem:

(a) Cambium (b) Bark

(c) Meristem (d) None of the above

125. Sal is a:

(a) Ring porous wood (b) Diffuse porous wood

(c) Intermediate wood (d) None of the above

126. The institute of wood science and technology is located in:

(a) Thrissur (b) Jodhpur

(c) Jabalpur (d) Bangalore

127. The basic structural and physiological unit of a plant:

(a) Cell (b) Fiber

(c) Tracheid (d) Cambium

128. Secondary thickening in wood is caused by:

(a) Xylem (b) Phloem

(c) Cambium (d) Meristem

129. The lignin content of hard woods range and between:

(a) 5 and 10 per cent (b) 8 and 18 per cent

(c) 18 and 25 per cent (d) 25 and 40 per cent

130. The region in which cells divide repeatedly and form new cells is termed as:

(a) Surface region (b) Meristematic region

(c) Cambial region (d) None of the above

131. The water held by capillary attraction inside the empty spaces in and between the cell is termed as:

(a) Free water (b) Absorbed water

(c) Bound water (d) None of the above

132. The formula for estimating the moisture content of wood is:

(a) (OD weight – Green wt/OD weight) X 100

(b) (Green wt-OD weight/Green wt) X 100

(c) (Green wt-OD weight/OD weight) X 100

(d) None of the above

133. The defects which show longitudinal distortion of wood in which one of the wood is cancave is called:

(a) Cup (b) Bow

(c) Twist (d) Spring

134. Urea formaldehyde (UF) is used as:

(a) Fire retardant (b) Wood adhesive

(c) Insect repellent (d) Chemical preservative

135. Tree with high cellulose content:

(a) *Eucalyptus* (b) *Acacia*

(c) *Albizia* (d) *Anthocephalus*

136. Tree with long fiber length:

(a) Eucalyptus (b) Bamboos

(c) Casuarinas (d) Leucaena

137. The ability of cell wall to reflect light is termed:

(a) Colour
(b) Iusture
(c) Odour
(d) None of the above

138. Heavy wood (Tick odd man out):

(a) *Acacia nilotica*
(b) *Dalbergia sissoo*
(c) *Tectona grandis*
(d) *Abies pindrow*

139. Pick out most refractory wood:

(a) *Sizygium cuminii*
(b) *Adina cordifolia*
(c) *Bombax ceiba*
(d) *Morus alba*

140. Choose the non refractory wood:

(a) *Tectona grandis*
(b) *Veteria indica*
(c) *Mangifera indica*
(d) *Pinus roxburghi*

141. Moluscan borer (Tick the odd man out):

(a) Xylotria
(b) Teredo
(c) Nausitoria
(d) Limnoria

142. Perishable wood:

(a) *Tectona grandis*
(b) *Albizia lebbeck*
(c) *Boswellia serrata*
(d) *Dalbergia paniculata*

143. The function of the phloem in the tree is to _______.

(a) Develop leaves from nutrients
(b) Conduct food manufactured in the leaves to the branches, trunk and roots
(c) Grow wood and bark cells
(d) Produce carbohydrates through photosynthesis

144. Epiblema of roots is equivalent to:

(a) Pericycle
(b) Endodermis
(c) Epidermis
(d) Stele

145. Phellogen and phellem respectively denote:

(a) Cork and cork cambium
(b) Cork cambium and cork
(c) Secondary cortex and cork
(d) Cork and secondary cortex

146. In which of the following pairs of parts of a flowering plant is epidermis absent?

(a) Root tip and shoot tip
(b) Shoot bud and floral bud
(c) Ovule and seed
(d) Petiole and pedicel

147. A piece of wood having no vessels trachea must be belong to:

(a) Teak
(b) Mango
(c) Pine
(d) Palm

148. A plant tissue, when stained, showed the presence of hemicelluloses and pectin in cell wall of its cells. The tissue represents:

(a) *Collenchymas*
(b) *Sclerenchyma*
(c) *Xylem*
(d) *Meristem*

149. Fibres are likely to be absent in:

(a) Secondary phloem
(b) Secondary xylem
(c) Primary phloem
(d) Leaves

150. A vesselless piece of stem possessing prominent sieve tubes would belong to:

(a) Pinus
(b) Eucalyptus
(c) Grass
(d) Trochodendron

151. Major components of wood:

(a) Cellulose
(b) Hemi cellulose
(c) Lignin
(d) All of the above

152. Vessels are present in:

(a) Hard wood
(b) Heart wood
(c) Sap wood
(d) Soft wood

153. In seasoning of wood, the rate of drying depends on:

(a) Circulation of the air around the timber
(b) Humidity or dryness of the air
(c) Heat
(d) All of the above

154. Based on seasoning behavior, the woods can be classified into:

(a) 3 categories
(b) 4 categories
(c) 5 categories
(d) 6 categories

155. In Air Seasoning, the function of the crossers is:

(a) Controls the dryness of air
(b) Allow uniform circulation of air around the timber
(c) Both (a) and (b)
(d) To maintain relative humidity

156. Serious wood destroying animals are:

(a) Elephant
(b) Crustaceans and mollusks
(c) Deers
(d) Borers

157. Ingredients of ASCU are:

(a) Arsenic Penta Oxide + copper sulphate + Na or K di chromate

(b) Arsenic Penta Oxide +Zinc sulphate + Na or K di chromate

(c) Arsenic Penta Oxide +Mercuric chloride + Na or K di chromate

(d) Arsenic Penta Oxide +Boric Acid + Na or K di chromate

158. In plywood, the grains of each veneers arranged:

(a) Parallel to adjacent veneer
(b) Right angle to adjacent veneer
(c) According to the end use
(d) None of the above

159. Compregnated wood is:

(a) Compressed wood
(b) Impregnated wood
(c) Compressed and impregnated wood
(d) None of the above

160. In Laminated Veneer Lumber manufacturing, the adhesives used:

(a) Phenol formaldehyde
(b) Urea formaldehyde
(c) Melamine formaldehyde
(d) All of the above

161. In pulp and paper manufacturing process, commonly used fillers:

(a) Soap
(b) Talc
(c) Rosin
(d) Aluminum sulphate

162. Radical growth in trees due to the activity of:

(a) Cambium
(b) Parenchyma cells
(c) Rays
(d) Sclerenchyma cells

163. The following microstructures affects the drying rate:

(a) Tyloses
(b) Rays
(c) Sclerenchyma
(d) None of the above

164. The lightest wood in the world is a:

(a) Soft wood
(b) Hard wood
(c) Gymnosperms
(d) All of the above

165. _______ is the reciprocal of resistivity and is evaluated in units m hos per centimeter:

(a) Conductivity
(b) Gravity
(c) Density
(d) None of the above

Answer Sheet

1.	(a)	31.	(b)	61.	(c)	91.	(d)	121.	(d)
2.	(a)	32.	(b)	62.	(b)	92.	(b)	122.	(d)
3.	(d)	33.	(b)	63.	(b)	93.	(b)	123.	(d)
4.	(a)	34.	(c)	64.	(b)	94.	(c)	124.	(a)
5.	(b)	35.	(c)	65.	(c)	95.	(a)	125.	(b)
6.	(a)	36.	(c)	66.	(d)	96.	(a)	126.	(d)
7.	(b)	37.	(b)	67.	(b)	97.	(d)	127.	(a)
8.	(b)	38.	(a)	68.	(b)	98.	(a)	128.	(c)
9.	(c)	39.	(c)	69.	(b)	99.	(b)	129.	(c)
10.	(a)	40.	(d)	70.	(b)	100.	(b)	130.	(b)
11.	(c)	41.	(d)	71.	(b)	101.	(c)	131.	(a)
12.	(c)	42.	(d)	72.	(d)	102.	(b)	132.	(c)
13.	(b)	43.	(a)	73.	(a)	103.	(a)	133.	(b)
14.	(b)	44.	(a)	74.	(a)	104.	(d)	134.	(b)
15.	(c)	45.	(d)	75.	(a)	105.	(d)	135.	(a)
16.	(b)	46.	(d)	76.	(b)	106.	(a)	136.	(b)
17.	(b)	47.	(c)	77.	(d)	107.	(b)	137.	(b)
18.	(c)	48.	(b)	78.	(a)	108.	(b)	138.	(d)
19.	(c)	49.	(c)	79.	(b)	109.	(b)	139.	(a)
20.	(d)	50.	(c)	80.	(d)	110.	(c)	140.	(a)
21.	(a)	51.	(b)	81.	(c)	111.	(a)	141.	(d)
22.	(a)	52.	(b)	82.	(c)	112.	(a)	142.	(d)
23.	(b)	53.	(c)	83.	(b)	113.	(a)	143.	(b)
24.	(b)	54.	(a)	84.	(b)	114.	(b)	144.	(d)
25.	(d)	55.	(c)	85.	(a)	115.	(c)	145.	(b)
26.	(a)	56.	(d)	86.	(b)	116.	(b)	146.	(a)
27.	(a)	57.	(c)	87.	(b)	117.	(d)	147.	(c)
28.	(b)	58.	(c)	88.	(a)	118.	(d)	148.	(a)
29.	(b)	59.	(b)	89.	(a)	119.	(d)	149.	(d)
30.	(a)	60.	(a)	90.	(b)	120.	(b)	150.	(d)

151.	**(d)**	**154.**	**(a)**	**157.**	**(a)**	**160.**	**(d)**	**163.**	**(a)**
152.	**(a)**	**155.**	**(b)**	**158.**	**(b)**	**161.**	**(b)**	**164.**	**(b)**
153.	**(d)**	**156.**	**(b)**	**159.**	**(c)**	**162.**	**(c)**	**165.**	**(a)**

Chapter 8

Forest Botany

1. **The science of botany is divided into branches. Which of these branches deals with the naming and classifying of plants?**
   (a) Plant taxonomy (b) Plant pathology
   (c) Plant morphology (d) Plant genetics

2. **The system used to name all living organisms is properly called _______.**
   (a) Binominal classification (b) Latin names
   (c) Scientific names (d) All of the above

3. **The basic unit of classification is:**
   (a) Genus (b) Species
   (c) Family (d) Taxon

4. **Unicellular plants found floating in oceans and freshwater are called:**
   (a) Algae (b) Zooplanktons
   (c) Phytoplanktons (d) Epiphytes

5. **Carolus Linnaeus proposed the following system of classification:**
   (a) Phylogenetic (b) Two kingdoms
   (c) Five kingdoms (d) Natural

6. **Who is known as the 'father of botany'?**
   (a) Theophrastus (b) Linnaeus
   (c) Aristotle (d) Mendel

7. **Thalloid and non flowering plants are known as _______:**
   (a) Angiosperm (b) Thallophyta
   (c) Gymnosperm (d) Dicotyledons

8. **Artificial system of classification of plants was proposed by a:**
   (a) British botanist (b) Swedish botanist
   (c) German botanist (d) Indina botanist

9. **Which of the following classification is a sexual system of classification ?**
   (a) Artificial system (b) Natural system
   (c) Phylogenetic system (d) Natural selection

10. **The botanist who introduced binomial system is:**
    (a) Carolus Linnaeus (b) Gaspard Bauhin
    (c) Sir Joseph Dalton Hooker (d) Adolf Engler

11. **Genera plantarum of Bentham and Hooker:**
    (a) A single volume (b) Two volumes
    (c) Three volumes (d) Four volumes

12. **In Bentham and Hooker classification of plants, the present day 'orders' were referred to by them as:**
    (a) Series (b) Cohorts
    (c) Orders (d) Families

13. **Plants having flowers with free petals are placed under:**
    (a) Monochlamydease (b) Monocotyledons
    (c) Gamopetalae (d) Polypetalae

14. **Inferae includes:**
    (a) 6 orders and 34 families (b) 4 orders and 23 families
    (c) 3 orders and 9 families (d) 5 orders and 27 families

15. **How many families were described by Bentham and Hooker in their classification?**
    (a) 204 (b) 212
    (c) 202 (d) 102

16. **In Bentham and Hooker's classification of plants, the present by "families" were referred to by them as:**
    (a) Families (b) Cohorts
    (c) Orders (d) Series

17. Thalamiflorae includes:

(a) 4 orders and 23 families
(b) 6 orders and 34 families
(c) 5 orders and 27 families
(d) 3 orders and 12 families

18. Which one of the following series includes the epigynous flowers?

(a) Thalamiflorae
(b) Disciflorae
(c) Inferae
(d) Heteromerae

19. The family included under the series Unisexuales is:

(a) Solanaceae
(b) Euphorbiaceae
(c) Malvaceae
(d) Musaceae

20. Thespesiapopulnea belongs to:

(a) Solanaceae
(b) Euphorbiaceae
(c) Malvaceae
(d) Musaceae

21. Malvaceae is placed in the series:

(a) Thalamiflorae
(b) Inferae
(c) Heteromerae
(d) Disciflorae

22. Anthers are monothecous in:

(a) Solanaceae
(b) Euphorbiaceae
(c) Malvaceae
(d) Musaceae

23. In Abelmoschus esulentus, the fruit is:

(a) Drupe
(b) Schizocarp
(c) Regma
(d) Loculicidal capsule

24. The change from meristematic tissue to permanent tissue is called:

(a) Differentiation
(b) Self perpetuating
(c) Photosynthesis
(d) Cell division

25. The type of tissue presents in the petioles of banana and Vanna, is:

(a) Stellate parenchyma
(b) Prosenchyma
(c) Aerenchyma
(d) Chlorenchyma

26. The tissue generally present in all organs of plants is:

(a) Parenchyma
(b) Chlorenchyma
(c) Collenchymas
(d) Sclerenchyma

27. The lamellar collenchyma is seen in the hypodermis of:

(a) Datura
(b) Helianthus
(c) Ipomoea
(d) Nicotiana

28. The root hairs are produced from:

(a) Rhizodermis (b) Trichomes

(c) Accessory cells (d) Trichoblasts

29. The osteosclereids are seen in:

(a) Seed coat of crotalaria (b) Seed coat of pisum

(c) Pulp of pyrus (d) Petioles of banana

30. Bicollateral vascular bundles are seen in the members of:

(a) Malvaceae (b) Musaceae

(c) Solanaceae (d) Cucurbitaceae

31. The root hairs originate from:

(a) Trichoblasts (b) Endodermis

(c) Hypodermis (d) Pericycle

32. The casparian strips are found in the endodermis of:

(a) Dicot stem (b) Dicot root

(c) Monocot stem (d) Dicot leaf.

33. The passage cells are found in endodermis of:

(a) Dicot stem (b) Monocot stem

(c) Dicot root (d) Dicot leaf

34. The polyarch condition is found in:

(a) Monocot leaf (b) Dicot leaf

(c) Dicot stem (d) Monocot root

35. The inner most layer of the cortex is:

(a) Epidermis (b) Hypodermis

(c) Endodermis (d) Pericycle

36. In angiosperms, the structures that produce male gametophyte are called the:

(a) Anthers (b) Sepals

(c) Pollen tubes (d) Stigmas

37. The outmermost circle of flower parts consists of several:

(a) Sepals (b) Petals

(c) Carpels (d) Corollas

38. Pollination occurs when pollen lands on the:

(a) Style (b) Stigma

(c) Filament (d) Anther

39. The seed leaves of flowering plants are known as:

(a) Endosperm
(b) Carpels
(c) Radicals
(d) Cotyledons

40. In seed plants, the structure that encloses the male gametophyte and transports it to another plant is called:

(a) Pollen grain
(b) Seed
(c) Flower
(d) Pollinator

41. If you examine an animal cell under the microscope you would not find a:

(a) Nucleus
(b) Plastid
(c) Cell membrane
(d) Centriole

42. This layer of parenchyma cells protects the vulnerable root tip where meristematic tissue is located:

(a) Cambium
(b) Rootcap
(c) Endodermis
(d) Pericycle

43. Cell sap provides turgor to plant cells and is found in the:

(a) Chloroplasts
(b) Mitochondrion
(c) Central vacuole
(d) Cytoplasm

44. This vascular tissue cell type is not found in gymnosperms:

(a) Tracheid
(b) Sieve tube element
(c) Vessels
(d) Phloem

45. This tissue is living at maturity and has unevenly thickened cell walls; acts to provide flexible support for growing stems.

(a) Parenchyma
(b) Xylem
(c) Phloem
(d) Collenchymas

46. In monocot roots which types of vascular bundles are found?

(a) Collateral, conjoint and closed
(b) Radial with exarch xylem
(c) Bicollateral, conjoint and closed
(d) Radial with endarch xylem

47. The modified stem of Opuntia is:

(a) Phyllode
(b) Phylloclade
(c) Cladode
(d) Stamroinode

48. Among plants "pheromones" are secreted by the cells of the following plants for given function:

(a) All plants for growth an development
(b) Yest for facilitating mating
(c) All fungi for sexual reproduction
(d) Rhizopus for formation of zygospore

49. Which are the two types of tissues?

(a) Meristematic and permanent
(b) Meristematic and temporary
(c) Meristems and temporary
(d) None of the above

50. Parenchyma is a type of _______:

(a) Complex tissue
(b) Simple t issue
(c) Xylem
(d) Phloem

51. Parenchyma: Simple: Phloem: _______.

(a) Simple
(b) Collenchymas
(c) Complex
(d) Xylem

52. Transpiration and exchange of gases are functions of _______.

(a) Stomata
(b) Xylem
(c) Both (a) and (b)
(d) None of the above

53. _______ tissue forms new cells in plants.

(a) Meristematic
(b) Permanent
(c) Simple
(d) None of the above

54. _______ tissues prevent loss of water in plants.

(a) Protective
(b) Xylem
(c) Phloem
(d) All of the above

55. A group of cells alike in form, function and origin are called _______:

(a) Tissue
(b) Organ
(c) Organelle
(d) None of the above

56. Plant length is increased by _______:

(a) Apical meristems
(b) Lateral meristems
(c) Periblem
(d) Parenchyma

57. The cell division is restricted to _______:

(a) Meristematic cells
(b) Permanent cells
(c) Secretory cells
(d) All of the above

58. Active division takes place in the cells of _______:

(a) Xylem
(b) Phloem
(c) Sclerenchyma
(d) Cambium

59. Collenchymas mainly forms _______:

(a) Hypodermis
(b) Epidermis
(c) Phloem
(d) Inner cortex

60. Which of the following tissues is composed of mainly dead cells ?

(a) Phloem (b) Epidermis

(c) Xylem (d) Endodermis

61. Phloem in the plants perform the function of ______:

(a) Conduction of food (b) Conduction of water

(c) Providing support (d) Photosynthesis

62. Which of the following characteristics is true of all protists?

(a) Contains a nucleus (b) Photosynthetic

(c) Has a cell wall (d) Very small

63. What organisms are most numerous on Earth?

(a) Eukaryotes (b) Archaea

(c) Prokaryotes (d) Plants

64. In an aqueous environment microscopic animals and plants are collectively known as:

(a) Herbivores (b) Fauna and flora

(c) Planktons (d) Symbionts

65. Which of the following types of plants is notable to self-pollinate?

(a) Dioecious (b) Monoecious

(c) Complete (d) Wind-pollinated

66. In flowering plants, pollen is released from the q:

(a) Anther (b) Stigma

(c) Carpel (d) Filament

67. What is the embryonic root called?

(a) Plumule (b) Hypocotyls

(c) Epicotyls (d) Radicle

68. Which of these structures is unique to the seed of a monocot?

(a) Cotyledon (b) Endosperm

(c) Coleoptiles (d) Radical

69. Fruits develop from:

(a) Microsporangia (b) Receptacles

(c) Fertilized eggs (d) Ovaries

70. What is the process by which the macrospores of a seed plant are transferred to another compatible plant of the same type?

(a) Pollination (b) Fertilization

(c) Germination (d) Reproduction

71. What is the botanical name for the male reproductive organ in flowering plants?

(a) Calyx (b) Frond
(c) Carpel (d) Stamen

72. Which of these terms is used to describe the means by which the male and female cells combine to form a new cell which then eventually develops into a new embryo?

(a) Fertilization (b) Fusion
(c) Respiration (d) Ptyxis

73. What is the name of the structure that is formed from the ovule of a plant ater fertilization has taken place?

(a) Xylem (b) Seed
(c) Scion (d) Nut

74. What is the name of the ripened ovary which contains the seeds in many plants?

(a) Fruit (b) Leaf
(c) Flower (d) Cellulose

75. What is a plant called that sheds its leaves before winter, or during a dry spell?

(a) Deciduous (b) Indehiscent
(c) Evergreen (d) Recessive

76. Which of the following items is a product of the possible product of the cambium of a plant?

(a) Cork (b) Wort
(c) Germ (d) Fruit

77. What substance produced from the plant is used to create its cell walls?

(a) Lignin (b) Cellulose
(c) Sap (d) Ligature

78. Which type of plant tissue is an active site of cell division?

(a) Permanent tissue (b) Meristematic tissue
(c) Dpithelial tissue (d) Connective tissue

79. Meristematic tissue is responsible for the growth of the plant. Which type of meristematic tissue is responsible for the increase in girth of the plant?

(a) Apical (b) Parenchyma
(c) Intercalary (d) Lateral

80. Which type of permanent tissue does carrot contain?

(a) Collenchymas (b) Aerenchyma

(c) Sclerenchyma (d) Parenchyma

81. Which type of parenchyma contains cholorophyll?

(a) Chlorenchyma (b) Aerenchyma

(c) Chlorosome (d) All types of parenchyma contain chlrophyll

82. Which tissue does the husk of a coconut contain?

(a) Sclerenchyma (b) Collenchymas

(c) Parenchyma (d) Aerenchyma

83. Which substance is responsible for thickening of sclerenchyma walls?

(a) Suberin (b) Cutin

(c) Lignin (d) Pectin

84. Epidermis in desert plants has a thin waxy coating of a chemical substance. Name the chemical:

(a) Suberin (b) Cutin

(c) Pectin (d) Rennin

85. Which are the tubular structures present in xylem?

(a) Xylem fibers (b) Sieve tubes

(c) Xylem parenchyma (d) Tracheids and vessels

86. Name the part of phloem with perforated walls.

(a) Phloem fibers (b) Companion cells

(c) Sieve tubes (d) Phloem parenchyma

87. Name the dead part of phloem.

(a) Companion cells (b) Parenchyma

(c) Fibers (d) Sieve tubes

88. What is a common function of modified roots:

(a) The storage of food (b) Photosynthesis

(c) Help plants walk (d) Gaseous exchange

89. What is *not* a kind of modified root ?

(a) Fascicles (b) Phylloclade

(c) Brace root (d) Stilt root

90. What are plants called if they grow one year, flower the next and die?

(a) Perennial (b) Annual

(c) Biennial (d) None of the above

91. All leaves are connected to the stem by a thin stalk. What is this stalk called?

(a) Stem (b) Vein

(c) Blade (d) Petiole

92. That does the word 'gymnosperm' mean?

(a) Naked seed (b) Lose leaves in the fall

(c) Covered seed (d) None of the above

93. What does the word 'angiosperm' mean?

(a) None of the above (b) Naked seed

(c) Covered seed (d) Lose leaves in the fall

94. How many seed leaves does a dicotyledon have?

(a) 0 (b) 2

(c) 1 (d) 3

95. The four basic parts of a plant are all of these except?

(a) Roots (b) Leaves

(c) Sepals (d) Stems

96. Monocot pollengrains are generally:

(a) Monocoplate (b) Bicolpate

(c) Tricolpate (d) Multicolpate

97. Exine of pollen-grain is made up of:

(a) Sporopollenin (b) Cellulo-pectin

(c) Callose (d) Cellulose

98. Ovule is:

(a) Megasporangium (b) Microsporangium

(c) Intgumented megasporangium (d) Megaspore

99. The tenuinucellate ovule has:

(a) Small amount of nucellus (b) Large amount of nucellus

(c) Chalazalnucellus (d) MIcropylenucellus

The point at which funicle is attached to body of ovule:

(a) Raphe (b) Hilum

(c) Endothelium (d) Chalaza

101. A typical angiospermic embryo sac is usually:

(a) One celled (b) Two celled
(c) Three celled (d) Seven celled

102. Continued self pollination results in:

(a) Better progeny (b) Weak progeny
(c) New varieties (d) Both (a) and (c)

103. Endosperm is generally:

(a) Haploid (b) Diploid
(c) Triploid (d) Polyploid

104. In monocot seeds, radical is covered by a protective covering called:

(a) Coleoptiles (b) Coleorrhiza
(c) Scutellum (d) Aleurone layer

105. Embryo axis is also called as:

(a) Tegman (b) Tiller
(c) Tigellum (d) Scutellum

106. The first node indicates point of attachment of _______. Embryo axis:

(a) Plumule (b) Radical
(c) Cotyledons (d) All of the above

107. The branch which deals with study of fungi is known as _______.

(a) Physiology (b) Ecology
(c) Mycology (d) Phycology

108. The type of seed in which endosperm is completely utilized by developing embryo:

(a) Endospermic seed (b) Non-endospermic seed
(c) Albuminous seed (d) None of the above

109. Agamospermy produces new plant through the formation of:

(a) Gamete (b) Bulbil
(c) Bud (d) Asexual embryo

110. Palynology deals with study of:

(a) DNA (b) Chromosomes
(c) Pollen grains (d) Flowers

111. How many pollen mother cells should undergo meiotic division to produce 64 pollen grains?

(a) 64 (b) 32
(c) 16 (d) 8

112. In a fully developed male gametophyte the number of nuclei is:

(a) One (b) Two

(c) Three (d) Four

113. If an endosperm cell of an angiosperm contains 24 chromosomes, the number of chromosomes in each cell of the root will be:

(a) 8 (b) 4

(c) 16 (d) 24

114. Which of the following is necessary for seed formation in angiosperm?

(a) Ovule (b) Pollination

(c) Double fertilization (d) All of the above

115. Anthesis is:

(a) Elongation of pollen tube is style (b) Emergence of anthers from corolla tube

(c) Dehiscence of anthers (d) The first opening of a flower

116. Karyogamy is:

(a) Fusion of zoospores (b) Fusion of cytoplasm

(c) Fusion of spores (d) Fusion of gametic nuclei

117. What is the simplest design of a plant body?

(a) Cytoplasm (b) Mud

(c) Cellulite (d) Single cell

118. Which of these trees is NOT an angiosperm?

(a) Siberian crabapple (b) Maidenhair tree (*Ginkgo biloba*)

(c) Crack willow (d) Common laburnum

119. The first step in the germination of a seed is usually:

(a) Pollination (b) Fertilization

(c) Imbibitions of water (d) Hydrolysis of starch and other food reserves

120. In flowering plants, pollen is released from the:

(a) Anther (b) Stigma

(c) Carpel (d) Filament

121. Which of the following types of plants is not able to self-pollinate?

(a) Dioecious (b) Monoecious

(c) Complete (d) Wind-pollinated

122. A perfect flower is fertile, but may be either complete or incomplete. Which of the following correctly describes a perfect flower?

(a) It has no sepals (b) It has fused carpels
(c) It is on a dioecious plant (d) It has both stamens and carpels

123. Biogeochemical cycle are also known as:

(a) Cycles of matter (b) Sedimentary cycles
(c) Material cycling (d) Gaseous cycles

124. In an aqueous environment microscopic animals and plants are collectively known as:

(a) Herbivores (b) Fauna and flora
(c) Planktons (d) Symbionts

125. A group of cells alike in form, function and origin are called _______.

(a) Tissue (b) Organ
(c) Organelle (d) None of the above

126. Grasses have _______ meristems:

(a) Intercalary (b) Apical
(c) Lateral (d) None of the above

127. This field of botany specializes in studying the internal structure of plant organs by using a microscope:

(a) Plant physiology (b) Plant anatomy
(c) Plant morphology (d) Plant taxonomy

128. The scientist credited with being the Father of Botany is:

(a) Darwin (b) Theophrastus
(c) Hinterthuer (d) Linnaeus

129. The Kingdom in which mushrooms are found is:

(a) Monera (b) Bacteria
(c) Protista (d) Fungi

130. Maximum absorption of rainfall water is done by:

(a) Tropical deciduous forest (b) Tropical evergreen forest
(c) Tropical savanna (d) Scrub forest

131. Basidiomycetes produces _______ type of fruiting bodies:

(a) Ascocarp (b) Cleistothecium
(c) Basidiocarp (d) Basidiospore

132. The study of fungi is known as. _______.

(a) Mycology (b) Phycology

(c) Ecology (d) Taxonomy

133. The algae growing in seawater is known as _______:

(a) Freshwater algae (b) Terrestrial algae

(c) Marine algae (d) Lithophytic algae

134. The study of algae is known as _______.

(a) Mycology (b) Phycology

(c) Taxonomy (d) Physiology

135. What is the scientific term for a leaf stalk?

(a) Petiole (b) Pedicel

(c) Peduncle (d)

136. Which of the following is an edible 'fungi'?

(a) Mucor (b) Penicillum

(c) Agaricus (d) Rhizopus

137. Energy transferred from one trophic level to another is:

(a) 5 per cent (b) 15 per cent

(c) 10 per cent (d) 20 per cent

138. Coir which is obtained from fruit of coconut is:

(a) Endocarp (b) Periderm

(c) Mesocarp (d) Epicarp

139. 'Pneumatophores' are present/common in:

(a) Xerophytes (b) Hygrophytes

(c) Mesophytes (d) Halophytes

140. Maximum absorption of rainfall water is done by:

(a) Tropical deciduous forest (b) Tropical evergreen forest

(c) Tropical savanna (d) Scrub forest

141. In which form does the food transported in plants?

(a) Sucrose (b) Fructose

(c) Glucose (d) Lactose

142. Largest flower is:

(a) Rafflesia (b) Lotus

(c) Rose (d) Lity

143. Which group of plants possesses both animal and plant characteristics?

(a) Myxomycetes (b) Basidiomycetes
(c) Phycomycetes (d) Rhodomycetes

144. Which type of plant tissue is an active site of cell division?

(a) Permanent tissue (b) Meristematic tissue
(c) Epithelial tissue (d) Connective tissue

145. Meristematic tissue isresponsible for the growth of the plant. Which type of meristematic tissue is responsible for the increase in girth of the plant?

(a) Apical (b) Parenchyma
(c) Intercalary (d) Lateral

146. Which type of permanent tissue does carrot contain?

(a) Collenchymas (b) Aerenchyma
(c) Sclerenchyma (d) Parenchyma

147. Which type of parenchymacontains chlorophyll?

(a) Chlorenchyma (b) Aerenchyma
(c) Chlorosome (d) All types of parenchyma contain chlorophyll

148. Which tissue does the husk of a coconut contain?

(a) Sclerenchyma (b) Collenchymas
(c) Parenchyma (d) Aerenchyma

149. Which substance is responsible for thickening of sclerenchyma walls?

(a) Suberin (b) Cutin
(c) Lignin (d) Pectin

150. Epidermis in desert plants has a thin waxy coating of a chemical substance. Name the chemical:

(a) Suberin (b) Cutin
(c) Pectin (d) Rennin

151. Which are the tubular structures present in xylem?

(a) Xylem fibers (b) Sieve tubes
(c) Xylem parenchyma (d) Tracheids and vessels

152. Name the part of phloem with perforated walls:

(a) Phloem fibers (b) Companion cells
(c) Sieve tubes (d) Phloem parenchyma

153. Name the dead part of phloem:

(a) Companion cells (b) Parenchyma

(c) Fibers (d) Sieve tubes

154. What is a common function of modified roots?

(a) The storage of food (b) Photosynthesis

(c) Help plants walk (d) Gaseous exchange

155. The cell was discovered by:

(a) Dioscorides (b) Robert hooke

(c) Rudolf Virchow (d) Louis Pasteur

156. The ability of an organism to change in response to the environment is called:

(a) Adaptation (b) Avoidance

(c) Response (d) None of the above

157. Scientific study on bryophytes is known as:

(a) Mycology (b) Taxonomy

(c) Bryology (d) Pharmacology

158. Scientific study on lichens is known as:

(a) Ecology (b) Lichenology

(c) Cytology (d) None of the above

159. Scientific study of a algae is called:

(a) Molecular biology (b) Phycology

(c) Physiology (d) None of the above

160. The structure of DNA was discovered by:

(a) Theodor Schwann (b) Matthias Schleiden

(c) James Watson and Francis Crick (d) Charles Darwin

161. The study of fossil plants is defined as:

(a) Palaeobotany (b) Taxonomy

(c) Ecology (d) Histology

162. The epidermis is made up of _______ cells:

(a) Collenchymas (b) Meristematic

(c) Paranchymatus (d) Sclerenchyma

163. Leaf is the chief organ of:

(a) Photosynthesis (b) Storage

(c) Conduction (d) All of the above

164. Leaves which have the stipules are called _______:

(a) Ex-stipulate
(b) Stipulate
(c) Lanceolate
(d) Palmate

165. Leaf blade is known as _______:

(a) Pulvinus
(b) Stipules
(c) Pinnae
(d) Lamina

166. _______ are mature, fertilized ovules:

(a) Seeds
(b) Fruit
(c) Anther
(d) All of the above

167. Radical is an embryonic _______ of the plant.

(a) Shoot
(b) Root
(c) Leaves
(d) None of the above

168. Stigma, style, ovary and ovule collectively called as _______.

(a) Stamen
(b) Calyx
(c) Pistil
(d) Corella

169. The individual flowering units of the monocot type of inflorescence is referred as _______.

(a) Spikelet
(b) Leaflet
(c) Raceme
(d) Cyme

170. Fungi which live within the living plant tissues are known as:

(a) Parasite
(b) Epiphyte
(c) Endophyte
(d) Halophyte

171. Basidiocarpis formed from:

(a) Primary mycelium
(b) Secondary mycelium
(c) Tertiary mycelium
(d) Quaternary mycelium

172. Anti-cancerous compound Lentinan and Eritadine are extracted from:

(a) Fruit bodies of *Agaricus bisporus*
(b) Fruit bodies of *Lentinus edodeds*
(c) Fruit bodies of *Pleurotus ostreatus*
(d) Fruit bodies of *Volvariel lavolvacae*

173. The number of cotyledon in a pinus seed is:

(a) One
(b) One or two
(c) Many
(d) None of the above

174. Which of the following description is correct for anisogamy?

(a) Fusing gametes are morphologically and physiologically similar

(b) Fusing gametes are morphologically and physiologically dissimilar

(c) Motile male gametes are produced in large number and female gamete produced singly

(d) Fusing gametes are formed by different parental thallus

175. Who classified the Gymnosperms on the basis of wood structure:

(a) Engler and Prantl (b) Chamberlain

(c) Seward (d) Pant

176. The seeds which of the following plants are consumed by human beings?

(a) *Pinus insularis* (b) *Pinus merkurii*

(c) *Ppinus gerandiana* (d) *Pinus roxburghii*

177. Which one among the gymnosperm plant is called a living fossil?

(a) *Gnetum* (b) *Ginkgo*

(c) *Ginkgodium* (d) *Ephedra*

178. Special type of inflorescence is:

(a) Spadix (b) Cyathium

(c) Corymb (d) Panicle

179. Capsule is a:

(a) Dry fruit (b) Dry dehiscent fruit

(c) Loculicidal capsule (d) Septicidal capsule

180. Cohorts and orders of Bentham and Hooker's classification is today known as:

(a) Orders and families (b) Families and orders

(c) Families and genera (d) Genera and species

181. Bentham and Hookers system of classification is based on:

(a) De condole and Lindley's system (b) De condole system

(c) Robert Brown's systems (d) Exclusive natural system

182. Presence of stomata on both surfaces of leaves is called:

(a) Epistomatic (b) Amphistomatic

(c) Hypsotomatic (d) Multistomatic

183. Robert H. Whittaker classified the organisms into how many kingdoms?

(a) Five (b) Four

(c) Three (d) Two

184. Gram positive cell wall is negatively charged because of the presence of:

(a) D-alanine (b) Teichoic acid
(c) Peptidoglycan (d) Glucose

185. Meiosis occurs in the:

(a) Somatic cells (b) Germ cells
(c) Vegetative cells (d) Meristem

186. Two or more crops grown together in the kitchen garden are called:

(a) Succession crop (b) Companion crop
(c) Annual crop (d) Biennial crop

187. In Monocots, grafting is not possible because:

(a) Cambium is absent (b) Vascular bundles are absent
(c) Parallel venation (d) Herbaceous

188. An example of the algal sulphated polysaccharide is:

(a) Fucoidin (b) Sodium alginate
(c) Agar-Agar (d) Carrageenan

189. Calyptras is seen in:

(a) Gametophyte (b) Sporophyte
(c) Archegoniophore (d) Antheridophore

190. The layer which is protective in function of the sporangium of ferns is called as:

(a) Annulus (b) False indusium
(c) Indusium (d) Stomium

191. Aggregate fruits are:

(a) Acorns (b) Borne in clusters
(c) Cones (d) Winged seeds

192. Gymnosperms bear naked seeds. They are the:

(a) Broad-leafed trees (b) Cone bearing trees
(c) Fruit trees (d) Oaks

193. The genus of the tree noted for its great size and long life is _______.

(a) Alnux (b) Betula
(c) Salix (d) Sequoia

194. Achene, samara, drupe and pome are types of _______:

(a) Bark (b) Flowers
(c) Fruit (d) Leaves

195. An achene _______.

(a) Describes twig texture
(b) Is a description of bud shape
(c) Is an unwinged seed
(d) Is a winged seed

196. An ovate leaf is _______:

(a) Egg shaped
(b) Heart shaped
(c) Kidney shaped
(d) Triangular in shape

197. A leaf margin described as entire is _______.

(a) Coarsely toothed
(b) Lobed
(c) Smooth
(d) Wavy

198. The plant kingdom division Magnoliophyta includes _______:

(a) Angiosperms
(b) Broadleaf trees
(c) Oaks and hickories
(d) All of the above

199. Open grown softwood trees (pines, *etc.* take on a _______. From:

(a) Triangular
(b) Bulb like
(c) Columnar
(d) Square

200. Open grown hardwood trees become _______ in form.

(a) Triangular
(b) Bulb like
(c) Columnar
(d) Square

201. If a flower lacks either a calyx or corolla, it is called a (n) _______. Flower.

(a) Imperfect
(b) Monoecious
(c) Dioecious
(d) Incomplete

202. If the male and female flower parts occur in separate flolwers on the same tree, the species is said to be:

(a) Monoecious
(b) Dioecious
(c) All of the above
(d) None of the above

203. Ovate, deltoid, lanceolate, star shaped are examples of:

(a) Leaf margins
(b) Leaf apexes
(c) Leaf bases
(d) All of the above

204. Entire, serrate, dentate, lobed and incised are examples of:

(a) Leaf margins
(b) Leaf apexes
(c) Bud types
(d) All of the above

205. Major characteristics of twigs used in tree identification are:

(a) Buds
(b) Pith
(c) Lenticels
(d) All of the above

206. Which of the following are female parts of a tree flower?

(a) Sepal (b) Pistil
(c) Pollen (d) All of the above

207. The outside edge of a leaf is called the _______.

(a) Lobe (b) Margin
(c) Petiole (d) Leaflet

208. The leaf stem is also called the _______.

(a) Petiole (b) Attach
(c) Joiner (d) Midrib

209. When a single leaf is attached to the twig it is said to be _______.

(a) Truncate (b) Pinnate
(c) Palmate (d) Simple

210. When many small leaflets are attached to the petiole to form a larger leaf that type of leaf is called _______.

(a) Multiple (b) Compound
(c) Acute (d) Oblique

211. Which is not a type of leaf arrangement?

(a) Alternate (b) Opposite
(c) Whorled (d) Oblique

212. As per Bentham and Hookers' classification, there are _______.non-standard exceptions to the family name ending 'aceae'.

(a) 8 (b) 6
(c) 4 (d) 2

213. The natural system of classification is based on:

(a) Numerical (b) Habit
(c) Form (d) None of the above

214. The flower lacking either stamen or pistil is called:

(a) Incomplete (b) Complete
(c) Intermediate (d) None of the above

215. The largest family among the Indian flora is:

(a) Fabaceae (b) Meliaceae
(c) Asteraceae (d) Myrtaceae

216. The origin of teak is:

(a) India (b) Burma

(c) Indonesia (d) Nepal

217. The arrangement of ovules within the ovary is:

(a) Aestivation (b) Phyllotaxy

(c) Placentation (d) None of the above

218. Ergrostisergrostis is an example for:

(a) Basionym (b) Homonym

(c) Autonym (d) Tautonym

219. A pair of appendages that present at the point of attachment of a leaf to the stem is:

(a) Bracts (b) Bracteoles

(c) Stipules (d) None of the above

220. Thick and fleshy root without any particular shape is:

(a) Tuberous (b) Rhizome

(c) Bulbs (d) None of the above

221. Sucking roots of parasites are:

(a) Prop roots (b) Haustoria

(c) Hanging roots (d) Tilt roots

222. The specimen or other element designated by the author or used by him as the nomenclatural type:

(a) Lectotype (b) Holotype

(c) Isotype (d) None of the above

223. Father of Botany is:

(a) Linnaeus (b) Aristotle

(c) Theophrastus (d) John Ray

224. If all the petals are free, the condition is called:

(a) Polypetale (b) Polysepalae

(c) Gamopetalae (d) Gamosepalae

225. The alternative name of Guttiferae is:

(a) Clusiaceae (b) Asteraceae

(c) Poaceae (d) None of the above

226. *Acacia nilotica* belongs to the family:

(a) Pappilionaceae (b) Mimoceae
(c) Caesalpinaceae (d) None of the above

227. Peeling of bark is a characteristic feature of:

(a) Magnoliaceae (b) Meliaceae
(c) Moraceae (d) Myrtaceae

228. *Santalum album* is found in _______. forests:

(a) Dry deciduous (b) Evergreen
(c) Semi-evergreen (d) Moist deciduous

229. Plants with male and female flowers separate are:

(a) Monoecious (b) Dioecious
(c) Polygamous (d) None of the above

230. Pellucid dots are characteristic of the family.

(a) Magnoliaceae (b) Rutaceae
(c) Myrtaceae (d) Asteraceae

231. _______ is an exotic:

(a) Eucalyptus (b) Casuarinas
(c) Simaruba (d) All of the above

Answer Sheet

1.	(a)	31.	(a)	61.	(a)	91.	(d)	121.	(a)
2.	(a)	32.	(b)	62.	(a)	92.	(a)	122.	(d)
3.	(b)	33.	(c)	63.	(c)	93.	(c)	123.	(a)
4.	(c)	34.	(d)	64.	(c)	94.	(b)	124.	(c)
5.	(b)	35.	(c)	65.	(a)	95.	(c)	125.	(a)
6.	(a)	36.	(d)	66.	(a)	96.	(a)	126.	(a)
7.	(b)	37.	(b)	67.	(d)	97.	(a)	127.	(b)
8.	(b)	38.	(b)	68.	(c)	98.	(c)	128.	(b)
9.	(a)	39.	(c)	69.	(d)	99.	(a)	129.	(d)
10.	(a)	40.	(a)	70.	(a)	100.	(b)	130.	(b)
11.	(c)	41.	(b)	71.	(d)	101.	(d)	131.	(d)
12.	(b)	42.	(b)	72.	(a)	102.	(b)	132.	(a)
13.	(d)	43.	(c)	73.	(b)	103.	(c)	133.	(b)
14.	(c)	44.	(c)	74.	(a)	104.	(b)	134.	(b)
15.	(c)	45.	(d)	75.	(a)	105.	(c)	135.	(a)
16.	(c)	46.	(b)	76.	(a)	106.	(c)	136.	(c)
17.	(b)	47.	(b)	77.	(b)	107.	(c)	137.	(b)
18.	(c)	48.	(b)	78.	(b)	108.	(b)	138.	(c)
19.	(b)	49.	(a)	79.	(d)	109.	(d)	139.	(d)
20.	(c)	50.	(b)	80.	(d)	110.	(c)	140.	(b)
21.	(a)	51.	(c)	81.	(a)	111.	(c)	141.	(a)
22.	(c)	52.	(a)	82.	(a)	112.	(c)	142.	(a)
23.	(d)	53.	(a)	83.	(c)	113.	(c)	143.	(a)
24.	(a)	54.	(a)	84.	(b)	114.	(d)	144.	(b)
25.	(a)	55.	(a)	85.	(d)	115.	(c)	145.	(d)
26.	(a)	56.	(a)	86.	(c)	116.	(d)	146.	(b)
27.	(b)	57.	(a)	87.	(c)	117.	(d)	147.	(a)
28.	(d)	58.	(d)	88.	(a)	118.	(b)	148.	(a)
29.	(b)	59.	(a)	89.	(b)	119.	(c)	149.	(c)
30.	(d)	60.	(c)	90.	(c)	120.	(a)	150.	(b)

151.	**(d)**	**168.**	**(c)**	**185.**	**(b)**	**202.**	**(a)**	**219.**	**(c)**
152.	**(c)**	**169.**	**(a)**	**186.**	**(b)**	**203.**	**(c)**	**220.**	**(a)**
153.	**(c)**	**170.**	**(c)**	**187.**	**(a)**	**204.**	**(a)**	**221.**	**(b)**
154.	**(a)**	**171.**	**(b)**	**188.**	**(d)**	**205.**	**(d)**	**222.**	**(b)**
155.	**(b)**	**172.**	**(b)**	**189.**	**(b)**	**206.**	**(b)**	**223.**	**(c)**
156.	**(a)**	**173.**	**(c)**	**190.**	**(c)**	**207.**	**(b)**	**224.**	**(a)**
157.	**(c)**	**174.**	**(b)**	**191.**	**(b)**	**208.**	**(a)**	**225.**	**(a)**
158.	**(b)**	**175.**	**(c)**	**192.**	**(b)**	**209.**	**(d)**	**226.**	**(b)**
159.	**(b)**	**176.**	**(c)**	**193.**	**(d)**	**210.**	**(b)**	**227.**	**(d)**
160.	**(c)**	**177.**	**(b)**	**194.**	**(c)**	**211.**	**(d)**	**228.**	**(a)**
161.	**(a)**	**178.**	**(b)**	**195.**	**(c)**	**212.**	**(a)**	**229.**	**(b)**
162.	**(c)**	**179.**	**(b)**	**196.**	**(a)**	**213.**	**(c)**	**230.**	**(c)**
163.	**(a)**	**180.**	**(a)**	**197.**	**(c)**	**214.**	**(a)**	**231.**	**(d)**
164.	**(b)**	**181.**	**(a)**	**198.**	**(d)**	**215.**	**(c)**		
165.	**(d)**	**182.**	**(b)**	**199.**	**(a)**	**216.**	**(b)**		
166.	**(a)**	**183.**	**(a)**	**200.**	**(b)**	**217.**	**(c)**		
167.	**(b)**	**184.**	**(b)**	**201.**	**(a)**	**218.**	**(d)**		

Chapter 9

Forest Genetics and Tree Breeding

1. **The DNA molecule is usually made up of how many strands?**
   (a) 1 (b) 2
   (c) 3 (d) 6

2. **The RNA molecule is made up of how many strands?**
   (a) 1 (b) 2
   (c) 3 (d) 6

3. **How many different types of RNA molecules are needed to make protein?**
   (a) 1 (b) 2
   (c) 3 (d) 4

4. **What is the form of RNA that carries the code from the DNA to the site where the protein is assembled?**
   (a) Messenger RNA (b) Nuclear RNA
   (c) Ribosomal RNA (d) Transfer RNA

5. **Which of the following carries amino acids to ribosomes, where amino acids are linked into the primary structure of a polypeptide?**
   (a) mRNA (b) tRNA
   (c) hnRNA (d) rRNA

6. **The nitrogenous base found in DNA but not in RNA is:**
   (a) Adenine (b) Cytosine
   (c) Uracil (d) Thymine

7. **Transfer RNA differs from other types of RNA because it:**
   (a) Transfers genetic instructions from the cell nucleus to the cytoplasm.
   (b) Specifies the amino acid sequence of a particular protein
   (c) Carries an amino acid at one end (d) Contains codons

8. **Which substance is found in RNA but not in DNA?**
   (a) Thymine (b) Deoxyribose
   (c) Ribose (d) Guanine

9. **The nitrogenous base found in RNA but not in DNA is:**
   (a) Adenine (b) Cytosine
   (c) Guanine (d) Uracil

10. **Uracil will pair with:**
   (a) Ribose (b) Adenine
   (c) Cytosine (d) Thymine

11. **The synthesis of an RNA molecule from a DNA template strand is:**
   (a) Replication (b) Translation
   (c) Transcription (d) DNA synthesis

12. **The relationship between RNA and the DNA that served as the template is:**
   (a) Antagonistic (b) Opposite
   (c) Complementary (d) An exact duplicate

13. **Transcription:**
   (a) Occurs on the surface of the ribosome
   (b) Is the final process in the assembly of a protein
   (c) Is the synthesis of any type of RNA from a DNA template
   (d) Is catalyzed by DNA polymerase

14. **Which of the following catalyzes the process of transcription?**
   (a) RNA polymerase (b) DNA polymerase
   (c) Ligase (d) Gyrase

15. **Three dimensional shape of tRNA is:**
   (a) L-shaped (b) Clover leaf-like
   (c) X-shaped (d) Y-shaped

16. **The portion of the DNA molecule that is translate is composed of:**
   (a) Introns (b) Anticodons
   (c) Exons (d) Transcriptions

17. In transcription:

(a) Several RNA molecules can be transcribed from the same DNA molecule

(b) Promoters are needed so that RNA polymerase can bind to DNA

(c) DNA produces messenger RNA

(d) All of the above

18. The portion of the DNA molecule that is not translated and is a non-coding portion:

(a) Introns (b) Anticodons

(c) Exons (d) Transcriptions

19. In transcription,:

(a) Several amino acids are assembled by the messenger RNA molecules at one time

(b) A special sequence called a promoter is necessary for transcription to begin

(c) Certain polypeptide sequences are governed by one ribosome, while other sequences are produced by other ribosomes

(d) The transfer RNA molecules arrange the messenger RNA codons into the appropriate sequence.

20. When a gene transcription occurs, which of the following is produced?

(a) Protein or polypeptide sequences (b) More DNA

(c) Messenger RNA (d) Enzymes

21. The genetic code is made up of units consisting of how many nucleotides?

(a) 2 (b) 3

(c) 5 (d) 6

22. There are how many different kinds of amino acids in proteins?

(a) 3 (b) 6

(c) 12 (d) 20

23. There are how many different RNA codons?

(a) 3 (b) 12

(c) 28 (d) 64

24. If the codon consisted of only two nucleotides, there would be how many different kinds of codons?

(a) 4 (b) 8

(c) 16 (d) 32

25. The concept that a set of three nucleotides specifies a particular amino acid provides the basis for:

(a) The one gene, one enzyme hypothesis

(b) The one gene, one polypeptide hypothesis

(c) The genetic code

(d) Biochemical reactions among nucleic acids

26. Of all the different codons that exist, three of them:

(a) Are involved in mutations

(b) Do not specify a particular amino acid

(c) Cannot be copied

(d) Don not specify a particular amino acid

27. Each "word" in the mRNA language consists of how many letters?

(a) Three (b) Four

(c) Five (d) More than five

28. If each nucleotide coded for a single amino acid, how many different types of amino acids could be combined to form proteins?

(a) 4 (b) 16

(c) 20 (d) 64

29. If the DNA triplets were ATG-CGT, the mRNA codons would be:

(a) AUGCGU (b) ATGCGT

(c) UACGCA (d) UAGCGU

30. If the DNA triplets were ATG-CGT, the tRNA anticodons would be:

(a) AUGCGU (b) ATGCGT

(c) UACGCA (d) UAGCGU

31. Ribosomes function as:

(a) A single unit (b) Two part units

(c) Three part units (d) Four part units

32. Which the following needs an anticodon to bind to the protein synthesizing machinery?

(a) DNA (b) mRNA

(c) rRNA (d) tRNA

33. The "start" codon is:

(a) UUU (b) CUC

(c) GUC (d) AUG

34. Amino acids are joined together in proteins by:

(a) Hydrogen bonds
(b) Peptide bonds
(c) Anticodons
(d) Wobble effects

35. Translation begins when the initiator:

(a) Large ribosomal subunit
(b) Small ribosomal subunit
(c) The second tRNA
(d) The terminator codon

36. Termination of protein translation is due to:

(a) The lack of amino acids
(b) Wobble effects
(c) The lack of messenger rna
(d) A stop

37. The difference between normal hemoglobin and sickle – cell hemoglobin is in the:

(a) Heme portion of the molecules
(b) Number of chains of amino acids
(c) Substitution of a specific amino acid for another specific amino acid
(d) Addition of one amino acid to the normal hemoglobin molecule

38. A gene mutation:

(a) Is a change in the nucleotide sequence of DNA
(b) May arise spontaneously
(c) Can occur in all organisms
(d) All of the above

39. Mutations can be:

(a) Beneficial
(b) Harmful
(c) Heritable
(d) All of the above

40. Which of the following can directly cause mutations?

(a) Ionizing radiation
(b) Non-ionizing radiation
(c) Alkylating agents
(d) Both ionizing and noninonizing radiation

41. Anticodon:

(a) Sites at which RNA polymerases can bind and initiate transcription
(b) Binds to small subunit platform of a ribosome
(c) Guided and catalyzed by RNA polymerases
(d) A tRNA triplet that binds to mRNA

42. Codon:

(a) Sites at which RNA polymerases can bind and initiate transcription
(b) Binds to small subunit platform of a ribosome
(c) Guided and catalyzed by RNA polymerases
(d) A set of three nucleotides

43. Messenger RNA:

(a) Sites at which RNA polymerases can bind and initiate transcription
(b) Binds to small subunit platform of a ribosome
(c) Guided and catalyzed by RNA polymerases
(d) A tRNA triplet that binds to mRNA

44. Promoter:

(a) Sites at which RNA polymerases can bind and initiate transcription
(b) Binds to small subunit platform of a ribosome
(c) Guided and catalyzed by RNA polymerases
(d) A tRNA triplet that binds to mRNA

45. Transcription:

(a) Sites at which RNA polymerases can bind and initiate transcription
(b) Binds to small subunit platform of a ribosome
(c) Guided and catalyzed by RNA polymerases
(d) A tRNA triplet that binds to mRNA

46. This codon terminates a coding region.

(a) AUG (b) UAA
(c) UUU (d) UUA

47. The anticodon AAA would pair with this codon.

(a) AUG (b) UAA
(c) UUU (d) UUA

48. A single mutation involving the second letter codon AUA would convert it to this codon.

(a) AUG (b) UAA
(c) UUU (d) AAA

49. A DNA codon of ATT would be complementary to this RNA codon.

(a) AUG (b) UAA
(c) UUU (d) UUA

50. This codon specifies an amino acid and indicates the beginning:

(a) AUG (b) UAA

(c) UUU (d) UUA

51. The process of making RNA from DNA is:

(a) Transcription (b) Translation

(c) Conjunction (d) Peptide synthesis

52. The process of making proteins from RNA is:

(a) Transcription (b) Translation

(c) Conjunction (d) Peptide synthesis

53. Which of the following is NOT one of the bases in a DNA molecule?

(a) Adenine (b) Cytosine

(c) Uracil (d) Thymine

54. Cytosine will form a base pair only with:

(a) Cytosine (b) Adenine

(c) Thymine (d) Guanine

55. What is a nucleotide?

(a) A phosphate, five – carbon sugar, and a nitrogen base

(b) A group of linked amino acids

(c) One half of a DNA molecule

(d) A protein, a sugar, and a phosphate

56. The weak bonds between complementary nitrogen bases involve:

(a) RNA (b) DNA

(c) Hydrogen (d) Five carbon sugars

57. DNA assumes a form termed what?

(a) A triple – linked chain (b) A six sided figure

(c) A step ladder (d) A double helix

58. DNA may be described functionally as:

(a) A source for RNA production (b) An enzyme

(c) An arrangement of mRNA molecules (d) A sink for disposal of RNA

59. Where is the messenger RNA synthesized in a cell?

(a) Nucleus (b) Mitochondria

(c) Ribosome (d) Endoplasmic reticulum

60. Which is the following correctly describe messenger RNA?

(a) Single stranded (b) Uracil replaces thymine

(c) Many sugar to phosphate bonds occur (d) All of the above

61. Which of the following are correct types of RNA?

(a) mRNA (b) qRNA

(c) tRNA (d) Both (a) and (c)

62. The directions used in protein synthesis are provided by:

(a) mRNA (b) Centromere

(c) RNA in the ribosomes (d) tRNA

63. The coding of amino acids by multiple sets of nucleotides is referred to as;:

(a) Base pairing specificity (b) Hydrogen bonding

(c) Chemical coding (d) Triplet coding

64. What is the role of ribosomes in protein synthesis?

(a) They provide a source of amino acids

(b) They provide a site for transfer RNAs to link to messenger RNAs

(c) They translate the basic DNA code using transfer RNA

(d) They carry the proteins to their site of action

65. Which of the following is not a major grouping of proteins in cells?

(a) Enzymes (b) Antibodies

(c) Energy storage units (d) Structural units

66. _______ is responsible for replacing RNA primers with new DNA in Okazaki fragments.

(a) RNA exonuclease (b) DNA polymerase III

(c) DNA ligase (d) DNA polymerase I

67. All the amino acids have multiple codons except?

(a) Met (b) Arg

(c) Cys (d) Leu

68. The _______ separate during anaphase of Mitosis.

(a) Homologous chromosomes (b) Alleles of a gene

(c) Daughter cells (d) Sister chromatids

69. Which of the following will give cell-specific regulation of gene expression?

(a) Estrogen receptor (b) TFIID

(c) TATA box (d) AP-1

70. The codon is found in:

(a) DNA
(b) rRNA
(c) tRNA
(d) mRNA

71. Developed first electron microscope:

(a) Oparin and Haldane
(b) Watson and crick
(c) Miller and Urey
(d) Knoll and Ruska

72. The stage of meiosis in which chromosomes pair and cross over is:

(a) Prophase I
(b) Metaphase I
(c) Prophase II
(d) Metaphase II

73. Polyploidy refers to:

(a) Extra copies of a gene adjacent to each other on a chromosome
(b) An individual with complete extra sets of chromosomes
(c) A chromosome which has replicated but not divided
(d) Multiple ribosomes present on a single mRNA

74. A gene showing co dominance:

(a) Has both alleles independently expressed in the heterozygote
(b) Has one allele dominant to the other
(c) Has alleles tightly linked on the same chromosome
(d) Has alleles expressed at the same time in development

75. DNA ligase is:

(a) An enzyme that joins fragments in normal DNA replication
(b) An enzyme involved in protein synthesis
(c) An enzyme of bacterial origin which cuts DNA at defined base sequences
(d) An enzyme that facilitates transcription of specific genes

76. Replication of DNA:

(a) Takes place in a "conservative" manner
(b) Takes place in a "dispersive" manner
(c) Takes place in a "semi-conservative" manner
(d) Usually involves one origin of replication per chromosome in eukaryotes

77. A mutation in a codon leads to the substitution of one amino acid with another. What is the name for this type of mutation?

(a) Nonsense mutation
(b) Missense mutation
(c) Frameshift mutation
(d) Promoter mttion

78. A duplication is:

(a) An exchange between non-homologous chromosomes, resulting in chromosomes with new genes adjacent to each other

(b) Loss of genes in part of a chromosome

(c) An extra copy of the genes on part of a chromosome

(d) A reversal of order of genes on a chromosome

79. Mapping of human chromosomes:

(a) Has been restricted to the sex chromosomes because of small family sizes

(b) Proceeded much more successfully as large numbers of DNA markers became available

(c) Has determined that the number of linkage groups is about twice the number of chromosomes

(d) Has demonstrated that almost all of the DNA is involved in coding for genes

80. Germ plasm theory was postulated by:

(a) Lamark
(b) Weisman
(c) Kolliker
(d) Hertwig

81. The ability of a given gene or gene combination to be expressed phenotypically to any degree:

(a) Penetrance
(b) Expressivity
(c) Pleiotropism
(d) Prepotency

82. Total number of genotypes in human ABO blood group system:

(a) 3
(b) 12
(c) 6
(d) 4

83. Eye colour in drosophila is an example of:

(a) Sex linked inheritance
(b) Sex limited inheritance
(c) Sex influenced inheritance
(d) Incomplete dominance

84. If the X/A ratio is 0.5, the individual will be:

(a) Intersex
(b) Normal female
(c) Super female
(d) Normal male

85. If a woman heterozygous for colour blindness marries a colour blind man, what is the probability that their first child will be colour blind daughter?

(a) 50 per cent
(b) 25 per cent
(c) 75 per cent
(d) 100 per cent

86. In traslocation, the exchange of chromosomal segments occurs:

(a) Within the same chromosome

(b) Between homologous chromosomes

(c) Between non homologous chromosomes

(d) Both (b) and (c)

87. Shortest phase in mitosis is:

(a) Prophase (b) Anaphase

(c) Metaphase (d) Telophase

88. Soluble RNA)sRN(a) is:

(a) tRNA (b) rRNA

(c) mRNA (d) Mitochondrial RNA

89. The theory of epigenesist was proposed by:

(a) K.F. Wolf (b) Charles Darwin

(c) Swammerdam (d) Koliker

90. Phenotypic F2 ratio of duplicative recessive epistasis (complementary gene interaction) is:

(a) 12:3:1 (b) 9:7

(c) 9:6:1 (d) 9:3:3:1

91. Source of energy in nucleus is:

(a) Glycolysis only (b) TCA cycle only

(c) Both (a) and (b) (d) None of the above

92. Crossing over taking place at:

(a) Mitosis (b) Meiosis II

(c) Meiosis I (d) All of the above

93. Xanthine is deaminated product of:

(a) Adenine (b) Thymine

(c) Cytosine (d) Guanine

94. The types of phenotypes in F2 generation is:

(a) 2n (b) 3n

(c) n2 (d) n3

95. Sum total of genes in a population is:

(a) Genotype (b) Karyotype

(c) Gene pool (d) Gene frequency

96. Rho factor is required for the termination of:

(a) Replication
(b) Transcription
(c) Translation
(d) Transversion

97. The types of histone proteins present in eukaryotic chromosome are:

(a) 4
(b) 5
(c) 6
(d) 7

98. The classical test cross ratio in dihybrid is:

(a) 7:1:1:7
(b) 1:7:7:1
(c) 1:1:1:1
(d) Both (a) and (b)

99. Chromatin consists of:

(a) DNA and protein only
(b) RNA and protein only
(c) RNA and protein only
(d) DNA, RNA and protein

100. Epistasis is _______ type of interaction:

(a) Inter allelic
(b) Intra allelic
(c) Allelic genetic
(d) Both (a) and (b)

101. If the number multiple alleles for a trait is 5, then what is the total number of genotypes:

(a) 9
(b) 16
(c) 12
(d) 15

102. Chromosome theory of linkage was proposed by:

(a) Beadle and Tatum
(b) Bateson and Punnett
(c) Morgan and Castle
(d) Sutton and Bovery

103. Quantitative traits are controlled by:

(a) Major genes
(b) Minor genes
(c) Poly genes
(d) Both (b) and (c)

104. Breeding value (BV) is:

(a) Equal to the TA
(b) Twice the TA
(c) ½ of TA
(d) of TA

105. Heritability in narrow sense is:

(a) VA/VP
(b) VE/VP
(c) VG/VP
(d) VA+VG/VP

106. Repeatability value set in _______ of heritability:

(a) Lower limit of h2
(b) Upper limit of h2
(c) Intermediate of h2
(d) Both (a) and (b)

107. The most effective method of selection:

(a) Mass selection
(b) Pedigree selection
(c) Progeny testing
(d) Selection index

108. Quantitative traits shows:

(a) Continuous variation
(b) Discontinuous variation
(c) Both (a) and (b)
(d) None of the above

109. Genotypic frequency of progeny depends on:

(a) Genotypic freq. of population
(b) Genotypic freq. of parent
(c) Gene freq. of parent
(d) Both (a) and (c)

110. In a population sex linked genes are carried by homogametic sex is _______. of the total sex linked genes in the population:

(a) ½
(b)
(c) 1/3
(d) 2/3

111. Type of gene action in general combining ability (GC(a) is:

(a) Additive
(b) Non additive
(c) Both (a) and (b)
(d) None of the above

112. Range of h2 and fitness is:

(a) 0 to 1
(b) -1 to +1
(c) 0 to infinity
(d) 0 to xn

113. The proportion of population which shows genetic death is:

(a) Genetic sterility
(b) Genetic linkage
(c) Genetic load
(d) Genetic drift

114. For inbreeding, mated individuals should have common ancestors with in:

(a) 2-3 generations
(b) 4-6 generations
(c) 6-7 generations
(d) More than 10 generations

115. Inbreeding coefficients of progenies produced by self mating, parent offspring mating, full sibs and half sibs are in the order:

(a) 1,0.5,0.25,0.125
(b) 0.5,0.5,0.25,0.125
(c) 1,0.5,0.5,0.25
(d) 05,0.25,0.25,0.125

116. In the second generation, hybrid vigor will be:

(a) Same as the first
(b) Doubled the first
(c) Half of the first
(d) None of the above

117. Crossing of two inbred lines of the same breed is:

(a) Incrossing (b) Incross breeding
(c) Top incrossing (d) Top incross breeding

118. The intensity of selection depends with number traits considered 'n' is:

(a) 1/n (b) 1/n2
(c) n (d) 1/vn

119. Non additive gene action include,:

(a) Epistasis (b) Dominance
(c) Interaction (d) All of the above

120. Mildest form of out breeding is:

(a) Cross breeding (b) Close breeding
(c) Out crossing (d) Rotational crossing

121. The dispersive process mainly occurs in small population in which,:

(a) Only direction can be predicted not amount
(b) Only amount can be predicted not direction
(c) Both amount and direction can be predicted
(d) Direction and amount cannot be predicted

122. The contribution of off spring to the next generation is called:

(a) Fitness (b) Adaptive value
(c) Selective value (d) All of the above

123. Genotypic correlation is due to:

(a) Polymorphism (b) Linkage
(c) Pleiotropism (d) Both (b) and (c)

124. A new breed can be evolved by:

(a) Out crossing (b) Cross breeding
(c) Grading up (d) None of the above

125. If the coefficient of selection is 0.25, then the fitness is:

(a) 1 (b) 0.25
(c) 0.75 (d) 0.5

126. Grading up produces pure breed in how many generations:

(a) 2-4 (b) 4-5
(c) 10-12 (d) 7-8

127. Which of the following would represent a mutation?

(a) A change in the base sequence of DNA (b) Crossing over in meiosis

(c) Both (a) and (b) (d) None of the above

128. Arrange the following structures/chemicals in order of size from smallest to largest.

1. codon 2. Gene 3. Base 4. Chromosome:

(a) 1,2,3 and 4 (b) 2,1,3 and 4

(c) 3,1,2 and 4 (d) 1,2,4 and 3

129. A change in which of the following would result in a mutation?

(a) Order of bases in DNA (b) Gene

(c) Chromosome (d) All of the above

130. The production of a protein from mRNA is called _______.

(a) Replication (b) Transformation

(c) Transcription (d) Translation

131. The role of mRNA is to:

(a) Carry the genetic instructions to the cytoplasm

(b) Identify and transport amino acids

(c) Provide nucleotides for RNA production

(d) Produce amino acids for protein

132. Transcription refers to:

(a) DNA replication

(b) Protein synthesis using mRNA and tRNA

(c) mRNA synthesis from DNA

(d) Converting An inactive protein into its active form

133. Which of the following molecules has are amino acid attached to it?

(a) DNA (b) tRNA

(c) mRNA (d) All of the above

134. Codons may:

(a) Act as start codons (b) Act as stop codons

(c) Code for a specific amino acid (d) All of the above

135. How many amino acids are coded for in this mRNA sequence GGUACCUUUACU:

(a) 6 (b) 4

(c) 1 (d) 12

136. Thirty percent of the bases in DNA extracted from a prokaryotic cell is adenine. What percentage of cytosine is present in this DNA?

(a) 20
(b) 30
(c) 40
(d) 10

137. All cells of a species (except for the sex cells) contain the same amount of DNA. However, the cells of multicellular organisms vary in both structure and function from one tissue type to another. This specialization is evidence that most of the DNA in a cell is:

(a) Inactive
(b) Removed as the cell matures
(c) Converted into protein
(d) Converted into glycoprotein

138. Which of the following is the role of the structural genes in an inducible enzyme system?

(a) To code for proteins that metabolize an inducer substance
(b) To produce a repressor that binds to the operator
(c) To stop the transcription of the operator
(d) To recognize and form a complex with the inducible enzyme

139. Substances that cause cells to increase production of an enzyme (or group of enzymes) are known as:

(a) Introns
(b) Repressors
(c) Inducers
(d) Promoters

140. New alleles arise by:

(a) Mutation
(b) Migration
(c) Genetic drift
(d) Random mating

141. New variations of genes in a population may be produced by:

(a) Mutation
(b) Crossing over
(c) Sexual reproduction
(d) All of the above

142. The Hardy-Weinberg formula:

(a) Is useful in determining the extent to which a sexually reproducing population is evolving
(b) Is used to predict when genetic drift will occur in a sexually reproducing population
(c) Is useful in determining the extent to which polyploidy is occurring in specific plant populations
(d) Is used to predict when specific groups of organisms will become extinct

143. The maintenance of Hardy-Weinberg equilibrium is encouraged:

(a) When sexual selection occurs
(b) When mutations occur
(c) In small populations
(d) When there is no gene flow among different populations

144. In the Hardy-Weinberg equation, the term q2 refers to the frequency of:

(a) A recessive allele of a given locus
(b) The homozygous recessive genotype at a given locus
(c) Recessive alleles in a population
(d) Heterozygotes in a population

145. The influence of genetic drift on allele frequencies increases as:

(a) Gene flow increases
(b) Population size decreases
(c) Mutation rate decreases
(d) The number of heterozygous loci increases

146. Two individuals are members of the same species if they:

(a) Possess the same number of chromosomes
(b) Breed at the same time
(c) Are phenotypically indistinguishable
(d) Can mate and produce fertile offspring

147. _______ is the practice of rapidly multiplying stock plant material to produce a large number of progeny plants:

(a) Micro propagation
(b) Macro propagation
(c) Seed propagation
(d) All of the above

148. Self incompatibility (SI) uses first reported by:

(a) Breese
(b) Koelreuter
(c) Davis
(d) Riddle

149. _______ is somewhat similar to runners but with shorter internodes:

(a) Suckers
(b) Rhizomes
(c) Corm
(d) Bulbs

150. The pollen sterility is caused by nuclear genes is termed as:

(a) Cytoplasmic male sterility
(b) Cytoplasmic genetic male sterility
(c) Genetic male sterility
(d) None of the above

151. Physical separation of anther and stigma is called:

(a) Cleistogamy
(b) Allogamy
(c) Autogamy
(d) Hergogamy

152. The respective stigma of *Eucalyptus tereticornis* shows exudation and change in colour blue to:

(a) Reddish white (b) White
(c) Creamish white (d) None of the above

153. Gametophytic system is described by:

(a) Higher (b) East and Manglsdorf
(c) Koelreuter (d) Hayward

154. Tropical trees are predominantly _______ trees.

(a) Selfed (b) Apomixes
(c) Bisexual (d) Outcrossed

155. Mechanism of self incompatibility:

(a) Pollen-stigma interaction (b) Pollen tube-style interaction
(c) Pollen tube-ovule interaction (d) All of the above

156. Transfer of pollen grains from the anther to the stigma of same flower is:

(a) Cleistogamy (b) Allogamy
(c) Autogamy (d) Homogamy

157. Eucalyptus is mainly pollinated through:

(a) Ants (b) Honey bee
(c) Water (d) Air

158. Plants which are not flowering plants:

(a) Green algae (b) Ferns
(c) Gymnosperms (d) All of the above

159. The development of embryo sac from a megaspore is known as:

(a) Megagametogenesis (b) Microgametogenesis
(c) Microphyle (d) Sporogenesis

160. Maturation of anthers and stigma of a flower at the same time is called:

(a) Cleistogamy (b) Allogamy
(c) Autogamy (d) Homogamy

161. Pollination occurs through wind is known as:

(a) Entomophylly (b) Zoophylly
(c) Anemophylly (d) Hydrophylly

162. Development of phenotypically stable population called:

(a) Vybrid (b) Hybrid
(c) Zybrid (d) None of the above

163. An organism which has acquired a heritable variation as a result of sudden change in the hereditary material is called as a:

(a) Hybrid (b) Biotype

(c) Mutant (d) Albino

164. A polyploid having chromosome sets from different sources such as different species is known as:

(a) Autopolyploid (b) Aneuploid

(c) Tetraploid (d) Allopolypoid

165. The increased performance of F1 hybrid over its parents is called:

(a) Dominance (b) Heterosis

(c) Crossing over (d) None of the above

166. Ratio of the genetic variance to the total variance which determines the relationship between phenotype and genotype is known as:

(a) Genetic advance (b) Penetrance

(c) Hetritability (d) Dominance

167. Mass pedigree selection method of breeding is particularly, useful for improvement of:

(a) Self compatible cross pollinated crops (b) Self pollinated crops

(c) Self incompatible crops (d) Diocious crops

168. Emasculation of flower is necessary for hybridization in:

(a) Male sterile plants (b) Self pollinated plants

(c) Self incompatible plants (d) Dioecious plants

169. Mendel did not face the problem of linkage, however, seven characters studied by him were located on how many chromosomes?

(a) Seven (b) Four

(c) Five (d) Three

170. Test cross ratio in case of duplicate gene action will be:

(a) 1:1 (b) 1:1:1:1

(c) 3:1 (d) 1:3

171. A:T ratio in one strand of DNA duplex is 2, what will be the ratio in the complementary strand?

(a) 1.0 (b) 0.5

(c) 2.5 (d) 2.0

172. The term gene was coined by:

(a) Mendel
(b) Bateson
(c) Shull
(d) Johannsen

173. Multiple effects of a single gene is known as:

(a) Pleiotropy
(b) Polyploidy
(c) Heteroploidy
(d) None of the above

174. An example of a heterozygous but homogeneous population is:

(a) Pureline
(b) Open pollinated variety
(c) Hybrid variety
(d) Synthetic variety

175. The cross of an inbred line with an open pollinated variety is known as:

(a) Back cross
(b) Top cross
(c) Three way cross
(d) Double cross

176. The first artificial hybrid was made by:

(a) Mendel
(b) Thomas Fairchild
(c) Koelreuter
(d) Beal

177. Polyploidy derived from two different species is called:

(a) Autopolyploid
(b) Triploid
(c) Allopolyploid
(d) Monoploid

178. DNA synthesis takes place in which stage?

(a) G1 phase
(b) G2 stage
(c) S Phase
(d) None of the above

179. The number of bases in the single turn of the DNA helix is:

(a) 5
(b) 7
(c) 10
(d) 12

180. The common form of DNA present in the living organisms is the:

(a) A form
(b) Z form
(c) C form
(d) B form

181. Okazaki fragments are formed due to synthesis of DNA in direction:

(a) 5′-3′
(b) 3′-5′
(c) Both
(d) None of the above

182. An amino acid which is related to drought tolerance in plants is:

(a) Glutamic acid
(b) Proline
(c) Glycine
(d) Hydroxyproline

183. The variance which is fixable is:

(a) Additive (b) Dominance

(c) Epistatic (d) All of the above

184. The term cybrid refers to the:

(a) Somatic hybrids (b) Gametic hybrids

(c) Both (a) and (b) (d) None of the above

185. During mitosis, chromosomal DNA duplication takes place in:

(a) Interphase (b) Prophase

(c) Metaphase (d) Anaphase

186. A mating among the following cannot lead to heterosis:

(a) AAbb x aaBB (b) aaBBcc x AABBCC

(c) AABBccdd x AABBccdd (d) AABBCCDD x aabbCCDD

187. IPGRI headquater is situated in:

(a) New Delhi, India (b) New York, USA

(c) Sydney, Australia (d) Rome, Italy

188. Law of homologue series is given by:

(a) Biffen (b) Debary

(c) Dar (d) Vavilov

189. Land races are:

(a) Advanced strains (b) Hybrid varieties

(c) Local varieties (d) All of the above

190. A condition in which F1 plants of an interspecific cross are vigorous and fertile but their F2 progeny is weak and sterile is called:

(a) Hybrid sterility (b) Hybrid in viability

(c) Hybrid breakdown (d) All of the above

191. Term heterosis was introduced by:

(a) Davenport (b) Shull

(c) Hull (d) East

192. Highest uniformity is present in:

(a) Double cross (b) Multiple cross

(c) Single cross (d) Three way cross

193. Progeny of a single plant obtained by asexual reproduction is:

(a) Pure line (b) Inbred

(c) Clone (d) Strain

194. The progeny of a single homozygous self pollinated plant is known as:

(a) Pure line (b) Inbred
(c) Clone (d) Land race

195. The term genetics is coined by:

(a) Johannsen (b) Bateson
(c) Punnet (d) Morgan

196. An individual with gametic chromosome number is:

(a) Haploid (b) Monoploid
(c) Dihaploid (d) Polyploidy

197. The individual with 2n+1 chromosome are commonly known as:

(a) Euploid (b) Aneuploid
(c) Heteroploids (d) None of the above

198. Pollination and fertilization in an unopened flower bud is the situation:

(a) Cleistogamy (b) Chasmogamy
(c) Homogamy (d) Dichogamy

199. The situation in which one gene hides the effect of a second gene when both are present in a chromosome is called:

(a) Dominance (b) Epistasis
(c) Both (a) and (b) (d) None of the above

200. An individual with 2n-1-1 chromosome constitution is called:

(a) Monosomic (b) Double monosomic
(c) Disomic (d) Double trisomic

201. The smallest unit of genetic material that cannot be divided but is interchangeable among units refers to:

(a) Cistron (b) Muton
(c) Recon (d) Plasmon

202. In apomixes, the progeny closely resembles:

(a) Male parent (b) Female parent
(c) Both male and female parent (d) None of the above

203. Superiority of F1 hybrid over a standard commercial variety is called:

(a) Heterobeltiosis (b) Luxuriance
(c) Economic heterosis (d) Relative heterosis

204. NBPGR is located at:

(a) Lucknow (b) Karnal
(c) Bikaner (d) New Delhi

205. Mutation due to change in gene because of:

(a) Loss
(b) Degeneration
(c) Addition
(d) All of the above

206. The chemical mutagens usually induce:

(a) Chromosomal breakages
(b) Gene mutation
(c) Chromosomal translocation
(d) None of the above

207. A cybrid is a:

(a) Hybrid formed by cell division
(b) Sexual hybrid
(c) Plant formed by tissue culture
(d) Synonymous to hybrid

208. Reversion occurs due to:

(a) Gene interaction
(b) Mutation
(c) Gene interaction and mutation
(d) None of the above

209. Physical basis of life is:

(a) Nucleus
(b) Cell
(c) Protoplasm
(d) Food

210. The capacity of a cell to develop into a complete plant is called:

(a) Tissue culture
(b) Fecundity
(c) Totipotency
(d) Haploidy

211. Colchicines interferes with:

(a) Replication
(b) Recombination
(c) Formation of spindle fibres
(d) None of the above

212. Colchicines is generally used for:

(a) Chromosome doubling
(b) Hybridization
(c) Mutation
(d) All of the above

213. Chromosomal theory of inheritance was given by:

(a) Beadle and Tatum
(b) Watson and Crick
(c) Bateson and Punnet
(d) Sutton and Boveri

214. RFLP is a:

(a) Genetic marker
(b) Molecular marker
(c) Morphological marker
(d) Physiological marker

215. Chromosome other than sex chromosomes are called:

(a) Allosomes
(b) Autosomes
(c) Heterosomes
(d) Autoallosomes

216. Improved seed distributed to farmers for commercial cultivation is:
(a) Nucleus (b) Breeder
(c) Foundation (d) Certified

217. Recommended concentration of EMS for induced mutagenesis is:
(a) 0.03 to 1.5 per cent (b) 03.0 to 3.0 per cent
(c) 1.00 to 3.5 per cent (d) >4 per cent

218. DNA can be denatured by heating at a temperature of:
(a) 720c (b) 860c
(c) 900c (d) 940c

219. The following amongst the molecular markers is co dominant in nature:
(a) RAPD (b) SSR and RFLP
(c) VNTR (d) None of the above

220. Genomic imprinting occurs in:
(a) Plants (b) Animals
(c) Humans (d) All of the above

221. In mitosis, longest phase is:
(a) Anaphase (b) Telophase
(c) Prophase (d) Metaphase

222. Collection of Germplasm from for foreign countries is called:
(a) Indigenous collection (b) Indirect introduction
(c) Exotic collection (d) Alien collection

223. The best method for developing disease resistant plants is:
(a) Mutation (b) Selection
(c) Hybridization (d) Back crossing

224. The growth of tissues of living plants on a suitable culture medium is called:
(a) Anther culture (b) Plant tissue culture
(c) Genetic engineering (d) None of the above

225. Clonal selection is used in:
(a) Flowering species (b) Non flowering species
(c) Seeds propagating species (d) None of the above

226. Population which intermate freely are called as:
(a) Random mating population (b) Mendalian population
(c) Panmicitic population (d) All of the above

227. **What is the most effective method for transfer of on oligogenic character in an otherwise agronomically superior genotype?**
(a) Pedigree
(b) Bulk population
(c) Recurrent selection
(d) Back cross

228. **Which of the following will NOT promote speciation?**
(a) Gamete differences
(b) Gene flow
(c) Seasonal fertility
(d) Natural selection

229. **When germination of all seeds occurs over a relatively brief period it is called:**
(a) Continous
(b) Intermittent
(c) Quasi simultaneous
(d) Periodic

230. **Physical dormancy is found in:**
(a) *Eucalyptus delegatensis*
(b) *Betula pendula*
(c) *Acacia nilotica*
(d) *Fraxinus* spp

231. **Water soaking seed treatment is best suited for:**
(a) *Cassia fistula*
(b) *Santalum album*
(c) *C. terminalia arjuna*
(d) *Grevillea robusta*

232. **Vertical antenna like branches often occurring in conifers outside the normal branch whorl:**
(a) Epicromic branch
(b) Lateral branch
(c) Ramicron
(d) None of the above

233. **For a wind pollinated species, the width of the pollen dilution zone is:**
(a) 500-600m
(b) 100-150m
(c) 200-400m
(d) 450-500m

234. **Clonal seed orchard has:**
(a) Broader genetic base and high selection differential
(b) Broader genetic base and less selection differential
(c) Narrow genetic base and less selection differential
(d) Narrow genetic base and high selection differential

235. **Pollen contamination in seed orchard can be reduced by:**
(a) Geographic isolation
(b) Supplemental mass pollination
(c) Both (a) and (b)
(d) None of the above

236. Example for closed panmictic population:

(a) Seed stand (b) Seed production area
(c) Seed orchard (d) None of the above

237. Minimum number of clones to be planted in the first generation seed orchard is:

(a) 10-20 (b) 15-25
(c) 25-35 (d) 25-40

238. Tumbling is the method for:

(a) Wet extraction of pulpy fruits (b) Dry extraction of fruits
(c) Extraction of cones (d) None of the above

239. Dry strobile is seen in:

(a) Dogwood (b) Redwood
(c) Casuarinas (d) Fir

240. A tree that has been selected for grading because of the desirable character is:

(a) Candidate tree (b) Check tree
(c) Elite tree (d) Plus tree

241. Factorial and diallel designs are commonly used in the seed orchards of:

(a) Broad leaved spp (b) Conifers
(c) Both (a) and (b) (d) None of the above

242. Systematic removal of individuals in orchards:

(a) Roughing (b) Thinning
(c) Clearing (d) None of the above

243. The species with special ordchard problems are:

(a) Monoecious species (b) Dioecious species
(c) Both (a) and (b) (d) None of the above

244. The unique functional position of organisms is:

(a) Niche (b) Habitat
(c) Species interaction (d) Ecotone

245. Speciation occurs through physical separation of population.

(a) Ecotype (b) Sympatry
(c) Cline (d) Allopatry

246. CVP index on forest productivity does not include the following:

(a) Climatic (b) Biotic
(c) edaphic (d) None of the above

247. Oxygen sag relates to:

(a) Soil pollution (b) Water pollution

(c) Air pollution (d) Noise

248. Trees that have proven to be genetically superior by means of progeny testing:

(a) Candidate tree (b) Plus tree

(c) Elite tree (d) Comparison tree

249. Most frequently used seed orchard design is:

(a) Completely randomized design (b) Randomized block design

(c) Pure rows (d) Chess board/Rotating block

250. Plus tree selection method used in an even aged stand:

(a) Baseline selection (b) Comparison tree selection

(c) Regression selection (d) Mother tree system

251. The original geographic area from which seed or other propagules were obtained is known as:

(a) Seed orchard (b) Seed production area

(c) Seed source (d) Provenance

252. In tree improvement programme, many cycles of selection and breeding refers to:

(a) Tandem selection (b) Independent culling

(c) Selection index (d) Recurrent selection

253. The book on applied tree improvement was written by:

(a) B.D. Singh (b) Namkoong

(c) Bruce Zobel and John Talbert (d) M. Gibson

254. The pioneering tree improvement studies were attempted by:

(a) Japanese (b) Chinese

(c) Swedish (d) None of the above

255. Successful tree improvement programme depends on:

(a) Parentage (b) efficient management

(c) Both (a) and (b) (d) None of the above

256. Variation within the clonal progenies is due to:

(a) Environment (b) genetics

(c) Both (a) and (b) (d) biotic

257. Selection differential can be calculated using:

(a) Selection intensity and phenotypic variance
(b) Selection intensity and genotypic variance
(c) Phenotypic and genotypic variance
(d) None of the above

258. The following mating designs are termed as incomplete mating design:

(a) Full diallel
(b) Polymix
(c) Nested
(d) Partial diallel

259. More genetic gain can be achieved in the following seed orchard:

(a) Seedling seed
(b) Clonal
(c) Extensive seedling seed
(d) All of the above

260. Pureline theory was proposed by:

(a) Jinks
(b) Vavilov
(c) Muller
(d) Johannsen

261. Homoclime approach suited to:

(a) Hybrids
(b) Exotics
(c) Clone
(d) Genetic depression

262. Self incompatibility hampers the development of:

(a) Inbreds
(b) Clone
(c) Hybrids
(d) Exotics

263. Progeny having both parent identities is known as:

(a) Halfsib
(b) Fullsib
(c) One parent family
(d) None of the above

264. Selection can be resorted for:

(a) Non heritable trait
(b) Environmental influenced trait
(c) Heritable trait
(d) All of the above

265. Provenance – hybrid seed orchard is used for:

(a) Hybridizing diverse and compatible provenance
(b) Hybridizing genetically similar provenance
(c) Hybridizing two different but compatible species
(d) Hybridizing diverse and non compatible provenances

266. The presence of male and female flowers in different plants is called:

(a) Dioecious (b) Monecious

(c) Halfsib (d) None of the above

267. Emasculation by suction method is more suitable:

(a) Big flower (b) Medium flower

(c) Small flower (d) None of the above

268. In trees, normally the width of pollen dilution zone is used:

(a) 100-200m (b) 200-300m

(c) 300-400m (d) 400-500m

269. IBPGR refers to:

(a) Indian Bureau of Plant Genetic Resources

(b) International Bureau of Plant Genetic Resources

(c) Indian Botanical Plant Genetic Resources

(d) International Botanical Plant Genetic Resources

270. _______ proposed the concept of hormonal control of organ formation:

(a) Skoog and Miller (b) Gautherect

(c) Caplin and Steward (d) Kohlenbeck

271. The virus free plant was first produced by:

(a) Morel and Martin (b) White

(c) Cocking and white (d) Stanford and Muir

272. _______ is a major carbon source in plant tissue culture medium:

(a) Glucose (b) Sucrose

(c) Mannose (d) Fructose

273. _______ is a method of sterilizing with water vapour under high pressure:

(a) Hot sterilization (b) Autoclaving

(c) Flame sterilization (d) Filter sterilization

274. Cybrid is a _______.

(a) Somatic embryo (b) Somatic hybrid

(c) Somaclonal variant (d) Virus free plant

275. _______ mediated transformation is an important indirect gene transfer technique:

(a) Agrobacterium (b) Protoplast transformation

(c) Pollen (d) Particle bombardment

276. The two polynucleotide strands of the DNA double helix are:

(a) Parallel
(b) Complementary
(c) Non-complementary
(d) Parallel and complementary

277. Tree improvement consists of:

(a) Silviculture
(b) Tree breeding
(c) Both (a) and (b)
(d) None of the above

278. The unorganized mass of tissues are called as:

(a) Callus
(b) Totipotency
(c) Dedifferentiation
(d) Shooting

279. The original tree (Donor tree) from which parts to be propagated are taken are called:

(a) Clone
(b) Ramet
(c) Ortet
(d) Progeny

280. Ramet is:

(a) Population of genetically similar plant
(b) Individual of clone
(c) Callus
(d) Clone

281. Cleistogamous flowers are always:

(a) Insect pollinated
(b) Wind pollinated
(c) Cross pollinated
(d) Self pollinated

282. Orthotropic growth refers to:

(a) Tree
(b) Shrub
(c) Herb
(d) None of the above

283. In tree improvement programme, many cycles of selection and breeding refers to:

(a) Tandem selection
(b) Independent culling
(c) Selection index
(d) Recurrent selection

284. The trees produced from the seed of a parent called:

(a) Selection
(b) Population
(c) Recurrent selection
(d) Progeny

285. Population of genetically similar plants obtained from same individual by vegatative method:

(a) Propagule
(b) Bud
(c) Clone
(d) Callus

286. Area where superior phenotypes or genotypes are established and managed intensively and entirely for seed production is referred:

(a) Provenance (b) Family

(c) Seed orchard (d) Seed source

287. A condition in which F1 plants of an interspecific cross are vigorous and fertile and their F1 progeny is weak and sterile:

(a) Sterility (b) Fully incompatibility

(c) Hybrid breakdown (d) Partial incompatibility

288. An individual having more than two basic sets of chromosomes:

(a) Autopolyploids (b) Allopolyploids

(c) Polyploids (d) Amphidiploids

289. Ability of a gene to affect more than one phenotypic characteristic:

(a) Pseudoalleles (b) Modifying factors

(c) Pleiotrophism (d) Penetrance

290. In cross pollinated species a true breeding line obtained by continuous inbreeding:

(a) Variety (b) Hybrid

(c) Inbred (d) Synthetic variety

291. A plant with functional pollen fails to set seed when self pollinated:

(a) Sterility (b) Incompatibility

(c) Gamete abortion (d) Genetic barrier

292. Regeneration of whole plant from tissues of an actively dividing plant part such as stem tip, is referred:

(a) Anther culture (b) Callus culture

(c) Meristem culture (d) Ovule culture

293. Variation t hat is observed among the plants which are regenerated from callus cultures of somatic explants such as meristem:

(a) Gametoclonal variation (b) Protoclonal variation

(c) Somoclonal variation (d) Zygotic variation

294. Regeneration of plants from isolated meristematic cell or tissues:

(a) Embryogenesis (b) Macro-propagation

(c) Micro-propagation (d) Protoplast culture

295. The plant part used in regeneration in tissue culture termed as:

(a) Callus (b) Explants

(c) Ovule (d) Embryo

296. Isolation, introduction and expression of foreign DNA in plants and animals:

(a) Gene cloning (b) Genetic engineering

(c) Transgenic breeding (d) Gene splicing

297. CVP index on forest productivity does not include the following:

(a) Climatic (b) Biotic

(c) Edaphic (d) None of the above

298. Trees that have proven to be genetically superior by means of progeny testing is:

(a) Candidate tree (b) Plus tree

(c) Elite tree (d) Comparison tree

299. Most frequently used seed orchard design is:

(a) Completely randomized design (b) Randomized block design

(c) Pure rows (d) Chess board/Rotating block

300. Plus tree selection method used in an even aged stand is:

(a) Baseline selection (b) Comparison tree selection

(c) Regression selection (d) Mother tree system

301. The original geographic area from which seed or other propagule were obtained is known as:

(a) Seed orchard (b) Seed production area

(c) Seed source (d) Provenance

302. Usually, seed production areas should contain a minimum area (h(a) of:

(a) 1ha (b) 2ha

(c) 3ha (d) 4ha

303. Breeding value (BV) is:

(a) Equal to the TA (b) Twice the TA

(c) ½ of TA (d) of TA

304. Heritability in narrow sense is:

(a) VA/VP (b) VE/VP

(c) VG/VP (d) VA+VG/VP

305. Repeatability value set in _______ of heritability:

(a) Lower limit of h^2 (b) Upper limit of h^2

(c) Intermediate of h^2 (d) Both (a) and (b)

306. The most effective method of selection:

(a) Mass selection (b) Pedigree selection

(c) Progeny testing (d) Selection index

307. Quantitative traits shows:

(a) Continuous variation (b) Discontinuous variation

(c) Both (a) and (b) (d) None of the above

308. Male sterility promotes:

(a) Inbreeding (b) Out breeding

(c) Cross breeding (d) None of the above

309. Wide crossing is also known as:

(a) Distant hybridization (b) Intergeneric hybridization

(c) Interspecific hybridization (d) None of the above

310. Character showing continuous variation and considerable effect of the environment is:

(a) Quantitative character (b) Qualitative character

(c) Heritable character (d) None of the above

311. Non-additive variation accounts for _______ phenotypic variation:

(a) 10 per cent (b) 25 per cent

(c) 75 per cent (d) 100 per cent

312. Genetic gain depends upon:

(a) Heritability (b) Phenotypic variance

(c) Selection intensity (d) All of the above

313. Larger units of similar vegetation are referred as:

(a) Community (b) Population

(c) Ecosystem (d) Biome

314. Lack of seed set on self pollination is:

(a) Male sterility (b) Self incompatibility

(c) Self fertility (d) None of the above

315. If all variation was due to genetics, the broad sense heritability would be equal to:

(a) 0 (b) 1

(c) -1 (d) 2

316. Emasculation means:

(a) Removal of female flowers/stigma

(b) Removal of male flowers/anthers

(c) Dusting of pollen grains on female flowers

(d) None of the above

317. Genetic gain in tree improvement programme is estimated by the formula:

(a) G=gxh2 (b) G=Sxh

(c) G=Sxh2 (d) G=E+GE

318. Which studies are performed in order to describe genetic variation within and between population?

(a) Population variation (b) Genetic variation

(c) Population dynamics (d) Genetic gene

319. Where from the best quality seeds would be obtained for producing quality planting stock?

(a) Seed stand (b) Seed production area

(c) Seed orchard (d) Clonal orchard

320. Selection in evenaged forest:

(a) Regression method (b) Check tree method

(c) Mass selection (d) Family solution

321. The dispersive process mainly occurs in small population in which,:

(a) Only direction can be predicted not amount

(b) Only amount can be predicted not direction

(c) Both amount and direction can be predicted

(d) Direction and amount cannot be predicted

322. The contribution of offspring to the next generation is called:

(a) Fitness (b) Adaptive value

(c) Selective value (d) All of the above

323. Genotypic correlation is due to:

(a) Polymorphism (b) Linkage

(c) Pleiotropism (d) Both (b) and (c)

324. A new breed can be evolved by:

(a) Out crossing (b) Cross breeding

(c) Gradiong up (d) None of the above

325. If the coefficient of selection is 0.25, then the fitness is:

(a) 1
(b) 0.25
(c) 0.75
(d) 0.5

326. Grading up produces pure breed in how many generations q:

(a) 2-4
(b) 4-5
(c) 10-12
(d) 7-8

327. Which is the only gaseous hormone?

(a) Abscisic acid
(b) Auxins
(c) Cytokinins
(d) Ethylene

328. Which hormone is primarily responsible for enhancing apical dominance?

(a) Gibberellins
(b) Auxins
(c) Abscisic acid
(d) Ethylene

329. Which of the following is responsible for promoting seed dormancy?

(a) Abscisic acid
(b) Auxins
(c) Cytokinins
(d) Ethylene

330. Which hormone promotes the ripening of fruit?

(a) Gibberellins
(b) Auxins
(c) Cytokinins
(d) Ethylene

331. Which of the hormones regulates cell division and differentiation?

(a) Gibberellins
(b) Auxins
(c) Cytokinins
(d) Ethylene

332. Which hormone instigates "the triple response?":

(a) Abscisic acid
(b) Auxins
(c) Cytokinins
(d) Ethylene

333. Which hormone is located in every major organ and living tissue?

(a) Abscisic acid
(b) Auxins
(c) Cytokinins
(d) Gibberellins

334. Which hormone has anti-aging effects?

(a) Gibberellins
(b) Auxins
(c) Cytokinins
(d) Ethylene

335. Which hormone is responsible to stem elongation ?

(a) Gibberellins
(b) Auxins
(c) Abscisic acid
(d) Ethylene

336. Which hormone allows seeds to ignore environmental conditions and germinate?

(a) Abscisic acid (b) Auxins

(c) Cytokinins (d) Gibberellins

337. Which of the following is the only known naturally occurring auxin?

(a) 2,4-D (b) IAA

(c) GA (d) ABA

338. Which of the following hormones would be most useful in causing the rooting of plant cuttings?

(a) Abscisic acid (b) Cytokinins

(c) Gibberellins (d) Auxins

339. Which plant hormones is are most closely associated with cell division?

(a) Ethylene (b) Cytokinins

(c) Abscisic acid (d) Phytochrome

340. The application of which of the following hormones would be a logical first choice in an attempt to produce normal growth in mutant dwarf plants?

(a) Cytokinins (b) Gibberellins

(c) Abscisic acid (d) Ethylene

341. One effect of gibberellins is to stimulate cereal seeds to produce:

(a) Lipids (b) Abscisic acid

(c) Starch (d) Amylase

342. While responses to plant hormones are normally slow, which of the following hormones has been shown to be involved in the rapid opening and closing of stomata?

(a) Auxin (b) Cytokinins

(c) Ethylene (d) Abscisic acid

343. If you were shipping green bananas to a supermarket in Boston, which of the following chemicals would you want to eliminate from the plants' environment?

(a) CO_2 (b) Cytokinins

(c) Ethylene (d) Auxin

344. Auxin is responsible for many plant growth responses, including all of the following except:

(a) Formation of adventitious roots (b) Phototropism

(c) Apical dominance (d) Promotion of abscission

345. Plants synthesize auxin from the amino acid:

(a) Cystine (b) Phenylalanine
(c) Ornithine (d) Tryptophan

346. Preserving germplasm in frozen state is:

(a) Cryopreservation (b) Cold storage
(c) *In situ* preservation (d) Vernalisation

347. An example of *ex-situ* conservation is:

(a) Seed bank (b) Sacred groves
(c) National parks (d) Wildlife sanctuary

348. Ex situ conservation is carried out in:

(a) Sanctuary (b) National park
(c) Biosphere reserve (d) Zoo

349. A condition in which pollination and fertilization take place in an unopened flower is termed as:

(a) Homogamy (b) Apogamy
(c) Polygamy (d) Cleistogamy

350. Cleistogamy encourages:

(a) Self pollination (b) Cross pollination
(c) Apomixes (d) Inbreeding depression

351. Clone in general is:

(a) Homozygous (b) Heterogenous
(c) Heterozygous (d) None of the above

352. Cross of F1 hybrid with either parents is called:

(a) Test cross (b) Single cross
(c) Back cross (d) None of the above

353. Development and formation of pollen grains in anther of the stamen is known as:

(a) Pollination (b) Fertilization
(c) Microsporogensis (d) Megasporogensis

354. Development of a fruit without fertilization is known as:

(a) Parthenogensis (b) Polyembryony
(c) Budding (d) Apomixes

355. Development of seed without-sexual process is:

(a) Apomixes
(b) Polyembryony
(c) Both (a) and (b)
(d) None of the above

356. Development of embryos without fertilization:

(a) Meiosis
(b) Apomixes
(c) Sporogenesis
(d) Gametogenesis

357. Heterosis is commercially exploited in:

(a) Often cross pollinated species
(b) Cross pollinated species
(c) Vegetatively propagated species
(d) Self pollinated species

358. Heterosis over the better parent is called as:

(a) Reklative heterosis
(b) Standard heterosis
(c) Pseudoheterosis
(d) Heterobeltiosis

359. In self incompatibility system:

(a) Pollen not produced
(b) Self pollination takes place
(c) Pollen does not germinate on self pollination
(d) All of the above

360. The basic difference between a synthetic and composite variety lies in:

(a) The difference in synthesis
(b) Testing for picking ability
(c) The mode of pollination
(d) The number of components

361. The best method to break the unfavourable linkage is:

(a) Single seed descent method
(b) Diallel selective mating
(c) Pedigree method
(d) Bulk method

362. The condition in which the anthers and stigmas of a bisexual flower mature at the same time is called:

(a) Cleistogamy
(b) Heterogamy
(c) Syngamy
(d) Homogamy

363. The condition in which the pollen o a flower has no fertilizing effect on the stigma of the same flower is called:

(a) Herkogamy
(b) Dichogamy
(c) Dicliny
(d) Self-sterility

364. The effect of the genotype of the pollen grain on the phenotype of the seed is termed as:

(a) Apospory
(b) Pollinia
(c) Endopolyploidy
(d) Xenia

365. When the male gametes are shed into embryo sac through micropyle is called:

(a) Porogamy (b) Chasmogamy

(c) Misogamy (d) Syngamy

366. The period from when the flower open to when seed is set called as:

(a) Anthesis (b) Apomixes

(c) Herkogamy (d) Dichogamy

367. Which is a common method used ion cross pollinated crops?

(a) Mass selection (b) Bulk method

(c) Pure line selection (d) Pedigree method

368. Which of the following procedures are not used for producing new varieties of self pollinated crop?

(a) Recurrent selection (b) Mass selection

(c) Pure line selection (d) Hybridization

369. The best method to carry forward the maximum number of genotypes/ genes to the advanced generations against the force of natural selection is:

(a) Pedigree method (b) Bulk method

(c) Single seed descent method (d) Back cross method

370. Pollen grains falling on the stigma of another flower of the same kind but on a different plant is called as:

(a) Cross-pollination (b) Self-pollination

(c) Often cross pollination (d) Mixed pollination

371. Pollen is dispersed by:

(a) Wind (b) Water

(c) Animal and Birds (d) All of the above

372. The diversity of place at the ecosystem level is termed as:

(a) Morphological diversity (b) Ecosystem diversity

(c) Species diversity (d) Genetic diversity

373. Pollen viability test can be done by:

(a) Baker's procedure (b) p-Phenylenediamine test

(c) None of the above (d) Both (a) and (b)

374. A measure of the possible choices of information provided by a gene is called as:

(a) Genetic diversity (b) Species diversity

(c) Alpha diversity (d) Morphological diversity

375. Assessment of Genetic Diversity can be done by:

(a) Species richness (b) Simpson Index

(c) Shannon-Wiener index (d) All of the above

376. The Shannon-Wiener index is defined as:

(a) E=H/log(S) (b) D= sum(Pi2)

(c) H=-sum(Pilog[Pi] (d) 1-D

377. A measure that accounts for both richness and proportion (percent) of each species:

(a) Simpson diversity index (b) Shannon-Wiener index

(c) All of the above (d) None of the above

378. Conservation of species in their natural habitat:

(a) *In situ* conservation (b) *Ex situ* conservation

(c) Habitat conservation (d) Natural conservation

379. Example for *Ex situ* Conservation:

(a) Natural parks (b) Nature reserves

(c) Sanctuary (d) Botanical gardens

380. Protection and maintenance of forest diversity through legal means is known as.

(a) Protected areas (b) Reserved area

(c) Botanical gardens (d) None of the above

381. First preservation plot in India:

(a) Bihar in 1906 (b) Gujarat in 1906

(c) Delhi in 1908 (d) Maharashtra in 1909

382. Preservation plot is maintained for conservation of _______.

(a) Phenotypic diversity (b) Genetic diversity

(c) Animal diversity (d) Morphology diversity

383. Tree showing the inferred evolutionary relationships among various biological species based upon similarities and differences in their physical and/or genetic characteristics.

(a) Phylogenetic tree (b) Evolutionary tree

(c) Commercial tree (d) Both (a) and (b)

384. The current method of choice to infer phylogenetic trees is:

(a) Cladistics (b) Osteometric data

(c) Phonetics (d) Synapomorphies

385. A_______ is a broad term for the diagrammatic representation of phylogenetic tree.

(a) Chronogram (b) Cladogram

(c) Dendrogram (d) Dendrology

386. Phylogenetics is richly informed _______.

(a) Taxonomy of organisms (b) Classification of organisms

(c) Identification of organisms (d) All of the above

387. A directed tree with a unique node corresponding to the (usually impute (d) most recent common ancestor of all the entities at the leaves of the tree is called as _______:

(a) Rooted phylogenetic tree (b) Un rooted phylogenetic tree

(c) Bifurcated phylogenetic tree (d) None of above

388. Interactions between two different species as selective forces on each other, resulting in adaptations that increase their interdependency.

(a) Covaluation (b) Coevolution

(c) Pollination (d) Reproduction

389. Most pollen falls within _______ from its sources.

(a) 150m (b) 250m

(c) 350m (d) 140m

390. Control pollen was killed by placing in an incubator at:

(a) 56°C for up to 2 week (b) 50°C for up to 1 week

(c) 60°C for up to 1 week (d) 66°C for up to 2 week

391. A sugary solution produced in special flower glands is called as _______.

(a) Nect ar (b) Honey

(c) Pollen (d) None of the above

392. Pollen was stored for periods ranging from 1 day to _______:

(a) 3 years (b) 10 years

(c) 5 years (d) 15 years

393. Some plants take advantage of the sex drive of certain insects, it is called as _______.

(a) Plant diversion (b) Plant duplication

(c) Plant Mimicry (d) Plant modification

394. _______ in which only one parent is known for any given progeny.

(a) Incomplete pedigree designs (b) Complete pedigree designs

(c) All of the above (d) None of the above

395. Poly cross mating designs is otherwise known as _______.

(a) Pollen mix mating designs (b) Total cross mating designs
(c) Open pollinated mating design (d) cross mating designs

396. Members of one sex are crossed in all combinations with several members of the other sex is called as _______.

(a) Nested design (b) Factorial design
(c) Disconnected factorial design (d) Random design

397. Each parent is mated to another member of the population only once, it is known as _______.

(a) Single pair mating design (b) Double pair mating design
(c) Three pair mating design (d) None of the above

398. Groups of parents of one sex are mated to members of the other sex is known as _______.

(a) Nested design (b) Group design
(c) Diallele design (d) Factorial design

399. A _______ is a molecular unit of heredity of a living organism.

(a) Allele (b) Gene
(c) Chromosome (d) Cell

400. The first interspecific hybridization between sweet William and Carnation was developed by:

(a) Thomas fairchild (b) Knight
(c) John le couteur (d) De vilmorin

401. Genes can be detected by _______.

(a) Transgene detection methods (b) Ligase-mediated gene detection technique
(c) Apoptotic etection methods (d) All of the above

402. Who discovered the Principles of Inheritance?

(a) Rimpu (b) Mendel
(c) Neilsson (d) Shull

403. The term heterosis for hybrid vigour is attributable to:

(a) Jones (b) Bruce
(c) Hardy (d) Shull

404. How many centers of origin are identified?

(a) 8 (b) 9
(c) 7 (d) 10

405. Auxins (pick the odd man out):

(a) IAA (b) BAP
(c) IBA (d) NAA

406. Auto vegetative propagation (pick the odd man out):

(a) Rooting of cuttings (b) Layering
(c) Budding (d) TC Plants

407. Heterosis is confined to:

(a) F1 (b) F2
(c) BC1 (d) BC2

408. The average performance of an individual in a particular series of crosses is termed as:

(a) General combining ability (b) Specific combining ability
(c) Epistasis (d) None of the above

409. The selection method predominantly followed in plantation:

(a) Regression method (b) Comparison tree method
(c) Base line method (d) All of the above

410. The term provenance means:

(a) Land race (b) Geographical source
(c) Seed source (d) None of the above

411. Ecotype refers to:

(a) Variation in the phenotypic characters
(b) Phenotypic variation of different species
(c) Differential response of different species to environmental factors
(d) Differential response of the same species to environmental factors

412. The selection method used for improvement of traits for disease resistance and cold hardiness is:

(a) Tandem selection (b) Independent culling
(c) Selection index (d) Recurrent selection

413. The seed orchard established using grafted plant is called:

(a) Clonal seed orchard (b) Seedling seed orchard
(c) Extensive seedling seed orchard (d) None of the above

414. The growth regulators used for rooting in vegetative propagation are collectively called as:

(a) Vitamins (b) Cytokinins
(c) Gibberllins (d) Auxins

415. The exotic tree which has natural hybrids in India is:

(a) Casuarinas (b) Eucalyptus

(c) Silver oak (d) Acacia

416. A plus tree whose genetic worth has been established is called:

(a) Candidate tree (b) Comparison tree

(c) Elite tree (d) Superior tree

417. The pistil matures before anthers is called:

(a) Protandry (b) Protogyny

(c) Herkogamy (d) Heterostyly

418. The stamens mature before pistil is called:

(a) Protandry (b) Protogyny

(c) Herkogamy (d) Heterostyly

419. The staminate and pistillate flowers occurs in the same plant is called:

(a) Monoecy (b) Dioecy

(c) Protogyny (d) Protandry

420. The male and female flowers are present on different plants is called:

(a) Monoecy (b) Dioecy

(c) Protogyny (d) Protandry

421. The transfer of pollen from the anthers to the stigma in the same flower is:

(a) Autogamy (b) Allogamy

(c) Geitonogamy (d) None of the above

422. The transfer of pollen from a flower of one plant falls on the stigmas of other flowers of the same plant is called:

(a) Autogamy (b) Allogamy

(c) Geitonogamy (d) None of the above

423. Single seed descent (SS(d) method is the modified form of _______.

(a) Bulk method (b) Mass selection method

(c) Clonal selection (d) Pedigree method

424. _______ is the Mating between phenotypic dissimilar individuals belonging to same populations:

(a) Phenotypic assortative mating system

(b) Phenotypic disassortative mating system

(c) Genotypic assortative mating system

(d) Genotypic disassortative mating system

425. _______ mating systems reduces frequency of other alleles:

(a) Random mating system

(b) Phenotypic assortative mating system

(c) Phenotypic disassortative mating system

(d) Genotypic assortative mating system

426. The average phenotypic value of the selected individuals, expressed as a deviation from the population mean is:

(a) Selection index (b) Selection differential

(c) Half sib selection (d) Full sib selection

427. Selection of parent trees based upon the performance of their progeny:

(a) Mass selection (b) Sib selection

(c) Recurrent selection (d) Progeny selection

428. In cryopreservation, cells are stored at temperature of:

(a) -196°C (b) -112°C

(c) -120°C (d) -100°C

429. Synthetic seed in sandal wood is attributable to:

(a) Nadgir (b) Bapat and Rao

(c) Lakshmi Sita (d) Mascarenhas

430. Tick the antiauxin:

(a) Azoindole (b) Indole acetic acid

(c) Pyridoxine (d) None of the above

431. The chemical basis of heredity is:

(a) Protein (b) DNA

(c) Carbohydrates (d) Amino acids

432. PCR invented by:

(a) Kary Muller (b) William Benting

(c) Karl Mullis (d) Totenzef

433. EDTA stands for:

(a) Enzyme derived testing analysis

(b) Ethylene diamine tetra acetic acid

(c) Ethylene dichloro tetra acetic acid

(d) Ethylene dinitro triacetic acid

434. The buffer used in Electrophoresis is:

(a) Tris acetate
(b) Tricloride
(c) Ethanol
(d) Bromine

435. RAPD stands for:

(a) Randomly amplified polymorphic DNA
(b) Restricted amplified polymorphic DNA
(c) Rapid denaturation of protein test
(d) Restriction and polymerization of DNA

436. Father of plant tissue culture:

(a) Murashige
(b) Nitsch
(c) Skoog
(d) Haberlandt

437. Anther culture technique was developed by:

(a) Guha and Maheswari
(b) Bajaj
(c) Haberlandt
(d) H.J. Paikar

438. Single cell culture first developed by:

(a) Muir
(b) Gautheret
(c) Van overbeek
(d) Morel

439. Cryopreservation technique was first developed by:

(a) Zimmerman
(b) Scibert
(c) Melchers
(d) Bergamann

440. Protoplast fusion was attributable to:

(a) Zimmerman
(b) Scibert
(c) Melchers
(d) Bergamann

441. The optimum temperature and Humidity in the culture room is normally:

(a) 25 °c and 50 per cent
(b) 20°c and 40 per cent
(c) 20 °c and 80 per cent
(d) 25°c and 80 per cent

442. The membrane filter used in tissue culture should contain pore size of:

(a) Less than 0.45 micro meter
(b) More than 0.45 micro meter
(c) Less than 1.00 micro meter
(d) More than 0.95 micro meter

443. The main energy source in Plant tissue culture technique is:

(a) Agarose
(b) Sucrose
(c) Fats
(d) Proteins

444. During culture process activated charcoal is added to adsorb:

(a) Sucrose
(b) Agarose
(c) Phenols
(d) Fungi bodies

445. Agar is basically a:

(a) Poly saccharide
(b) Lipid
(c) Protein
(d) Phenolic compound

446. Pick the odd man out:

(a) Cytokinins
(b) Gibberellins
(c) Abscisic acid
(d) Thiamine

447. Auxins can dissolve in:

(a) Ethanol
(b) Water
(c) Oil
(d) Petrol

448. In tissue culture IAA induces:

(a) Cell division
(b) Root induction
(c) Apical dominance
(d) All of the above

449. Pick the odd man out:

(a) Benzyl aminopurine
(b) Benzyl adenine
(c) Kinetin
(d) Auxin

450. The potential to development of an adult organism from a single cell is referred as:

(a) Totipotency
(b) Apomixes
(c) Parthenocarpy
(d) Callus induction

451. From a single cell, development of reproducible non zygotic embryos capable of germinating to form complete plants is called:

(a) Somatic embryogenesis
(b) Shoot tip culture
(c) Meristem culture
(d) Zygote differentiation

452. Pick the odd man out:

(a) Suspension culture
(b) Batch culture
(c) Continuous culture
(d) Anther culture

453. Haemocytometer is used to:

(a) Cell counting
(b) Electrolyting callus culture
(c) Differentiation inducer
(d) Measure O_2 level in growing media

454. Morel and Martin (1952) developed techniques for:

(a) Induced somatic embryogenesis
(b) Disease free plants
(c) Isolation of specific pathogens
(d) Callogenesis

455. The formation of callus from explants is called:

(a) Redifferentiation
(b) De differentiation
(c) Regeneration
(d) Culturing

456. Caulogenesis refers to:

(a) Root formation
(b) Shoot formation
(c) Brach formation
(d) Callus formation

457. Rhizogenesis in plant tissue culture involves:

(a) Root formation
(b) Shoot formation
(c) Brach formation
(d) Callus formation

458. NOA stands for:

(a) Napthoxy acetic acid
(b) Nicotinamide oxalo acetate
(c) Non oxalic acetones
(d) None of the above

459. What is HEPA?

(a) A polyurathene sheet
(b) a particulate media
(c) A high density aluminium sheet
(d) A filter used in Laminar chamber

460. The commonly used medium in tissue culture (Pick the odd man out):

(a) Rose Bengal agar medium
(b) MS medium
(c) LS medium
(d) N6 medium

461. The lignifications of cell wall is a characteristic feature of:

(a) Non-Embryogenic calli
(b) Embryogenic calli
(c) Both of the above
(d) None of the above

462. Aseptic isolation and growth of sexually produced embryos in vitro to produce viable plant is called:

(a) Somatic embryogenesis
(b) Embryo culture
(c) Anther culture
(d) Suspension culture

463. TMV stands for:

(a) Tomographic molecular virulent
(b) Tobacco mosaic virus
(c) Tissues modified for vitality
(d) Tomato monomeric virus

464. FDA stands for:

(a) Fluorescein di acetate
(b) Flavin di acetate
(c) Ferridoxin dissolved auxin
(d) None of the above

465. PEG used in tissue culture for:

(a) Separation of tissues
(b) Fusion of protoplast
(c) Organ development
(d) Organ dissociation

466. Protoplast fusion is to develop the following (Pick the odd man out):

(a) Aneuploids
(b) Cybrids
(c) Somatic hybrids
(d) Transfer of genomes

467. National chemical laboratory is located:

(a) Chennai
(b) Delhi
(c) Pune
(d) Mumbai

468. The genotypic clusters separated by largest distance show:

(a) Maximum divergence
(b) Minimum divergence
(c) Equal divergence
(d) None of the above

469. D2 statistics measures the forces of differentiation:

(a) Inter-cluster
(b) Intra-cluster
(c) Both (a) and (b)
(d) Intermediate cluster

470. Base population includes:

(a) Natural stands
(b) Unimproved plantations
(c) Both (a) and (b)
(d) Segregated population

471. The concept of D2 statistics was proposed by:

(a) Mahalanobis
(b) Rao
(c) Peterson
(d) Zobel

472. In recurrent mass selection the genetic gain for the traits of medium heritability is:

(a) 12.0 per cent
(b) 12.5 per cent
(c) 13.0 per cent
(d) 13.5 per cent

473. Heterosis is:

(a) Hybrid vigour
(b) Increased expression
(c) Back cross
(d) Inbreeding depression

474. Nucleus population is:

(a) Highly productive individuals
(b) Control crossed
(c) Both (a) and (b)
(d) Assertive mating population

475. Selection in clonal orchard from forwards and backwards selection is executed:

(a) Within the open pollinated breeding population
(b) Best trees from best families
(c) Both (a) and (b)
(d) Recurrent selection

476. In clonal orchard from forwards and backwards selection gain achieved is:

(a) Equal to seedling orchard
(b) Greater than seedling orchard
(c) Less than seedling orchard
(d) None of the above

477. The methods by which trees are combined to produce progeny is called:

(a) Progeny testing
(b) Seedling seed orchard
(c) Mating design
(d) Provenances

478. Incomplete pedigree design include:

(a) Open pollinated matting design
(b) Polymix matting design
(c) Both (a) and (b)
(d) Single pair matting design

479. General combining ability is:

(a) Average performance of the progeny of individual parent
(b) Average performance of the progeny of multiple parent
(c) Average performance of the parent
(d) Both (a) and (b)

480. In open pollinated mating design estimates of genetic parameters is biased because of:

(a) Random mating
(b) Non random mating
(c) Both (a) and (b)
(d) Not biased

481. Singe pair mating design ensures complete avoidance of:

(a) In breeding
(b) Out breeding
(c) Back cross
(d) Multiple cross

482. In polymix mating design the number of male parents selected should not be less than:

(a) 5
(b) 10
(c) 15
(d) 20

483. An operon is most closely associated with:

(a) DNA replication
(b) Point mutations
(c) Translation
(d) Transcription

Answer Sheet

1.	(b)	31.	(b)	61.	(d)	91.	(a)	121.	(b)
2.	(a)	32.	(d)	62.	(a)	92.	(c)	122.	(d)
3.	(c)	33.	(d)	63.	(d)	93.	(d)	123.	(d)
4.	(a)	34.	(b)	64.	(b)	94.	(a)	124.	(b)
5.	(b)	35.	(b)	65.	(c)	95.	(c)	125.	(c)
6.	(d)	36.	(d)	66.	(d)	96.	(b)	126.	(d)
7.	(c)	37.	(c)	67.	(a)	97.	(b)	127.	(a)
8.	(c)	38.	(d)	68.	(d)	98.	(c)	128.	(c)
9.	(d)	39.	(d)	69.	(a)	99.	(d)	129.	(d)
10.	(b)	40.	(b)	70.	(d)	100.	(a)	130.	(d)
11.	(c)	41.	(d)	71.	(d)	101.	(d)	131.	(a)
12.	(c)	42.	(d)	72.	(a)	102.	(c)	132.	(c)
13.	(c)	43.	(b)	73.	(b)	103.	(d)	133.	(b)
14.	(a)	44.	(a)	74.	(a)	104.	(b)	134.	(d)
15.	(b)	45.	(c)	75.	(a)	105.	(a)	135.	(b)
16.	(c)	46.	(b)	76.	(c)	106.	(b)	136.	(a)
17.	(d)	47.	(c)	77.	(b)	107.	(c)	137.	(a)
18.	(a)	48.	(d)	78.	(c)	108.	(a)	138.	(a)
19.	(b)	49.	(b)	79.	(b)	109.	(c)	139.	(c)
20.	(c)	50.	(a)	80.	(b)	110.	(d)	140.	(a)
21.	(b)	51.	(a)	81.	(a)	111.	(a)	141.	(d)
22.	(d)	52.	(b)	82.	(c)	112.	(a)	142.	(a)
23.	(d)	53.	(c)	83.	(a)	113.	(c)	143.	(d)
24.	(c)	54.	(d)	84.	(d)	114.	(b)	144.	(b)
25.	(c)	55.	(a)	85.	(b)	115.	(d)	145.	(b)
26.	(d)	56.	(c)	86.	(a)	116.	(c)	146.	(d)
27.	(a)	57.	(d)	87.	(b)	117.	(a)	147.	(a)
28.	(a)	58.	(a)	88.	(a)	118.	(d)	148.	(b)
29.	(c)	59.	(a)	89.	(a)	119.	(d)	149.	(a)
30.	(a)	60.	(d)	90.	(b)	120.	(c)	150.	(a)

151.	(d)	185.	(a)	219.	(b)	253.	(c)	287.	(c)
152.	(c)	186.	(c)	220.	(d)	254.	(a)	288.	(c)
153.	(b)	187.	(d)	221.	(c)	255.	(c)	289.	(c)
154.	(d)	188.	(d)	222.	(c)	256.	(a)	290.	(c)
155.	(d)	189.	(c)	223.	(d)	257.	(a)	291.	(b)
156.	(c)	190.	(a)	224.	(b)	258.	(b)	292.	(c)
157.	(b)	191.	(b)	225.	(b)	259.	(b)	293.	(c)
158.	(d)	192.	(a)	226.	(d)	260.	(d)	294.	(c)
159.	(a)	193.	(c)	227.	(d)	261.	(b)	295.	(b)
160.	(d)	194.	(a)	228.	(b)	262.	(a)	296.	(b)
161.	(d)	195.	(b)	229.	(c)	263.	(b)	297.	(c)
162.	(b)	196.	(a)	230.	(c)	264.	(c)	298.	(c)
163.	(c)	197.	(b)	231.	(d)	265.	(a)	299.	(b)
164.	(d)	198.	(a)	232.	(c)	266.	(a)	300.	(b)
165.	(b)	199.	(b)	233.	(c)	267.	(c)	301.	(d)
166.	(c)	200.	(b)	234.	(b)	268.	(d)	302.	(d)
167.	(a)	201.	(c)	235.	(c)	269.	(b)	303.	(b)
168.	(b)	202.	(b)	236.	(c)	270.	(a)	304.	(a)
169.	(b)	203.	(c)	237.	(d)	271.	(c)	305.	(b)
170.	(c)	204.	(d)	238.	(c)	272.	(a)	306.	(c)
171.	(b)	205.	(d)	239.	(d)	273.	(b)	307.	(a)
172.	(d)	206.	(b)	240.	(a)	274.	(a)	308.	(b)
173.	(a)	207.	(a)	241.	(b)	275.	(a)	309.	(a)
174.	(c)	208.	(b)	242.	(a)	276.	(d)	310.	(a)
175.	(b)	209.	(c)	243.	(b)	277.	(c)	311.	(b)
176.	(b)	210.	(c)	244.	(a)	278.	(a)	312.	(d)
177.	(c)	211.	(c)	245.	(d)	279.	(c)	313.	(d)
178.	(c)	212.	(a)	246.	(c)	280.	(b)	314.	(b)
179.	(c)	213.	(d)	247.	(b)	281.	(d)	315.	(b)
180.	(d)	214.	(b)	248.	(c)	282.	(a)	316.	(b)
181.	(a)	215.	(b)	249.	(b)	283.	(d)	317.	(c)
182.	(b)	216.	(c)	250.	(b)	284.	(d)	318.	(a)
183.	(a)	217.	(a)	251.	(d)	285.	(c)	319.	(d)
184.	(a)	218.	(d)	252.	(d)	286.	(c)	320.	(b)

321.	(b)	354.	(a)	387.	(a)	420.	(b)	453.	(a)
322.	(d)	355.	(a)	388.	(b)	421.	(a)	454.	(b)
323.	(d)	356.	(b)	389.	(a)	422.	(c)	455.	(b)
324.	(b)	357.	(b)	390.	(c)	423.	(a)	456.	(b)
325.	(c)	358.	(d)	391.	(a)	424.	(b)	457.	(a)
326.	(d)	359.	(c)	392.	(a)	425.	(a)	458.	(a)
327.	(d)	360.	(d)	393.	(c)	426.	(b)	459.	(d)
328.	(b)	361.	(b)	394.	(a)	427.	(d)	460.	(a)
329.	(a)	362.	(d)	395.	(a)	428.	(a)	461.	(a)
330.	(d)	363.	(d)	396.	(b)	429.	(b)	462.	(b)
331.	(c)	364.	(d)	397.	(a)	430.	(a)	463.	(b)
332.	(d)	365.	(a)	398.	(a)	431.	(b)	464.	(a)
333.	(a)	366.	(a)	399.	(b)	432.	(a)	465.	(b)
334.	(c)	367.	(d)	400.	(a)	433.	(b)	466.	(a)
335.	(a)	368.	(b)	401.	(d)	434.	(a)	467.	(c)
336.	(d)	369.	(c)	402.	(b)	435.	(a)	468.	(a)
337.	(b)	370.	(a)	403.	(d)	436.	(d)	469.	(c)
338.	(d)	371.	(d)	404.	(a)	437.	(a)	470.	(c)
339.	(b)	372.	(b)	405.	(b)	438.	(a)	471.	(a)
340.	(b)	373.	(d)	406.	(c)	439.	(b)	472.	(b)
341.	(d)	374.	(a)	407.	(a)	440.	(a)	473.	(a)
342.	(d)	375.	(a)	408.	(a)	441.	(d)	474.	(c)
343.	(c)	376.	(c)	409.	(b)	442.	(a)	475.	(c)
344.	(d)	377.	(a)	410.	(b)	443.	(b)	476.	(b)
345.	(d)	378.	(a)	411.	(d)	444.	(c)	477.	(c)
346.	(a)	379.	(d)	412.	(a)	445.	(a)	478.	(c)
347.	(a)	380.	(a)	413.	(a)	446.	(d)	479.	(a)
348.	(d)	381.	(a)	414.	(d)	447.	(a)	480.	(b)
349.	(d)	382.	(b)	415.	(b)	448.	(d)	481.	(a)
350.	(a)	383.	(d)	416.	(c)	449.	(d)	482.	(b)
351.	(c)	384.	(a)	417.	(b)	450.	(a)	483.	(d)
352.	(a)	385.	(c)	418.	(a)	451.	(a)		
353.	(c)	386.	(d)	419.	(a)	452.	(a)		

Chapter 10

Tree Seed Technology

1. **Seed are formed, but the embryos developed without fertilization:**
   (a) Parthenogenesis (b) Apomixis
   (c) Sporogenesis (d) Gametogenesis
2. **Embryos develop from the embryo sac without pollination:**
   (a) Parthenogenesis (b) Pseudogamy
   (c) Androgenesis (d) Apomixes
3. **Chromosome number of *Jatropha curcus* is:**
   (a) 2n =22 (b) 2n=24
   (c) N=46 (d) None of the above
4. **Which one of the following tree is unisexual:**
   (a) Golden shower tree (b) Dung oil tree
   (c) Paradise tree (d) None of the above
5. **The most important factor which determines stability of seed is:**
   (a) Light (b) Temperature
   (c) Moisture (d) Nutrient
6. **In seed testing, the grow out test is conducted to determine:**
   (a) Physical purity (b) Seed viability
   (c) Genetic purity (d) None of the above

7. **The stalk of the ovules is called:**
   (a) Pedicel (b) Petiole
   (c) Funicle (d) Micropyle

8. **A fertilized mature ovule, which possesses am embryonic plant, food reserve and protective coating in a plant species is:**
   (a) Embryo sac (b) Embryonic callus
   (c) Embryo (d) Seed

9. **The visible scar (spot) observed on a seed at the place of the detachment from seed:**
   (a) Micropyle (b) Stome
   (c) Pole (d) Hillum

10. **Triploids are normally:**
   (a) Sterile (b) Fertile
   (c) Male sterile (d) None of the above

11. **Genetically most pure seed is:**
   (a) Breeder seed (b) Nucleus seed
   (c) Foundation seed (d) Certified seed

12. **Usually, seed production areas should contain a minimum area (ha) of:**
   (a) 1 ha (b) 2 ha
   (c) 3 ha (d) 4 ha

13. **Notification of seed control order by Government of India in the year:**
   (a) 1985 (b) 1989
   (c) 1983 (d) 1954

14. **The seed Act was enacted in the year:**
   (a) 1965 (b) 1966
   (c) 1968 (d) 1969

15. **The original tree (Donor tree) from which parts to be propagated are taken are called:**
   (a) Clone (b) Ramet
   (c) Ortet (d) Progeny

16. **Orthotropic growth refers to:**
   (a) Tree (b) Shrub
   (c) Herb (d) None of the above

17. The most common plant growth regulators compound used in rooting of tree species is:

(a) IAA (b) IBA

(c) NAA (d) 2, 4-D

18. Pure Live Seed is related to:

(a) Physical purity (b) Genetic purity

(c) Germination percentage (d) Defective purity

19. Seed lot certificate colour is:

(a) Orange (b) Green

(c) Purple (d) Orange and Green

20. Seed sample certificate colour is:

(a) Blue (b) Orange

(c) Purple (d) White

21. Difference between seed and grain is:

(a) Seed processing (b) Genetic purity

(c) Seed quality (d) Cost

22. Embryo exised test_______ used for surface disinfestation:

(a) NaOCl (b) Lactophenal

(c) $ALK(SO_4)$ (d) All of the above

23. Factors that affect tree seed demand and supply:

(a) Plantation area (b) Plantation types

(c) Species selection (d) All of the above

24. Among the gymnosperm, the seed demand is more for:

(a) *Pinus* spp. (b) *Quercus* spp.

(c) *Podocarpus* spp. (d) *Araucaria* spp.

25. Seeds are unique in natural regeneration and propagation because:

(a) Seed constitute unique genetic composition

(b) Seeds are produced in large number

(c) Seeds can be stored for long period

(d) All of the above

26. The "Embryon" is published by:

(a) ISTA (b) AOSA

(c) ASEAN Forest Tree Seed Centre (d) CAB International

27. The principle mode of pollination in Gymnosperm is by:

(a) Insects (b) Wind

(c) Birds (d) Bats

28. Seeds which maintain high moisture content during maturation:

(a) Recalcitrant seeds (b) Orthodox seeds

(c) Hard seeds (d) All of the above

29. The scar on the seed coat left by the funiculus is:

(a) Micropyle (b) Hilum

(c) Chalaza (d) Caruncle

30. Dalbergia seeds are dispersed by:

(a) Wind (b) Animals

(c) Water (d) None of the above

31. Ovule integument gets transformed into:

(a) Seed (b) Fruit wall

(c) Seed coat (d) Cotyledons

32. Innate seed dormancy is:

(a) Caused due to internal factors (b) Different from quiescence

(c) Called primary dormancy (d) All of the above

33. A seed which is just waiting for favourable environmental condition to germinate is called:

(a) Dormant seed (b) Quiescent seed

(c) Non-viable seed (d) Dead seed

34. Seed are subjected to hydraulic pressure at 2000 atm to:

(a) Increase permeability of seed coat to water

(b) Break seed coat

(c) Completion of after ripening period

(d) Leach out the germination inhibitors

35. Failure in pollination and fertilization in tree crops is due to:

(a) Shortage of pollen production (b) Closed flowers or cones

(c) Inbreeding barrier (d) All of the above

36. If most of the trees in the stand have abundant flowers, the reproduction capacity is:

(a) Very good (b) Good

(c) Poor (d) Very poor

37. Collection of immature fruits may:

(a) Minimize seed loss
(b) Avoid dormancy
(c) Reduce pre-processing damage
(d) All of the above

38. Labeling of seed lot in tree crops is:

(a) Not important
(b) Very difficult
(c) Important for maintaining the identity
(d) Not followed elsewhere

39. Speed of seed deterioration depends on:

(a) Condition of seed at collection
(b) Species
(c) Environment
(d) All of the above

40. Bulk reduction in fleshly fruits can be achieved by:

(a) Drying
(b) Removing the fruit pulp
(c) Soaking
(d) Threshing

41. Which of the following is recommended to maintain the viability in dry fruits of recalcitrant species:

(a) Moist storage
(b) Depulping
(c) Sun drying
(d) Hot air drying

42. Removal of wings from seed/fruit during processing is called:

(a) Grading
(b) Dewinging
(c) Pre-cleaning
(d) Pre-curing

43. Keeping the fruits at moist condition for prolonged period before extraction is called:

(a) Bulking
(b) Pre-cleaning
(c) Pre-curing
(d) None of the above

44. Seeds of tree species can be graded based on its density or weight by using:

(a) Specific gravity separator
(b) Indented cylinder
(c) Hand sieves
(d) All of the above

45. Orthodox seeds can be stored safely at a seed moisture content of:

(a) 0-4 per cent
(b) 8-10 per cent
(c) 18-20 per cent
(d) 28-30 per cent

46. While processing of tree seeds, seed quality may be affected by:

(a) Mechanical damage
(b) Heat damage
(c) Chemical damage
(d) All of the above

47. The colour of breeder seed tag is:

(a) Golden yellow (b) Blue

(c) Green (d) White

48. Which of the following is a recalcitrant seed:

(a) *Acacia nilotica* (b) *Melia dubia*

(c) *Syzygium cuminii* (d) *Delonix regia*

49. Characteristics of Good quality seed includes:

(a) Physical quality (b) Physiological quality

(c) Genetic quality (d) All the three

50. The stage at which the maximum dry matter accumulation in seed is:

(a) Harvestable maturity (b) Physical maturity

(c) Physiological maturity (d) Full maturity

51. Foundation seed is the progeny of:

(a) Breeder seed (b) Certified seed

(c) Labeled seed (d) Registered seed

52. Breeder seed is the progeny of:

(a) Foundation seed (b) Registered seed

(c) Nucleus seed (d) Certified seed

53. Certification is not required for:

(a) Nucleus seed (b) Breeder seed

(c) Foundation seed (d) Certified seed

54. Improved seed includes:

(a) Nucleus seed (b) Breeder seed

(c) Foundation seed (d) All of the above

55. Middle layer of fruit wall is called:

(a) Mesocarp (b) Mericarp

(c) Meristem (d) Micropyle

56. The Rudimentary root of the seed or seedling that forms the primary root of the young plant is known as:

(a) Rachis (b) Radical

(c) Rachilla (d) Racerne

57. New genes are created by:

(a) Mutation (b) Recombination

(c) Inversion (d) None of the above

58. Mode of pollination in maize is:

(a) Self-pollination
(b) Vegetative propagation
(c) Cross-pollination
(d) None of the above

59. The genome contained by bread wheat are:

(a) ABC
(b) XYZ
(c) ABD
(d) None of the above

60. Seed drying is very important to maintain its:

(a) Viability and vigour
(b) Protein content
(c) Oil content
(d) Chemical composition

61. The two most important factors influencing the life span of seeds under storage are:

(a) Relative humidity and temperature
(b) Rainfall and RH
(c) Climate and temperature
(d) Seed size and moisture content

62. The chemical seed treatment is given to control:

(a) Fungi and bacteria
(b) Soil insects
(c) Storage insects
(d) All of the above

63. In India, normally how many generation system seed are produced?

(a) 1
(b) 2
(c) 3
(d) 4

64. Off types are removed by:

(a) +ve selection
(b) -ve selection
(c) Pure line selection
(d) Mass selection

65. Seed certification involves:

(a) 1 phase
(b) 3 phase
(c) 5 phase
(d) 7 phase

66. Interaction between alleles of two different loci is called as:

(a) Epistasis
(b) Dominance
(c) Over-dominance
(d) None of the above

67. Seedlessnes in fruits is called as:

(a) Parthenogensis
(b) Parthenocarpy
(c) Apomixes
(d) None of the above

68. Solar treatment is used for:

(a) Loose smut (b) Stem rust

(c) Powdery mildew (d) Blast

69. Which of the following is not the part of seed embryo:

(a) Scutellum (b) Endosperm

(c) Cotyledon (d) Radical

70. Presence of inhibitors in pericarp seeds to:

(a) Mechanical dormancy (b) Physiological dormancy

(c) Chemical dormancy (d) Morphological dormancy

71. The area of seed stand treated and managed for abundant seed production is:

(a) Seed production area (b) Seed orchard

(c) Seed source (d) Provenance

72. Minimum area suggested to establish seed production area is:

(a) 20 Ha (b) 15 Ha

(c) 10 Ha (d) 4 Ha

73. Pollen dilution zone recommended for seed production area is:

(a) 400 m (b) 300 m

(c) 200 m (d) 100 m

74. Flower induction can be enhanced by:

(a) Girdling (b) Hormone application

(c) Inducing stress (d) All the three

75. Seeds can be graded according to its length using:

(a) Specific gravity separator (b) Spiral separator

(c) Indented cylinder (d) Magnetic separator

76. Allowing the seed to attain moisture equilibrium with in the seed during artificial drying process is:

(a) Cooling (b) Aeration

(c) Stratification (d) Tempering

77. Which of the following is a pre- storage seed treatment:

(a) Dormancy breaking (b) Seed coating

(c) Dydration-dehydration (d) Halogenations

78. Seed certification requires:

(a) Improved variety
(b) Genetic purity
(c) Physical purity
(d) All of the above

79. Plants which cotyledons do not come above the soil surface are called:

(a) Hypogeal germination
(b) Epigeal germination
(c) Both (a) and (b)
(d) None of the above

80. The first private seed company came into existence in:

(a) Mumbai
(b) Lucknow
(c) New Delhi
(d) Kolkata

81. The flowering plants a second seed coat is known as:

(a) Integument
(b) Aleurone layer
(c) Tegamen
(d) Inner ventral scale

82. In India, foundation seed is produced by:

(a) Registered growers
(b) National seed corporation
(c) SSCA
(d) All of the above

83. Failure in pollination and fertilization in tree crops is due to:

(a) Shortage of pollen production
(b) Closed flowers or cones
(c) Inbreeding barrier
(d) All of the above

84. If most of the trees in the stand have abundant flowers, the reproduction capacity is:

(a) Very good
(b) Good
(c) Poor very
(d) Very poor

85. Collection of immature fruits may:

(a) Minimize seed loss
(b) Avoid dormancy
(c) Reduce pre-processing damage
(d) All of the above

86. Labeling of seed lot in tree crops is:

(a) Not important
(b) Very difficult
(c) Important for maintain the identity
(d) Not followed elsewhere

87. Speed of seed deterioration depends on:

(a) Condition of seed at collection
(b) Species
(c) Environment
(d) All of the above

88. Bulk reduction in fleshly fruits can be achieved by:

(a) Drying
(b) Removing the fruit pulp
(c) Soaking
(d) Threshing

89. Which of the following is recommended to maintain the viability in dry fruits of recalcitrant species:

(a) Moist storage (b) Depulping

(c) Sun drying (d) Hot air drying

90. A set standards used to regulate own or community activity in relation to biological world is:

(a) Biopotency (b) Biopiracy

(c) Biowar (d) Bioethics

91. In most of monocot plants, germination is:

(a) Epigeal (b) Hypogeal

(c) Viviparous (d) Hydrogeal

92. Emergence of radical and plumule from seed is called:

(a) Pregermination (b) Post germination

(c) Germination (d) Dormancy

93. In stratification, dormant seeds are kept at:

(a) Very high temp (b) Low temperature (0-10°C)

(c) Under aerated condition (d) Optimum temperature

94. Longevity of seed can be prolonged by keeping them at:

(a) Dry condition and low temperature (b) Humidity and low temperature

(c) Low temperature and dryness (d) High temperature

95. The resumption of active growth of embryo is called as:

(a) Germination (b) Growth

(c) Quiescence (d) Viability

96. Biopiracy means:

(a) Use of biopatents

(b) Thefts of plants and animals

(c) Stealing of bioresources

(d) Exploitation of bioresources without authentic permission

97. Bioethics is r elated to:

(a) Preventing biopiracy

(b) Regulation of unethical activities like gene cloning in animals

(c) Preventing theft of living materials

(d) Moral guidance to the problems in biology

98. Tetrazolium test for testing viability of seeds is attributable to:

(a) Clement (b) Lakon

(c) Bonner (d) Skoog

99. Pick the odd man out – Orthodox seed:

(a) Acacia (b) Eucalyptus

(c) Terminalia (d) Shorea

100. The growth hormones used for shoot multiplication under in vitro condition are collectively called as:

(a) Auxins (b) Cytokinins

(c) Giberellins (d) Vitamins

101. Type of pre treatment required for teak is:

(a) Acid scarification

(b) Hot water treatment

(c) Soaking in ambient for 24 hrs

(d) Alternate wetting and drying for 15 days

102. Which among the following tree seeds has the lowest viability period?

(a) *Acacia nilotica* (b) *Cassia fistula*

(c) *Casuarnia equisetifolia* (d) *Albizzia lebbek*

103. Initial seed of an improved variety is called:

(a) Nucleus seed (b) Breeder seed

(c) Foundation seed (d) Certified seed

104. The seed which farms generally used are called:

(a) Breeder seed (b) Foundation seed

(c) Certified seed (d) Nucleus seed

Answer Sheet

1.	(b)	22.	(a)	43.	(c)	64.	(b)	85.	(d)
2.	(a)	23.	(d)	44.	(a)	65.	(c)	86.	(c)
3.	(a)	24.	(a)	45.	(b)	66.	(a)	87.	(d)
4.	(c)	25.	(d)	46.	(d)	67.	(b)	88.	(b)
5.	(c)	26.	(c)	47.	(a)	68.	(a)	89.	(a)
6.	(c)	27.	(b)	48.	(c)	69.	(b)	90.	(d)
7.	(c)	28.	(a)	49.	(d)	70.	(c)	91.	(b)
8.	(b)	29.	(b)	50.	(c)	71.	(a)	92.	(c)
9.	(d)	30.	(a)	51.	(a)	72.	(d)	93.	(b)
10.	(a)	31.	(c)	52.	(c)	73.	(c)	94.	(b)
11.	(b)	32.	(d)	53.	(b)	74.	(d)	95.	(a)
12.	(d)	33.	(b)	54.	(d)	75.	(c)	96.	(d)
13.	(c)	34.	(a)	55.	(a)	76.	(d)	97.	(b)
14.	(b)	35.	(d)	56.	(b)	77.	(d)	98.	(b)
15.	(c)	36.	(a)	57.	(a)	78.	(d)	99.	(d)
16.	(a)	37.	(d)	58.	(c)	79.	(a)	100.	(b)
17.	(b)	38.	(c)	59.	(c)	80.	(c)	101.	(d)
18.	(c)	39.	(d)	60.	(a)	81.	(c)	102.	(c)
19.	(d)	40.	(b)	61.	(a)	82.	(b)	103.	(a)
20.	(a)	41.	(a)	62.	(d)	83.	(d)	104.	(c)
21.	(b)	42.	(b)	63.	(c)	84.	(a)		

Chapter 11

Tree Physiology

1. **Wilting of plants results from excessive:**
   (a) Absorption (b) Photosynthesis
   (c) Respiration (d) Transpiration

2. **Process of transpiration plants helps in a:**
   (a) Absorption of CO_2 (b) Upward conduction
   (c) Opening of stomata (d) None of the above

3. **Plasmolysis will occur when the cell is placed in a:**
   (a) Acidic solution (b) Isotonic solution
   (c) Hypotonic solution (d) Hypertonic solution

4. **Blue green algae are:**
   (a) Actinomy cetes (b) Acellular
   (c) Prokaryotes (d) Eukaryotes

5. **The solute present into the cells is strongly dependent on and ability of cells to respire aerobically and to produce it:**
   (a) ADP (b) ATP
   (c) RNA (d) DNA

6. **Simplest amino acid is:**
   (a) Alanine (b) Glycine
   (c) Leucine (d) Aspartic acid

7. **Fermentation is conducted by:**
   (a) All fungi
   (b) All bacteria
   (c) Some fungi and some bacteria
   (d) All micro-organisms

8. **Peptide linkages are formed among:**
   (a) Amino acids
   (b) Gluco molecules
   (c) Sucrose molecules
   (d) Glycerol molecules

9. **Proteins are made up of:**
   (a) Amino acids
   (b) Nucleic acids
   (c) Sugars
   (d) Fatty acids

10. **Carbon dioxide is liberated during:**
    (a) Ascent of sap
    (b) Transpiration
    (c) Photosynthesis
    (d) Respiration

11. **Which is a micronutrient?**
    (a) Ca
    (b) Zn
    (c) P
    (d) Mg

12. **Which of the following effects of auxins on plants is the basis of commercial application?**
    (a) Callus formation
    (b) Curvature of stem
    (c) Induction of root formation
    (d) In stem cuttings

(e). All of the above:

13. **During the photosynthesis the oxygen in glucose comes from:**
    (a) Carbon dioxide
    (b) Water
    (c) Both from carbon dioxide and water
    (d) Oxygen in the air

14. **When yeast and certain bacteria carry on aerobic respiration they produce:**
    (a) Carbon dioxide
    (b) Nitrogen
    (c) Water
    (d) Oxygen

15. **One of the following is a water soluble pigment:**
    (a) Carotene
    (b) Chlorophyll
    (c) Anthocyanin
    (d) Xanthophylls

16. **Photosynthesis is:**
    (a) Catabolic process
    (b) Anabolic process
    (c) Exothermic process
    (d) Exergonic process

17. Which of the following structures controls the transport of the material into and out of living cells?

(a) Centrosome (b) Cell membrane
(c) Cell wall (d) Ribosome

18. Functions and vital activities of the plants are studied under:

(a) Ecology (b) Physiology
(c) Cytology (d) Morphology

19. Excessive use of fertilizer causes death of plants due to:

(a) Endosmosis (b) Exosmosis
(c) Imbibitions (d) Turgidity

20. Swelling of wooden door during rainy season due to:

(a) Endosmosis (b) Imbibitions
(c) Capillary (d) Deplasmolysis

21. Stomata are bound by a pair of guard cells and open guard cells are:

(a) Flaccid (b) Large
(c) Turgid (d) Green

22. Osmosis means:

(a) Flow of solute from low concentration to higher concentration
(b) Flow on solute from higher concentration to low concentration
(c) Flow of solvent from low concentration to higher concentration
(d) Flow of solvent from high concentration to low concentration

23. The rate of transpiration will greatly depend upon:

(a) Frequency of stomata (b) Position of stomata
(c) State of stomata (d) None of the above

24. Photochemical reactions of photosynthesis take place in:

(a) Frets (b) Stroma
(c) Thylakoids (d) All of the above

25. Which of the following is an essential part of photosynthesis?

(a) Glycolsis (b) Photorespiration
(c) Plasmolysis (d) Photolysis

26. Hydroponics are:

(a) Growing of plants in water culture (b) Growing of aquatic plants
(c) Growing of floating aquatic plants (d) Soil less cultivation of plants

27. When a seed is placed in a suitable medium for germination, by which of the following first process water enter into the seed coat?

(a) Imbibitions (b) Absorption

(c) Osmosis (d) Root pressure

28. Plasma membrane:

(a) Impermeable (b) A selective barrier

(c) A non-selective barrier (d) Made of up cellulose

29. Anaerobic respiration is also known as:

(a) Intra-molecular respiration (b) Inter-molecular respiration

(c) Extra-molecular respiration (d) Molecular respiration

30. Which of the following is not a part of chlorophyll?

(a) Iron (b) Nitrogen

(c) Carbon (d) Hydrogen

31. Wilting of leaves is hot weather is due to:

(a) Iron and calcium (b) Sodium and copper

(c) Calcium and potassium (d) Iron and Magnesium

32. Leaves with yellow or colorless patches are termed as:

(a) Lecorine (b) Decolourized

(c) Etiolated (d) Variegated

33. Oxygen evolved in photosynthesis is:

(a) Less that CO_2 consumed (b) More that CO_2 consumed

(c) Equal to CO_2 consumed (d) None of the above

34. The dry weight of a green leaf will be:

(a) Lesser in evening (b) Grater in the afternoon

(c) Greater in the morning (d) Almost same throughout the whole day

35. Etiolated plants are formed due to lack of:

(a) Light (b) Hg

(c) Fe (d) Mg

36. For photosynthesis (*i.e.* for the synthesis of organic matter) the green plants need only:

(a) Light (b) Chlorophyll

(c) CO_2 and water (d) All of the above

37. Growth is primarily affected by two climatic factors which are:

(a) Light and temperature
(b) Temperature and relative humidity
(c) Light and wind
(d) Rainfall and temperature

38. One of the following is not found in intact cells;:

(a) Lomasomes
(b) Dictyosomes
(c) Microsomes
(d) Ribosomes

39. A plant which needs not less than 10 hours of light to flower is:

(a) Long day plant
(b) Short day plant
(c) Day neutral plant
(d) None of the above

40. If cells are broken up and sedimented by centrifugation, the new structure formed in one of the fraction is:

(a) Lysosome
(b) Ribosome
(c) Microsome
(d) Centrosome

41. Probably the most useful measurable property in the soil-plant-air system is:

(a) Matric potential
(b) Water potential
(c) Osmotic potential
(d) Redox potential

42. DPD is equal to:

(a) OP x TP
(b) OP – TP
(c) TP – OP
(d) OP + TP

43. Seeds swell when placed in water is due to:

(a) Osmosis
(b) Imbibitions
(c) Hydrolysis
(d) Plasmolysis

44. The loss of water in the form of vapour from aerial plant parts is known as:

(a) Osmosis
(b) Respiration
(c) Photosynthesis
(d) Transpiration

45. Semi-permeability is the property of:

(a) Plasma membrane
(b) Cell wall
(c) Golgibody
(d) Endoplasmic recticulum

46. Protoplasm is:

(a) Polyphasic colloidal system
(b) Colloidal solution
(c) Suspension
(d) True solution

47. Photometer is use to measure:

(a) Rate of transpiration (b) Rate of respiration
(c) Growth (d) Osmosis

48. Photosynthesis is a:

(a) Reductive, endermic, catabolic process
(b) Reductive, endergonic, anabolic process
(c) Reductive, exergonic, anabolic process
(d) Oxidative, Exergonic, catabolic process

49. Plants synthesis protein from:

(a) Starch (b) Sugar
(c) Amino acids (d) Fatty acids

50. The end products of respiration in plants are:

(a) Starch and O_2 (b) H_2O and energy
(c) CO_2, H_2O and energy (d) Sugar and O_2

51. The process by which proteins are synthesized in a cell is called:

(a) Translation (b) Transduction
(c) Translocation (d) Transcription

52. Process of endosmosis stops:

(a) When water concentration becomes equal
(b) When solution becomes isotonic
(c) When the leaves fall
(d) When there is no light

53. Pigment system is concerned with:

(a) Photolysis of water (b) Reduction of CO2
(c) Flowering (d) None of the above

54. Respiration:

(a) Exothermic process (b) Endothermic process
(c) Endergonic process (d) Anabolic process

55. Photosynthesis is to the chloroplast, as _______ is to the mitochondrion.

(a) Respiration (b) Microtubule
(c) Extracellular matrix (d) Microfilament

56. The smooth endoplasmic reticulum functions in synthesis of:

(a) DNA (b) Lipids
(c) Pollysaccharides (d) Proteins

57. Which of the following is an eukaryotic characteristic?

(a) Cell fission
(b) Little processing of RNA
(c) 70s ribosomes
(d) Membrane-bound nucleus

58. Lignin is an important component in:

(a) Bacterial cell walls
(b) Plasmodesmata
(c) Tight junctions
(d) Secondary plant cell walls

59. Which of the following is true of a plant cell?

(a) Plant cells are often much larger than animal cells
(b) Plant cells may contain chloroplasts
(c) Only plant cells have a cell membrane
(d) Chromosomes are found only in plant cells

60. Which organelle has only a single membrane around it ?

(a) Peroxisome
(b) Nucleus
(c) Chloroplast
(d) Ribosome

61. Which of the following is a phototroph?

(a) A mushroom
(b) Bacteria growing on decomposing matter
(c) An earthworm
(d) A tree

62. To enter or leave a cell, substances must pass through:

(a) A microtubule
(b) The golgi apparatus
(c) A ribosome
(d) The plasma membrane

63. Which of the following correctly matches an organelle with its function?

(a) Mitochondrion – photosynthesis
(b) Ribosome – Manufacture of lipids
(c) Lysosome – Movement
(d) Central vacuole – Storage

64. Oxidation in biological systems is accompanied by:

(a) Hydrogenation
(b) Resonance stabilization
(c) Repulsion
(d) Dehydrogenation

65. During the first phase of glycolysis, ATP is used to form what type of bond with glucose?

(a) Dehydrogenation
(b) Hydrogenation
(c) Substrate – level phosphorylation
(d) Phosphoester

66. The enzyme used to break down fructose 1,6-bisphosphate into two trioses is called:

(a) Phosphoglucoisomerase (b) Enolase

(c) Aldolase (d) Triose phosphate isomerase

67. Beta oxidation:

(a) Is a process used to oxidize fatty acids

(b) Is the second step in t eh oxidation of glucose

(c) Is a process used to digest some polysaccharides

(d) Occurs in the cytoplasm.

68. The sequential process of fatty acid catabolism to acetyl CoA is called:

(a) Oxidative deamination (b) B-oxidation

(c) The Kerb cycle (d) The glyoxylate cycle

69. Transamination:

(a) Can involve only a small number of amino acids

(b) Involves the amination of an alpha keto acid

(c) Liberates a free ammonia molecule

(d) Involves transfer of amino groups from an amino acid to another amino acid

70. Which scientist is credited with coining the term "cell" from the Latin *cellulae*?

(a) Robert Hooke (b) Robert Brown

(c) Theodor Schwann (d) Antonie van Leeuwenhoek

71. The Hatch-Slack cycle occurs in which of the following types of plants?

(a) C3 plants (b) Tropical grasses

(c) C4 plants (d) Both (b) and (c)

72. The relationship between photosystem I (PSI) and photosystem II (PSII) is correctly described by which of the following statements?

(a) PSI absorbs only short wavelengths of light

(b) Electrons are excited by PSI or PSII, but not both

(c) The absorption maximum of PSII is 700 nm

(d) PSI is designated P 700, whereas PSII is designated P680

73. Which organelle is round and derives its name from the Latin word for "Kernel"?

(a) Nucleus (b) Lysosome

(c) Golgi complex (d) Mitochondrion

74. Photolysis is _______:

(a) The absorption of light by chlorophyll

(b) The assimilation of carbon dioxide

(c) The splitting of water

(d) None of the above

75. The molecules known as the energy currency of the cell are _______.

(a) NAD (b) NADP

(c) ATP (d) ADP

76. In addition to ATP, what are the end products of glycolysis?

(a) CO_2 and ethyl alcohol (b) CO_2 and H_2O

(c) NADH and pyruvate (d) H_2O and ethyl alcohol

77. All of the following are functions of the Krebs cycle except:

(a) Production of NADH

(b) Production of FADH

(c) Adding electrons and protons to oxygen to form water

(d) Production of ATP

78. Which of the following are products of the light reactions of photosynthesis that are utilized in the Calvin cycle?

(a) Electrons and H+ (b) ADP, Pi, and NADP+

(c) CO2 and glucose (d) ATP and NADP

79. The color of light least effective in driving photosynthesis is:

(a) Yellow (b) Red

(c) Green (d) Blue

80. What is the primary function of the light reactions of photosynthesis?

(a) To use ATP to make glucose

(b) To produce ATP and NADPH

(c) To convert light energy to the chemical energy of PGAL

(d) To produce NADPH used in respiration

81. Where do the microtubules of the spindle originated during mitosis in both plant and animal cells?

(a) Centriole (b) Centrosome

(c) Centromere (d) Kinetochore

82. Plants appear green because:
(a) They contain carotenoids
(b) They undergo photosynthesis
(c) They reflect green light
(d) They absorb green light

83. The primary source of energy for Earth is:
(a) Solar energy
(b) Thermal energy
(c) Hydroelectric energy
(d) None of the above

84. Land plants are composed of all the following tissue types except:
(a) Vascular
(b) Mesoderm
(c) Epidermal
(d) Meristematic

85. Which functional plant cells lack a nucleus?
(a) Xylem only
(b) Sieve-tube cell only
(c) Both companion an parenchyma cells
(d) Both xylem and sieve-tube cells

86. _______. Light is the most effective in producing a phototropic response.
(a) Green
(b) Yellow
(c) Orange
(d) Blue

87. Which of the following factors causes an increase in transpiration r/re?
(a) A hot dry windy day
(b) A cool damp windy day
(c) A hot damp windy day
(d) None of the above

88. An apparatus for measuring transpiration in plants is called:
(a) Auxanometer
(b) Photometer
(c) Hygrometer
(d) Lactometer

89. Sialic acid is a constituent of:
(a) Cell membrane
(b) Cell wall
(c) Chromosomes
(d) Dictyosomes

90. Which pair of structures are usually found in both plant and animal cell:
(a) Cell membrane and nucleolus
(b) Cell membrane and cell wall
(c) Nucleolus and chloroplast
(d) Nucleolus and cell wall

91. Unit membrane or (plasma membrane) is made of:
(a) Cellulose
(b) Lipoproteins
(c) Phosphor proteins
(d) Phospholipids

92. The mole of molybdenum is:

(a) Chromosome contraction (b) Carbon assimilation

(c) Nitrogen fixation (d) Flowering induction

93. The real force causing entry of water into a cell from other cell is:

(a) Diffusion pressure deficit (b) Tugor pressure

(c) Osmotic pressure (d) Wall pressure

94. Turgidity of a cell is maintained by:

(a) Diffusion pressure (b) Wall pressure

(c) Osmotic pressure (d) Turgor pressure

95. Mineral salt which are absorbed by the roots for the soil are in the form of:

(a) Dilute solution (b) Vary concentrated solution

(c) Concentrated solution (d) Very dilute solution

96. When a cell is fully turgid, which of the following will be zero?

(a) Osmotic pressure (b) Turgor pressure

(c) Wall pressure (d) Suction pressure

97. In aerobic respiration in plants:

(a) Oxygen is taken in (b) Oxygen is given out

(c) Carbon dioxide is given out (d) Carbon dioxide is taken out

98. Water potential concept was formulated by:

(a) Peter ray 1960 (b) M.H Mathew 1954

(c) Otto Renner 1915 (d) W.C. Hoffer 1920

99. Alcohol is produced during the process of:

(a) Photosynthesis (b) Aerobic respiration

(c) Anaerobic respiration (d) None of the above

100. Auxins promote growth:

(a) Rarely (b) Always

(c) Mostly (d) Never

101. The first event in photosynthesis is:

(a) Photoexcitation of cholorophyll all (b) Synthesis in ATP ejection of electron

(c) Photolysis (d) Release of oxygen

102. Carbohydrates are present in the plasmalemma in the form of:

(a) Starch cellulose (b) Cellulose

(c) Phospholipids and phosphoproteins (d) Hemicelluloses

Answer Sheet

1.	(d)	22.	(c)	43.	(b)	64.	(d)	85.	(d)
2.	(b)	23.	(b)	44.	(d)	65.	(d)	86.	(d)
3.	(d)	24.	(c)	45.	(a)	66.	(c)	87.	(a)
4.	(c)	25.	(d)	46.	(a)	67.	(a)	88.	(b)
5.	(b)	26.	(a)	47.	(b)	68.	(b)	89.	(a)
6.	(b)	27.	(a)	48.	(b)	69.	(b)	90.	(a)
7.	(c)	28.	(b)	49.	(c)	70.	(a)	91.	(b)
8.	(a)	29.	(a)	50.	(c)	71.	(d)	92.	(c)
9.	(a)	30.	(a)	51.	(a)	72.	(d)	93.	(a)
10.	(d)	31.	(a)	52.	(b)	73.	(a)	94.	(d)
11.	(b)	32.	(d)	53.	(a)	74.	(c)	95.	(c)
12	E	33.	(c)	54.	(a)	75.	(c)	96.	(d)
13.	(a)	34.	(b)	55.	(a)	76.	(c)	97.	(c)
14.	(a)	35.	(a)	56.	(b)	77.	(c)	98.	(c)
15.	(b)	36.	(d)	57.	(d)	78.	(d)	99.	(c)
16.	(b)	37.	(a)	58.	(d)	79.	(c)	100.	(c)
17.	(b)	38.	(b)	59.	(b)	80.	(b)	101.	(c)
18.	(c)	39.	(c)	60.	(a)	81.	(b)	102.	(c)
19.	(b)	40.	(b)	61.	(d)	82.	(c)		
20.	(b)	41.	(b)	62.	(d)	83.	(a)		
21.	(c)	42.	(b)	63.	(d)	84.	(b)		

Chapter 12

Soil Science

1. **Who is the Father of soil physics?**

   (a) Hillel (b) Richards

   (c) Dokuchaiev (d) Thomas way

2. **What is the number of classification in size groups of soil particles by International Society of Soil Science (ISSS)?**

   (a) Three (b) Four

   (c) Six (d) Seven

3. **What is the size of fine clay particles in USDA classification of soil texture?**

   (a) <2mm (b) <0.2mm

   (c) <0.02mm (d) <0.002mm

4. **Particle smaller than _______. Microns cannot be separated by sizing:**

   (a) <2 (b) <20

   (c) <50 (d) <100

5. **In which of the soils, water infiltration is more rapid?**

   (a) Sandy soils (b) Loamy soils

   (c) Clay soils (d) All of the above

6. **The arrangement of primary and secondary soil particles is called as:**

   (a) Soil structure (b) Soil texture

   (c) Soil consistency (d) Soil plasticity

7. **The relative proportion of soil particles is called as:**
   (a) Soil structure
   (b) Soil texture
   (c) Soil consistency
   (d) Soil plasticity

8. **Wet bulk density g cm^{-3} =:**
   (a) Weight of soil/volume of soil
   (b) Weight of dry soil/volume of soil
   (c) Weight of dry soil/volume of soil particles
   (d) Volume of soil/weight of moist soil

9. **Dry bulk density g cm^{-3} =:**
   (a) Weight of soil/volume of soil
   (b) Weight of dry soil/volume of soil
   (c) Weight of dry soil/volume of soil particles
   (d) Volume of soil/weight of moist soil

10. **Particle density g cm^{-3} =:**
    (a) Weight of soil/volume of soil
    (b) Weight of dry soil/volume of soil
    (c) Weight of dry soil/volume of soil particles
    (d) Volume of soil/weight of moist soil

11. **What is the bulk density of fine textured soils?**
    (a) 1.1-1.3g cm^{-3}
    (b) 2.65 g cm^{-3}
    (c) 1.4-1.8 g cm^{-3}
    (d) 2.4-2.6 g cm^{-3}

12. **What is the bulk density of coarse textured soils?**
    (a) 2.4-2.6 g cm^{-3}
    (b) 2.65 g cm^{-3}
    (c) 1.1-1.3g cm^{-3}
    (d) 1.4-1.8 g cm^{-3}

13. **Value of bulk density is:**
    (a) Higher than particle density
    (b) Lower than particle density
    (c) Equal to particle density
    (d) Unity

14. **Element essential for the flocculation of soil particles is:**
    (a) Sodium
    (b) Potassium
    (c) Calcium
    (d) Magnesium

15. **Element essential for the deflocculation of soil particles is:**
    (a) Potassium
    (b) Calcium
    (c) Magnesium
    (d) Sodium

16. More stable aggregates are formed by:

(a) Humus substances (b) Earthworms

(c) Plant roots decay (d) All of the above

17. The soil structural classification followed in India is:

(a) ISSS classification (b) USDA classification

(c) Atterberg classification (d) All of the above

18. Which soil is most desirable for agriculture ?

(a) Platy (b) Blocky

(c) Crumb (d) Granular

19. What are the total number of soil classes?

(a) Four (b) Five

(c) Six (d) Seven

20. What is the size of the non-capillary pore:

(a) >6_______. (b) >10_______.

(c) >15_______. (d) >20_______.

21. What is the size of micro or capillary pores?

(a) >6_______. (b) >10_______.

(c) >15_______. (d) >20_______.

22. Arrange the following soil texture in order of decreasing water retention capacity,:

(a) Loam>Clay>Sandy (b) Sandy>Loam>Clay

(c) Clay>Loam>Sandy (d) Sandy>Clay>Loam

23. The per cent of carbon dioxide in soil air?

(a) 0.03 (b) 0.003

(c) 0.3 (d) 3

24. More than 90 per cent of gaseous exchange in soil takes place by:

(a) Mass flow (b) Diffusion

(c) Particle diffusion (d) All of the above

25. The dominant spectral colour in Munsell colour chart indicates:

(a) Hue (b) Value

(c) Chroma (d) All of the above

26. Black soil cover almost _______ per cent of the geographical area of India.

(a) 9 (b) 20

(c) 21 (d) 22

27. Red soil cover almost _______ per cent of the geographical area of India.

(a) 9 (b) 20

(c) 21 (d) 22

28. Bulk density in soils can be estimated by:

(a) Core method (b) Excavation method

(c) Both (d) None of the above

29. Gravitational water in soil move through the:

(a) Microscopes (b) Macro pores

(c) Both (a) and (b) (d) None of the above

30. Water held against the force of gravity in microscopes is:

(a) Heavy water (b) Gravitational water

(c) Capillary water (d) Hygroscopic water

31. Capillary rise of water is due to the force of:

(a) Cohesion (b) Adhesion

(c) Cohesion and adhesion (d) Matric potential

32. Which type of water includes that which is available to plants?

(a) Hygroscopic water (b) Gravitational water

(c) Capillary water (d) All of the above

33. Capillary water in the soil is that which is found in the:

(a) Microscopes (b) Macro pores

(c) Root channels (d) Both (a) and (b)

34. The water held between -1/3 and -1/5 bar is called:

(a) Gravitational water (b) Plant available water

(c) Wilting point (d) Field capacity

35. A soil textural triangle measures:

(a) Percentage of sand, silt, clay

(b) Minerals lives in a soil sample

(c) Percent slope of a land class

(d) Percentage of soil erosion on a given location

36. Soil water potential always has:

(a) Positive sign (b) Negative sign

(c) Non sign (d) None of the above

37. Soil water suction always has:

(a) Positive sign (b) Negative sign

(c) Non sign (d) None of the above

38. Since that deals with soil genesis classification is called as:

(a) Petrology (b) Pedology

(c) Pedagogy (d) Petrography

39. Sandstone and limestone are:

(a) Metamorphic rocks (b) Sedimentary rocks

(c) Primary rocks (d) Igneous rocks

40. Major component of igneous rocks are:

(a) Quarts (b) Amphiboles and pyroxenes

(c) Feldspars (d) Micas

41. Of the following, which is an important chemical weathering process:

(a) Carbonation (b) Oxidation

(c) Hydrolysis (d) Hydration

42. Weathered products of rocks and rock minerals are called as:

(a) Pedon (b) Parent materials

(c) Solum (d) Ply pedons

43. The unconsolidated materials from which the solum develops:

(a) Pedon (b) Regolith

(c) Parent materials (d) Both (b) and (c)

44. Materials transported and depoisited by water are called as:

(a) Alluvium (b) Collivium

(c) Loess (d) Aeolian

45. Materials transported by the action of gravity are called as:

(a) Alluvium (b) Lacustrine

(c) Glacial drift (d) Colluviums

46. Materials transported by water and settled out in lakes are called as:

(a) Alluvium (b) Lacustrine

(c) Glacial drift (d) Colluviums

47. Silt material transported by wind are called as:

(a) Delta (b) Meander

(c) Loess (d) Aeolian

48. **Sand materials transported by water are called as:**
   (a) Delta (b) Meander
   (c) Loess (d) Aeolian

49. **Azonal soils are:**
   (a) Black soil (b) Red soil
   (c) Lateriate soil (d) Alluvial soil

50. **Process of eluviations of oxides of iron and aluminum, humus and carbonates from the upper horizons and their accumulation in subsurface horizons is called as:**
   (a) Laterization (b) Podsolization
   (c) Decalcification (d) Calcification

51. **Process o removal of silica and accumulation of sesquioxides is called as:**
   (a) Laterization (b) Podsolization
   (c) Decalcification (d) Calcification

52. **Cool and humid climatic conditions are favourable for:**
   (a) Podsolization (b) Laterization
   (c) Humification (d) Pedoturbation

53. **Warm and humid climatic conditions are favourable for:**
   (a) Podsolization (b) Laterization
   (c) Humification (d) Pedoturbation

54. **Bed rocks are obeviously absent in:**
   (a) Red soils (b) Black soils
   (c) Laterite soils (d) Alluvial soils

55. **Mater horizons are:**
   (a) O and C (b) O,A,E,B and C
   (c) O,A,B and C (d) A,B and C

56. **B horizon is a:**
   (a) Illuvial horizon (b) Elluvial horizon
   (c) Organic horizon (d) Bed rock

57. **Elluvial horizon is:**
   (a) A horizon (b) E horizon
   (c) B horizon (d) O horizon

58. Soils under the soil order Alfish is/are:

(a) Red soil (b) Laterite soil

(c) Black soil (d) Both (a) and (b)

59. Fertile soils of India are locally called as:

(a) Urvara (b) Anurvara

(c) Usara (d) Kari

60. Barren soils of India are locally called as:

(a) Usara (b) Anurvara

(c) Urvara (d) Both (a) and (b)

61. Total area of salt affected soils in India is:

(a) 25mha (b) 28mha

(c) 29mha (d) 10mha

62. Total area of saline soils in India is:

(a) 10mah (b) 7.2mha

(c) 2.8mha (d) 6.1mha

63. Total area of sodic or alkali soils in India:

(a) 10mah (b) 7.2mha

(c) 2.8mha (d) 6.1mha

64. What is the pH of saline-sodic soils ?

(a) Less than 8.5 (b) More than 8.5

(c) 8.5-10.0 (d) 8.5

65. Who is the father of soil science?

(a) Dokuchaiev (b) Justin von Liebig

(c) J.W. Leather (d) Marbut

66. Density of the earth:

(a) 4.0 g cm^{-3} (b) 5.5 g cm^{-3}

(c) 2.6-2.7 g cm^{-3} (d) 3.0-4.5 g cm^{-3}

67. Example(s) of light minerals is/are:

(a) Quartz (b) Feldspars

(c) Muscovite (d) All of the above

68. Example(s) of heavy minerals is/are:

(a) Olivines (b) Heamatite

(c) Pyrite and Augite (d) All of the above

69. Most important weathering of soil formation is/are:

(a) Physical weathering
(b) Chemical weathering
(c) Biological weathering
(d) All of the above

70. Active soil forming factors are:

(a) Parent material
(b) Relief or Topography
(c) Time
(d) Climate and vegetation

(e). All of the above:

71. Passive soil forming factors are:

(a) Parent material
(b) Relief or Topography
(c) Time
(d) All of the above

72. The igneous rock basalt, in humid tropics may give rise to:

(a) Black soil
(b) Red soil
(c) Laterite soil
(d) Alluvial soil

73. Soil that are formed from the same kind of parent materials but differing in their topographic positions within the landscape are called as:

(a) Soil catena
(b) Climosquence
(c) Toposquence
(d) Clinosquence

74. Zonal soil forming process is:

(a) Calcification
(b) Salinization
(c) Solinization
(d) Gleization

75. Saline soils are called as:

(a) Solonetz
(b) Solonchalks
(c) Soloth
(d) Both (b) and (c)

76. Sodic soils are also called as:

(a) Soloth
(b) Solonetz
(c) Solonchalks
(d) Both (a) and (b)

77. Moor soils are:

(a) Normal soil
(b) Transitional soil
(c) Abnormal soil
(d) None of the above

78. Tundra soils are:

(a) Normal soil
(b) Transitional soil
(c) Abnormal soil
(d) None of the above

79. **What is the total number of soil order?**
   (a) 10 (b) 11
   (c) 12 (d) 13

80. **Land capability class was given by:**
   (a) Sys (b) Klingebiel and Montgomery
   (c) Sys and Verhegs (d) Storie and Ricquier

81. **Total number of categories in the land capability classification is:**
   (a) Three (b) Five
   (c) Eight (d) Twelve

82. **Colour of the land capability class-V on LCC Maps is:**
   (a) Yellow (b) Pink
   (c) Green (d) Red

83. **The dominant soil order in India is:**
   (a) Vertisols (b) Inceptisols
   (c) Alfisols (d) Entisols

84. **Soil erosion is mostly affected by:**
   (a) Rainfall intensity (b) Rainfall variability
   (c) Rainfall erosivity (d) Total annual rainfall

85. **Clay minerals are called as _______.**
   (a) Primary minerals (b) Secondary minerals
   (c) Tertiary minerals (d) Accessory minerals

86. **The last layer in the soil profile is the:**
   (a) A Horizon (b) B Horizon
   (c) C Horizon (d) R Horizon

87. **The soil orders which have highest sub (7) are:**
   (a) Mollisols and Andosols (b) Mollisols and Vertisols
   (c) Vertisols and Andosols (d) Inceptisols and Alfisols

88. **The dominant soil order in India is:**
   (a) Entisols (b) Alfisols
   (c) Inceptisols (d) Vertisols

89. **The soil horizon of maximum biological activity is:**
   (a) A (b) B
   (c) C (d) E

90. The C horizon is also called as:

(a) Solum (b) Top soil

(c) Zone of loss (d) Parent material

91. Buffering capacity is high for:

(a) Sand (b) Silt

(c) Clay (d) Humus

92. Red colour of the soil is due to the presence of:

(a) Gibbsite (b) Magnetite

(c) Hematite (d) Goethite

93. The capability of soil to produce a plant or plant parts is called as:

(a) Soil fertility (b) Soil quality

(c) Soil productivity (d) All of the above

94. The inherent capacity of soil to provide nutrients is called as:

(a) Soil fertility (b) Soil quality

(c) Soil productivity (d) All of the above

95. The total number of essential nutrient elements required for plant growth is:

(a) 16 (b) 17

(c) 20 (d) 21

96. N,P and K are called as:

(a) Macro nutrients (b) Micro nutrients

(c) Secondary nutrients (d) Beneficial nutrients

97. Ca, Mg and S are called as:

(a) Macro-nutrients (b) Micro-nutrients

(c) Secondary nutrients (d) Beneficial nutrients

98. Iron, zinc, boron and molybdenum are called as:

(a) Macro nutrients (b) Micro nutrients

(c) Secondary nutrients (d) Beneficial nutrients

99. Na, Ve, Ni, Si, Co, Se and I are the examples of:

(a) Macro nutrients (b) Micro nutrients

(c) Secondary nutrients (d) Beneficial nutrients

100. The Law Dimnishing Return was given by:

(a) Justin von Liebig (b) Spill man

(c) Mitcherlich (d) Schofield

101. Total number of classes in nutrient index is:

(a) Three (b) Four

(c) Five (d) Six

102. What is the percentage of nitrogen in urea?

(a) 48 (b) 46

(c) 25 (d) 42

103. The percentage of K_2O in muriate of potash is (MOP) IS:

(a) 58.0 (b) 48.0

(c) 23.0 (d) 15.0

104. Which form of nitrogen is present in urea?

(a) Nitrate form (b) Ammonical form

(c) Amide form (d) Both (a) and (b)

105. The most deficient nutrient in Indian soil is:

(a) Nitrogen (b) Zinc

(c) Copper (d) Boron

106. Soils of India are generally:

(a) High in fertility (b) Poor in fertility

(c) Moderate in fertility (d) None of the above

107. Erosion created by the activities of man and sometimes by animals is called as:

(a) Natural geological erosion (b) Geologic erosion

(c) Accelerated soil erosion (d) All of the above

108. Scattering of detachated soil particles by the impact of rain drops is called:

(a) Splash erosion (b) Sheet erosion

(c) Gully erosion (d) Ravines

109. Removal by of a very thin layer of soil from the entire surface of large area is called:

(a) Splash erosion (b) Sheet erosion

(c) Rill erosion (d) Ravines

110. Sheet erosion is observed on:

(a) Deeply sloping land (b) Moderately sloping land

(c) Sloppy land (d) Gently sloping land

111. In Universal soil equation, A = RKLSCP, K denotes:

(a) Soil conservation practices
(b) Soil roughness
(c) Soil erodability
(d) Rainfall factor

112. Size range of particles in saltation type of wind erosion is:

(a) Less than 0.05 mm
(b) 0.5-1.0mm
(c) 0.5-0.2mm
(d) 0.05-0.5mm

113. Size range of particles in surface creep type of wind erosion is:

(a) Less than 0.05 mm
(b) 0.5-1.0mm
(c) 0.5-0.2mm
(d) 0.05-0.5mm

114. Size range of particles in suspension type of wind erosion is:

(a) Less than 0.05 mm
(b) 0.5-1.0mm
(c) 0.5-0.2mm
(d) 0.05-0.5mm

115. Wind erosion is calculated by:

(a) Ramer's formula
(b) Chepil equation
(c) Chepil and Woodruff equation
(d) All of the above

116. Diameter of particles that is most susceptibles to wind erosion is:

(a) 0.1mm
(b) 0.2mm
(c) 0.02mm
(d) 0.01mm

117. Dominant type of wind erosion in soil is:

(a) Saltation
(b) Surface creep
(c) Suspension
(d) All of the above

118. Total number of classes in the group, land suitable for cultivation is:

(a) Eight
(b) Four
(c) Six
(d) Two

119. Rills with more than _______ cm depth are generally called as gullies:

(a) 15
(b) 20
(c) 25
(d) 30

120. The most serious sheet erosion occurs in:

(a) Red and black soils
(b) Red and alluvial soils
(c) Black and alluvial soils
(d) Red and laterite soils

Answer Sheet

1.	(a)	25.	(a)	49.	(d)	73.	(c)	97.	(c)
2.	(b)	26.	(d)	50.	(b)	74.	(a)	98.	(d)
3.	(d)	27.	(b)	51.	(a)	75.	(b)	99.	(d)
4.	(b)	28.	(c)	52.	(a)	76.	(b)	100.	(c)
5.	(a)	29.	(b)	53.	(b)	77.	(c)	101.	(a)
6.	(a)	30.	(c)	54.	(d)	78.	(a)	102.	(b)
7.	(b)	31.	(c)	55.	(c)	79.	(c)	103.	(a)
8.	(a)	32.	(c)	56.	(a)	80.	(b)	104.	(c)
9.	(b)	33.	(b)	57.	(b)	81.	(a)	105.	(a)
10.	(c)	34.	(b)	58.	(d)	82.	(c)	106.	(a)
11.	(a)	35.	(a)	59.	(a)	83.	(b)	107.	(c)
12.	(d)	36.	(b)	60.	(d)	84.	(a)	108.	(a)
13.	(b)	37.	(a)	61.	(b)	85.	(b)	109.	(b)
14.	(c)	38.	(b)	62.	(b)	86.	(d)	110.	(d)
15.	(d)	39.	(b)	63.	(c)	87.	(a)	111.	(c)
16.	(c)	40.	(c)	64.	(a)	88.	(c)	112.	(d)
17.	(b)	41.	(c)	65.	(a)	89.	(a)	113.	(c)
18.	(c)	42.	(b)	66.	(b)	90.	(d)	114.	(a)
19.	(a)	43.	(c)	67.	(d)	91.	(d)	115.	(c)
20.	(c)	44.	(a)	68.	(d)	92.	(c)	116.	(a)
21.	(a)	45.	(d)	69.	(b)	93.	(c)	117.	(a)
22.	(c)	46.	(b)	70.	(d)	94.	(a)	118.	(b)
23.	(c)	47.	(c)	71.	(d)	95.	(b)	119.	(d)
24.	(b)	48.	(d)	72.	(b)	96.	(a)	120.	(a)

Chapter 13

Forest Microbiology

1. **The term 'rhizosphere' was coined by:**
   (a) Beijernich (b) Hiltner
   (c) Winogradsky (d) Wilfarth

2. **The total number of 'nif' genes present in Klepsiella pneumonia is:**
   (a) 21 (b) 16
   (c) 28 (d) 35

3. **The bacterial genes responsible for nodule formation in Rhizobium _______ legume symbiosis are termed as:**
   (a) 'Sym' genes (b) 'nif' genes
   (c) 'nod' genes (d) Nodulins

4. **The nitrogen fixing potential of a nodulating plant can be assessed by:**
   (a) Nitrite reductase activity
   (b) Acetylene reduction assay
   (c) Nitrate reductase activity
   (d) Estimation of ammonia formation

5. **An example of phosphate solubilizing microorganisms is:**
   (a) *Pseudomonas striata* (b) *Aspergillus niger*
   (c) *Thiobacillus ferroxidans* (d) *Nitrobacter* sp.

6. **A group of microbes capable of utilizing CO_2 as a carbon source is termed as:**
   (a) Autotrophs (b) Heterotrophs
   (c) Organotrophs (d) Lithotrophs

7. **The kind of microflora dominate in acidic environment is:**
   (a) Bacteria (b) Actinomycetes
   (c) Protozoa (d) Fungi

8. **The endosymbiont responsible for nodulation in actinorhizal plants is an:**
   (a) Actinomycetes (b) Fungi
   (c) Bacteria (d) Algae

9. **The plant genes responsible for nodule formation are termed as:**
   (a) 'nod' genes (b) Nodulins
   (c) 'ntr' genes (d) 'fix' genes

10. **VAM fungi develop characteristic structures called:**
   (a) Arbuscules (b) Vesicles and arbuscules
   (c) Mushrooms (d) Truffles

11. **Parasponis is a non-leguminous plant nodulated by:**
   (a) *Azotobacter* (b) *Rhizobium*
   (c) *Frankia* (d) *Azospirillum*

12. **The optimum temperature for the growth of mesophilic microbes ranges between:**
   (a) 0° and 15°C (b) 30° and 40°C
   (c) 40° and 60°C (d) 60° and 80°C

13. **A group of microbes with symbiotic relationship between fungi and algae is termed as:**
   (a) Actinorhizal plants (b) Lichens
   (c) Mycorrhizae (d) Cyanobacteria

14. **The site of nitrogen fixation in Frankia actinorhizal plants is:**
   (a) Spores (b) Bacteriodes
   (c) Vesicles (d) Conidia

15. **An example of non-nodulating leguminous tree is:**
   (a) *Acacia nilotica* (b) *Albizia lebbeck*
   (c) *Hardwickia binata* (d) *Acacia mellifera*

16. **Ectendomycorrhizas found in:**
    (a) Amaranthaceae (b) Ericaceae
    (c) Monotrophaceae (d) Orchidaceae

17. **The quantum of nitrogen fixed by *Leucaena leucocephala* is:**
    (a) 100 kg/ha/yr (b) 500 kg/ha/yr
    (c) 50 kg/ha/yr (d) 300 kg/ha/yr

18. **The microbes associated with seed surface as termed as:**
    (a) Phyllospdhere (b) Spermosphere
    (c) Phylloplane (d) Rhizosphere

19. **The organism associated with termite gut that is responsible for cellulose degradation is:**
    (a) Bacteria (b) Fungi
    (c) Algae (d) Protozoa

20. **An ectomycorrhizal fungus forming symbiotic relationship with most of the temperate trees is:**
    (a) *Glomus* (b) *Pisolithus*
    (c) *Gigaspora* (d) *Acalospera*

21. **An example of an associate diazotroph is:**
    (a) *Azotobacter* (b) *Azopirillum*
    (c) *Beijierinckia* (d) *Derxia*

22. **The substances produced by microbes which can kill other microbes at low concentrations is known as:**
    (a) Antibiotics (b) Antibody
    (c) Immunogen (d) Immunoglobulins

23. **The covalently closed circular DNA present in bacterium is:**
    (a) Tranposons (b) Insertional sequences
    (c) Plasmids (d) Cosmids

24. **An example of cellulose and lignin degrading organisms is:**
    (a) *Fusarium* sp. (b) *Pleurotus* sp.
    (c) *Amanita* sp. (d) *Aspergillus* sp.

25. **The basic unit of lignin is:**
    (a) Alpha 1-4 glucose (b) Amylopectin
    (c) Pectin (d) Phenyl proponaid

26. An example of plant growth promoting rhizobacteria is:

(a) *Bacillus coagulans* (b) *Pseudomonas fluorescens*

(c) *Rhodospirillum rubram* (d) *Acetobacter diazotrophicus*

27. One of the mechanisms by which plant growth promoting rhizobacteria improve plant growth:

(a) Siderophore (b) Drought resistance

(c) Heavy metal tolerance (d) Salt resistance

28. An example for biological insecticide is:

(a) *Trichoderma viride* (b) *Pseudomonas fluorescens*

(c) *Bacillus thuringiensis* (d) *Nosem bombyci*

29. Rhizobium from leguminous root nodules was first isolated by:

(a) Winogradsky (b) Hellriegel

(c) Beijerinck (d) Wilfarth

30. Mycorrhizae help in mobilizing:

(a) Potassium (b) Phosphorus

(c) Nitrate (d) Boron

31. The enzyme involved in lignin degradation is:

(a) Alpha 1-4 glycosidase enzymes (b) Beta 1-4 glycolytic enzymes

(c) Chitinase (d) Laccase

32. The dominant insoluble form of phosphorus present in acidic soils is:

(a) Tricalcium phosphate (b) Iron phosphate

(c) Magnesium phosphate (d) Dicalcium phosphate

33. An example of sulfur reducing bacterium is:

(a) *Thiobacillus* (b) *Desulfa tomaculum*

(c) *Ferrobacillus* (d) *Rhodo pseudomonas*

34. ATP generating system in clostridium:

(a) EMP pathway (b) TCA cycle

(c) Phosphoroclastic reaction (d) Oxidative electron transport system

35. Total number of ATP molecules required for nitrogen fixation is:

(a) 32 (b) 16

(c) 8 (d) 38

36. The nitrogenase of Azotobacter overcomes oxygen toxicity by:

(a) Gum production
(b) Conformational production
(c) Leghaemoglobin
(d) Vesicle formation

37. The organism involved in iron transformation is:

(a) Ferrobacillus
(b) Thiobacillus
(c) Thiothrix
(d) Chromatium

38. The process of conversion of nitrate into elemental nitrogen is termed as:

(a) Nitrate reduction
(b) Denitrification
(c) Nitrification
(d) Ammonification

39. The acid preferring nitrogen fixing bacterium that dominates forest floor is:

(a) Derxia
(b) Thiobacillus
(c) Thiothrix
(d) Chromaium

40. The rhizobial population of carrier based culture should be:

(a) X x 10^8 cfus/g carrier material
(b) X x 10^{10} cfus/g carrier material
(c) X x 10^5 cfus/g carrier material
(d) X x 10^2 cfus/g carrier material

41. The expiry period of any carrier based bioinculants in:

(a) 4 months
(b) 6 months
(c) 8 months
(d) 1 year

42. The tree species having association with both ecto and endomycorrhizae is:

(a) *Acacia nilotica*
(b) Casuarinas
(c) *Albizia lebbeck*
(d) Teak

43. The total number of tree families having association with Frankia:

(a) 14
(b) 8
(c) 6
(d) 24

44. The rate of organic matter decomposition under tropical to subtropical condition over temperate is:

(a) Slow
(b) Slow and steady
(c) Moderate
(d) Faster

45. One of the common mechanisms by which phosphate solubilizing microbes solubilize insoluble form of phosphorous:

(a) Organic acid production
(b) Enzymatic hydrolysis
(c) H_2S
(d) Soil respiration

46. Fermentation carried out in a limited volume of closed container is termed as:

(a) Continuous (b) Batch

(c) Fed batch (d) Air lift fermentation

47. The host plant that is normally used to multiply VAM culture:

(a) Sorghum (b) Maize

(c) Chilli (d) Cenchrus grass

48. Azotobacter fixes nitrogen in:

(a) Infertile soils (b) Organic matter rich soil

(c) Alkali soils (d) Acidic soils

49. The alkali preferring Azospirillum is:

(a) *A. amazonense* (b) *A. irakens*

(c) *A. halopraeferens* (d) *A. brasilense*

50. The mycrorhizae is _______. microflora:

(a) Ectorhizosphere (b) Endorhizosphere

(c) Rhizosplane (d) Rhizosphere

51. The Azospirillum was first isolated and developed by:

(a) Beijerinck (b) Dobereiner

(c) Omeliansky (d) Frank

52. The native microflora are termed as:

(a) Fermentative (b) Autochthonous

(c) Spermosphere (d) Rhizosphere

53. One of the quality control laboratories of National Biofertilizer Development Corporation is located at:

(a) Chennai (b) Coimbatore

(c) Bangalore (d) Hyderabad

54. VAM spores are recovered from soil by:

(a) Ultra centrification (b) Wet sieving and decantation

(c) Serial dilution and planting technique (d) Most probable number technique

55. The microbial process by which contaminants are removed from soil is termed as:

(a) Phytoremediation (b) Bioremediation

(c) Microbial remediation (d) Decomposition

56. The carrier material normally used for multiplication of VAM is:

(a) Lignite (b) Pest

(c) Vermiculite (d) Charcoal

57. The mycorrhizal infection is normally high in:

(a) Phosphorous poor soil (b) 'P' rich soil

(c) Molybdenum rich soils (d) Calcium rich soil

58. A compound that regulates entry of O_2 into the site of nitrogen fixation in leguminous plants is:

(a) Leghaemoglobin (b) Nodulin

(c) Ferridoxin (d) Flavodoxin

59. The maximum N fixing genera found in the sub-family of Leguminaceae:

(a) Papilionoideae (b) Mimosoideae

(c) Cesalpinoideae (d) Rosaceae

60. The total microbial load of soil can be assessed by:

(a) Soil phosphatase activity (b) Urease activity

(c) Dehydrogenase (d) Protease activity

61. Diatoms are:

(a) Soil algae (b) Soil bacteria

(c) Parent rocks (d) Minerals

62. Blue-green algae grows best in flooded soils exposed to sun and occupied by:

(a) Sun hemp (b) Rice crop

(c) Jute (d) Sorghum

63. Volume of soil under the influence of roots of growing plants is referred to as:

(a) Phyto-biosphere (b) Phyto-biozone

(c) Rhizosphere (d) Phytozone

64. Symbiotic association of fungal hyphae with plant roots is referred to as:

(a) Mycorrhiza (b) Mutual infection

(c) Lichen (d) Co-parasitism

65. VM fungi have reported to increase the uptake of:

(a) Phosphorus (b) Nitrogen

(c) Potash (d) None of the above

66. Which of the followings acts as storage organ for VAM fungi?

(a) Arbuscule (b) Mycelium

(c) Hypha (d) Vesicle

67. Actinomycetes provide nutrients to the plant by:

(a) Decomposing soil organic matter

(b) Breaking down complex clays

(c) Fixing from atmosphere secreting as metabolites

(d) None of the above

68. Soil organisms depending on dead tissue of both plants and animals as food source is called as:

(a) Herbivores (b) Carnivores

(c) Detritovores (d) None of the above

69. For their carbon and energy needs, heterotrophs reply on:

(a) Organic compounds (b) Inorganic compounds

(c) Photosynthesis (d) None of the above

70. In soil, action of the microflora is mostly:

(a) Physical (b) Both (a) and (b)

(c) Chemical (d) None of the above

71. Which of the followings is used as food source by microphytic feeders?

(a) Microfauna (b) Macrofauna

(c) Microflora (d) Macroflora

72. Which of the following groups of soil organisms are termed as 'ultimate decomposers'?

(a) Microflora (b) Macroflora

(c) Microfauna (d) Mesofauna

73. As energy source, photoautotrophs depend on:

(a) Organic compounds (b) Inorganic compounds

(c) Solar energy (d) None of the above

74. Chemoautotrophs dose not derive energy from:

(a) CO_2 (b) N_2

(c) S (d) Fe

75. The size range of soil inhabiting protozoa is:

(a) 0.1-1.5............m (b) 1-2............m

(c) 5-100............m (d) 1000-5000............m

76. The rhizosphere zone extends from root surface within:

(a) 1-2mm (b) 5-10mm

(c) 15-20mm (d) 50-100mm

77. Which of the following groups of soil algae are prokaryotes?

(a) Green algae (b) Blue-Green algae

(c) Yellow green algae (d) Diatoms

78. Fungi are:

(a) Photoautotrophs (b) Chemoautotrophs

(c) Heterotrophs (d) None of the above

79. Which of the following groups represent single celled fungi?

(a) Molds (b) Mushrooms

(c) Yeasts (d) None of the above

80. Individual fungal filaments are called:

(a) Hyphae (b) Mycelium

(c) Pseudopodia (d) Flagella

81. *Penicillum, Mucor, Fusarium and aspergillus* are genera of:

(a) Yeasts (b) Molds

(c) Mushroom fungi (d) None of the above

82. Mycrorhizae is a symbiotic association between:

(a) Bacteria and plant roots (b) Fungi and bacteria

(c) Fungi and roots of higher plants (d) Two species of fungi

83. Mycotoxin is produced by:

(a) Bacteria (b) Protozoa

(c) Fungi (d) Nematodes

84. Aflatoxin is a mycotoxin produced by:

(a) *Bacillus polymyxa* (b) *Aspergillus awamorii*

(c) *Pseudomonas striata* (d) *Aspergillus flavus*

85. Which of the following statement is not true regarding arbuscules of VAM fungi?

(a) Transfer mineral nutrients from the fungi to host plant

(b) Transfer sugar from host plant to the fungi

(c) Work as storage organ for the fungi

(d) A highly branched structure

86. Which of the followings acts as storage organ for VAM fungi?

(a) Hypha (b) Arbuscule

(c) Mycelium (d) Vesicle

87. Which of the following groups of plants does not form mycorrhizae?

(a) Solanaceae (b) Poaceae

(c) Leguminosae (d) Cruciferae

88. Which of the following products are not originated from actinomycetes?

(a) Actinomycin (b) Pencillum

(c) Neomycin (d) Streptomycin

89. Which of the following get oxidized during the first step of nitrification?

(a) Nitrate (b) Nitrite

(c) Ammonium (d) None of the above

90. The nitrogen fixing organism living inside the leaves of the aquatic fern azolla is a:

(a) Fungi (b) Cyanobacteria

(c) Actinomycetes (d) None of the above

91. Siderophores show strongest affinity to bind:

(a) Iron (b) Nitrogen

(c) Aluminum (d) Phosphorus

92. Who for the first time showed that legumes can obtain nitrogen from air?

(a) M.W. Beijerinck (b) J.B. Bousssingault

(c) Robbert Koch (d) Louis Pasteur

93. Bacteria from root nodules of legume was isolated for the first time by:

(a) J.B. Bousssingault (b) M.W. Beijerinck

(c) Robbert Koch (d) Louis Pasteur

94. Who for the first time obtained pure culture of bacteria by serial dilution in liquid media?

(a) Louis Pasteur (b) John Needham

(c) Joseph Lister (d) M.W. Beijerinck

95. Cellulose degradation by anaerobic bacteria was first discovered by:

(a) S.N. Wingradsky (b) Omeliansky

(c) J.B. Bousssingault (d) Joseph Lister

96. Study of rhizophere was initiated by:

(a) Hiltner (b) Omeliansky

(c) Lipman and Brown (d) S.N. Winogradsky

97. Penicillum was discovered by:

(a) Robert Koch (b) Alexander Fleming

(c) Waksman (d) Winogradsky

98. Streptomycin was discovered by:

(a) Robert Koch (b) Winogradsky

(c) Alexander Fleming (d) Waksman

99. Glomus and Acaulospora are examples of:

(a) Ectomycorrhiza (b) Actinomycetes

(c) Endomycorrhiza (d) None of the above

100. Rhizobium from root nodules of a non legume was first isolated by:

(a) Quipsel (b) Nutman

(c) Bond (d) Trinick

101. Azobacterin and Phosphobacterin were first used in:

(a) USA (b) U.K.

(c) Australia (d) Russia

102. Bacteria thriving well within a temperature range of 15-45°C are called:

(a) Thermophillic (b) Psychrophillic

(c) Mesophillic (d) None of the above

103. The bacteria thiobacillus converts:

(a) Inorganic sulphur to sulphate (b) Ammonia to nitrite

(c) Ferrous iron to ferric iron (d) None of the above

104. The most common genera of actinomycetes in soil is:

(a) Nocardia (b) Micromonospora

(c) Actinomyces (d) Streptomyces

105. The smallest inhabitants of soil is:

(a) Protozoa (b) Bacteria

(c) Actinomycetes (d) Bacteriophages

106. Which of the following does not represent anaerobic nitrogen fixing bacteria?

(a) Achromobacter (b) Clostridium

(c) Chlorobium (d) Chromatium

107. Acetylene Reduction Essay (AR(a) is done to determine:

(a) Nitrogenase activity
(b) Dehydrogenase activity
(c) Phosphatase activity
(d) None of the above

108. The 15N technique for quantitative estimation of nitrogenase activity was developed by:

(a) Hiltner
(b) Beijerinck
(c) Burris and Wilson
(d) None of the above

109. In cellulose miolecule, individual glucose units are bound together by:

(a) ....-1-4
(b) ....-1-4
(c) 1-6
(d) 1-6

110. Hemicelluloses represent various polymers:

(a) Hexoses
(b) Pentoses
(c) Uronic acid
(d) All of the above

111. Pecting is an example of:

(a) Cellulose
(b) Lignin
(c) Hemicelluloses
(d) None of the above

112. Which of the followings represents genera of lignin degrading fungi?

(a) Mycena and Marasmius
(b) Polyporus and pleurotus
(c) Citocybe and Collybia
(d) None of the above

113. Fulvic acid is:

(a) Acid soluble
(b) Alkali soluble
(c) Acid insoluble
(d) Acid and alkali soluble

114. Humin is:

(a) Resistant too cold alkali
(b) Soluble in cold alkali
(c) Soluble in hot water
(d) Soluble in cold water

115. Which of the followings is a well known synthetic chelator?

(a) EBT
(b) EDTA
(c) TTC
(d) TPF

116. Iron is an important component of enzymes:

(a) Nitrogenase
(b) Nitrate reductase
(c) Nitrite reductase
(d) All of the above

117. Which of the followings is microaerobic that fixes nitrogen freely or in association with grasses?

(a) *Azotobacter*
(b) *Azospirillum*
(c) *Rhizobium*
(d) *Frankia*

118. Which of the following bacteria genera take part in nitrification?

(a) *Nitrosomanas* and *Nitrobacter*
(b) *Nitrosococcus* and *Nitrosopira*
(c) *Nitrosocystis* and *Nitrosogloea*
(d) All of the above

119. Which of the followings plays a role in ripening of fruits?

(a) Auxin
(b) Cytokinin
(c) Ethylene
(d) None of the above

120. The insecticide Aldrin is converted to Dieldrin by:

(a) *Trichoderma*
(b) *Fusarium*
(c) *Penicillium*
(d) All of the above

121. The first organic fungicide is:

(a) Zineb
(b) Thiram
(c) Captan
(d) None of the above

122. Hartig net is developed by:

(a) Endomycorrhizae
(b) Ectomycorrhizae
(c) Nodulating bacteria
(d) None of the above

123. The under fined constituents of root exudates required for growth/culture of ectomycorrhizae are called:

(a) P-factor
(b) G-factor
(c) M-Factor
(d) None of the above

124. Which of the followings represents the liason tissue between the fungal sheath and the host cells?

(a) Cortex layer
(b) Casparian strip
(c) Hartig net
(d) None of the above

125. Which of the following structures of VAMF connect the fungal ramifications inside roots with the mycelium of the fungus outside roots?

(a) Vesicles
(b) Appressoria
(c) Arbuscules
(d) None of the above

126. The VAM fungi with spores bearing straight or angular stalks are grouped as:

(a) Glomus
(b) Gigaspora
(c) Acaulospora
(d) Endogone

127. The VAM fungi with spores bearing bulbous stalks are grouped as:

(a) Glomus
(b) Sclerocystis
(c) Gigaspora
(d) None of the above

128. The VAM fungi with sessile spores are grouped as:

(a) Sclerocystis (b) Acaulospora

(c) Gigaspora (d) Glomus

129. The VAM fungi with spores regularly arranged on a central core are grouped as:

(a) Sclerocystis (b) Acaulospora

(c) Gigaspora (d) Glomus

130. Decomposition of larger polysaccharides are carried out by:

(a) Intracellular enzymes (b) Extracellular enzymes

(c) Both intracellular and extracellular enzymes (d) None of the above

131. Biological oxidation actually does not proceed by:

(a) Addition of oxygen (b) Removal of hydrogen

(c) Removal of electron (d) All of the above

132. The group of enzymes always produced in microbial cell irrespective of presence or absence of a substrate is called:

(a) Inducible enzyme (b) Intracellular enzyme

(c) Extracellular enzyme (d) Constitutive enzyme

133. The term used to denote conversion of organic complexes of an element to inorganic state is:

(a) Mineralization (b) Immobilization

(c) Assimilation (d) All of the above

134. The phenomenon of enhancement in native humus decomposition owing to external addition of organic matter is termed.

(a) Mineralization (b) Assimilation

(c) Immobilization (d) Priming

135. In woody perennials, the bulk of the organic matter lost is derived from:

(a) Hemicelluloses (b) Cellulose

(c) Lignin (d) All of the above

136. As the clay content of a soil increases, the decay rate:

(a) Increases (b) Decreases

(c) Remains unchanged (d) Shows variable response

137. The most abundant organic compound in nature is:

(a) Hemicelluloses (b) Cellulose

(c) Lignin (d) None of the above

138. Beyond which of the following ratios between inorganic N and cellulose, the cellulose decomposition does not respond to further addition of inorganic N:

(a) 1:10 (b) 1:20

(c) 1:30 (d) 1:35

139. Which of the following chemicals are formed during anaerobic decomposition of cellulose:

(a) Ethanol (b) Acetic and Formic acid

(c) Lactic and Butyric acid (d) All of the above

140. Aerobic bacteria generally convert cellulose into:

(a) CO_2 and cell substance (b) H_2 and ethanol

(c) Acetic and Formic acid (d) Succinic and Butyric acid

141. Hemicelluloses are polymers of:

(a) Simple sugars (b) Uronic acids

(c) Both simple sugars and uronic acids (d) None of the above

142. In the breakdown of xylan, the action of endo enzyme yields:

(a) Cellobiose (b) Xylobiose

(c) Xylose (d) None of the above

143. In the breakdown of xylan, the action of exo enzyme yields:

(a) Cellobiose (b) Xylobiose

(c) Xylose (d) Glucose

144. The components of plant starches are:

(a) Amylase (b) Amylopectin

(c) Both (a) and (b) (d) None of the above

145. The anaerobic fermentation of starches yields:

(a) Lactic acid (b) Acetic acid

(c) Butyric acid (d) All of the above

146. The different enzymes catalyzing starch breakdown are:

(a) _______ amylase (b) _______ amylase

(c) Glucoamylase (d) All of the above

147. The main building block of pectin is:

(a) Galactan (b) Unronic acid

(c) Galactouronic acid (d) None of the above

148. Pectic substances are of types:

(a) Protopectin and pectin
(b) Pectinic acid and pectic acid
(c) Both (a) and (b)
(d) None of the above

149. Which of the following groups of bacterial genera represent pectin degrades?

(a) *Clostridium* and *Flavobacterium*
(b) *Arthrobacter* and *Bacillus*
(c) *Micrococcus* and *Pseudomonas*
(d) All of the above

150. The building block of chitin is:

(a) Glucose
(b) N-aceetylglucosamine
(c) Fructose
(d) None of the above

151. The enzymes responsible for breakdown of chitin are:

(a) Chitinase
(b) Chitobiase
(c) Both (a) and (b)
(d) None of the above

152. The anaerobic breakdown of protein rich materials is termed.

(a) Putrefaction
(b) Denitrification
(c) Deploymerization
(d) None of the above

153. Which of the following groups represents final products of anaerobic protein breakdown?

(a) Ammonium, amines and CO_2
(b) Organic acids, mercaptans and hydrogen sulphide
(c) Both (a) and (b)
(d) None of the above

154. Which of the following fungal genera are protein decomposers?

(a) Alternaria and Aspergillus
(b) Mucor and Penicillium
(c) Rhizopus
(d) All of the above

155. Which of the following soil moisture conditions favour nitrogen mineralization the most?

(a) Continuous wetting
(b) Alternate drying and wetting
(c) Continuous drying
(d) None of the above

156. The rise in pH near the urea particles applied to soil is caused by production of:

(a) Ammonia
(b) Nitrate
(c) Nitrite
(d) None of the above

157. Which of the following bacterial genera produce enzyme urease?

(a) *Bacillus* and *Micrococcus*
(b) *Pseudomonas* and Klebsiella
(c) *Cyanobacterium* and *Clostridium*
(d) All of the above

158. 'Nitrogen factor' defines:

(a) Number of units of inorganic nitrogen immobilized for each hundred units
(b) The amount of nitrogen need to be added to prevent a net immobilization from the environment
(c) Both (a) and (b)
(d) None of the above

159. Which of the followings are the most dominant microorganisms in nitrification?

(a) *Nitrosomonas*
(b) *Nitrobacter*
(c) Both (a) and (b)
(d) *Nitrosospira*

160. In methemoglobinemia, the haemoglobin is converted to:

(a) Methemoglobin
(b) Leghemoglobin
(c) Phytin
(d) None of the above

161. Denitrification may also be termed as:

(a) Nitrate assimilation
(b) Nitrate respiration
(c) Denitrification
(d) All of the above

162. Which of the followings represents dominant anaerobic N_2 fixer?

(a) *Azotobacter*
(b) *Clostridium*
(c) *Beizerinckia*
(d) None of the above

163. Which of the followings does not represent non leguminous N_2 – fixing tree?

(a) *Alnus*
(b) *Casuarinas*
(c) *Gmelina*
(d) *Myrica*

164. The rhizosphere effect can be quantitatively represented by:

(a) C:N ration
(b) R:S ratio
(c) C:P ratio
(d) None of the above

165. The principle of gram staining is based on differences in:

(a) Composition of bacterial cell wall
(b) Size of bacteria
(c) Shape of bacteria
(d) Colour of bacteria

166. YEMA medium was introduced for rhizobium by:

(a) Hiltner
(b) Nobbe
(c) Fred
(d) Joseph Lister

167. Which of the following genus under rhizobium contributes to stem nodulation:

(a) *Azorhizobium*
(b) *Bradyrhizobium*
(c) *Rhizobium*
(d) None of the above

168. Which of the following pigments is present in BGA?

(a) Carotene
(b) Xanthophylls
(c) Phycocyanine
(d) All of the above

169. The minimum shelf of Rhizobium biofertilizer should be:

(a) 4 months
(b) 6 months
(c) 12 months
(d) None of the above

170. The term autocolony stands for:

(a) Any microbial colony
(b) Microbial colony formed in the cells of parent colony
(c) Both (a) and (b)
(d) None of the above

171. 'Nif' gene is responsible for:

(a) Synthesis and regulation of nitrogenase
(b) Nitrification
(c) Dentrification
(d) None of the above

172. 'Biofertiliser Newsletter' is published by:

(a) IARI
(b) NifTAL
(c) National Biofertilizer Development Centre
(d) None of the above

173. Green plants are also called as:

(a) Primary producers
(b) Secondary producers
(c) Consumers
(d) Decomposers

174. Of the following, which is soil micro fauna?

(a) Protozoa
(b) Nematodes
(c) Earthworms
(d) Diatoms

175. Penicillium is an example of:

(a) Cellulose decomposer
(b) Protein decomposer
(c) Lignin decomposer
(d) Humus former

176. Cellulose is mainly decomposed by:

(a) Penicillium (b) Aspergillus

(c) Mucar (d) All of the above

177. Of the following, which is/are soil macro fauna?

(a) Earthworms (b) Moles

(c) Ants (d) All of the above

178. Arrange the following microorganisms based on their population in soil.

(a) Fungi>Bacteria>Actinomycetes

(b) Bacteria> Fungi> Actinomycetes

(c) Bacteria>Actinomycetes> Fungi

(d) Fungi> Actinomycetes>Bacteria

179. Who is the 'Father of field experiment'?

(a) M.W. Beijerink (b) J.B. Bousingault

(c) C. Theoder de Saucer (d) Arnon and Stout

180. The bacterial population per gram of rhizosphere of soil is:

(a) 6778×10^6 (b) 918×10^3

(c) 83×10^6 (d) 68×10^6

181. What is the temperature range of mesophiles?

(a) <10°C (b) 10-20°C

(c) 20-40°C (d) >40°C

182. All the autotrophic organisms get carbon source from:

(a) Organic matter (b) Carbon dioxide

(c) Both (a) and (b) (d) Minerals

183. All the heterotrophic organisms get carbon source from:

(a) Organic matter (b) Carbon dioxide

(c) Both (a) and (b) (d) Minerals

184. Phototrophic organisms derive their energy from:

(a) Carbon dioxide (b) Organic matter

(c) Sun light (d) Organic matter

185. What is the optimum temperature for the mesophyllic organisms?

(a) 30°C (b) 37°C

(c) 27°C (d) 17°C

186. The ideal pH range for the growth of fungi in soil is:

(a) 6.5 and 8.0 (b) 5.5 and 7.0

(c) 6.0 and 7.0 (d) 4.5 and 6.5

187. The ideal pH range for the growth of bacteria in soil is:

(a) 6.5 and 8.0 (b) 5.5 and 7.0

(c) 6.0 and 7.0 (d) 4.5 and 6.5

188. Soil protozoa are the predators of:

(a) Bacteria (b) Actinomycetes

(c) Fungi (d) All of the above

189. Lignin is mainly decomposed by:

(a) Bacteria (b) Actinomycetes

(c) Fungi (d) All of the above

190. Humus is primarily formed by:

(a) Bacteria (b) Actinomycetes

(c) Fungi (d) Protozoa

191. VAM fungi are:

(a) Obligate symbiont (b) Facultative symbiont

(c) Free-Living (d) Both (a) and (b)

192. Earthworms casts are rich in nutrients and:

(a) Bacteria (b) Fungi

(c) Viruses (d) All of the above

193. Earthworms are more abundant in:

(a) Coarse textured soils (b) Fine textured soils

(c) Medium textured soils (d) Both (a) and (b)

194. Which is an example of green sulphur bacteria?

(a) *Chlorobium* (b) *Rhodomicrobium*

(c) *Rhodosprillum* (d) All of the above

195. Propagules of VAM fungi consist of:

(a) Clamydospores (b) Azygospores

(c) Soil-borne vesicle (d) All of the above

196. What is the optimum C/N ratio for organic matter?

(a) <20:1 (b) 20:30-1

(c) >30:1 (d) None of the above

197. Phytin is a type of organic:

(a) Nitrogen
(b) Phosphorus
(c) Sulphur
(d) Molybdenum

198. Solidifying agent in a medium is called as:

(a) Agar-agar
(b) Starch
(c) Fats
(d) Sugars

199. Biofertilizer inoculation technique was started during:

(a) 1880
(b) 1885
(c) 1887
(d) 1895

200. 'Herting net' is produced by:

(a) N-fixing rhizobium
(b)' *Ecototrophic mycorrhiza*
(c) Both (a) and (b)
(d) None of the above

201. Extra chromosomal genetic materials is:

(a) Nucleoid
(b) Nucleus
(c) Nucleolus
(d) Plasmid

202. Structure involved in the transfer of genetic material during conjugation of bacteria:

(a) Cilia
(b) Flagella
(c) Pilus
(d) Cytoplasmic membrane

203. Infecting a bacterium:

(a) Bacteriophage
(b) Viroid
(c) Viroid
(d) Mycoplasma

204. Small, circular, self-replicating DNA found in *E.coli*:

(a) Chromosomes
(b) Plastids
(c) Plasmids
(d) Ribosomes

205. A segment of DNA that is transcribed but removed from the DNA sequences:

(a) Introns
(b) Exons
(c) Split genes
(d) Jumping genes

206. Total genetic material in a cell constitutes:

(a) Genome
(b) Gene Bank
(c) Gene
(d) Genome DNA

207. Which of the following is not a prokaryote?

(a) Bacterium
(b) Amoeba
(c) Blue Green Algae
(d) Archaeobacteria

208. Thermophiles grow in conditions of high:

(a) Temperature
(b) Salt
(c) CO_2
(d) Food

209. Halophiles grow in conditions of high:

(a) Salt
(b) Temperature
(c) Acidity
(d) Oxygen

210. Photoautotrophic metabolism involves:

(a) 1,2
(b) 1
(c) 2,3,4
(d) 4,5

211. Bacteria multiplying in the dead sea are mostly likely to be:

(a) Aerobic
(b) Anaerobic
(c) Thermophiles
(d) Holophiles

212. An organism that can grow either in the presence or in the absence of mol. O_2 is called a (n):

(a) Obligate aerobe
(b) Facultative anaerobe
(c) Obligate anaerobe
(d) Acidophiles

213. Bacteria which grow at a temperature around 10° are called as:

(a) Phycophites
(b) Thermophiles
(c) Mesophiles
(d) Acidophiles

214. How many generation did the population go through, when a bacterial culture began with 2 cells and ended with 128 cells?

(a) 12
(b) 10
(c) 6
(d) 5

215. Bacterial broth culture stops growing and enters a stationary phase, because it:

(a) Depletes an essential nutrient
(b) Accumulates essential nutrients
(c) Becomes too crowded
(d) Produces plenty of oxygen

216. Which of the following processes does not result in recombination?

(a) Conjugation
(b) Transformation
(c) Transduction
(d) Translation

217. Which of the following can now be produced in microorganisms?

(a) 1.2.4.5
(b) 1.2.3.4.5.
(c) 1.2.4
(d) 1.2.5

218. Five Kingdom scheme of classification was proposed by:

(a) E.H. Haeckel
(b) Carolus Linnaeus
(c) Dobzhansky
(d) R.H. Whittaker

219. Yeast is a:

(a) Bacteria
(b) Virus
(c) Fungi
(d) Algae

220. Scientist who for the first time demonstrated the principle of immunization:

(a) Louis Pasteur
(b) Edward Jenner
(c) John Tyndall
(d) Robert Koch

221. Name of the scientist who observed micro-organisms for the first time:

(a) Griffith
(b) A.V. Leeuwenhoek
(c) Lazaro Spallanzani
(d) Robert Hooke

222. Germ theory of disease was first proposed by:

(a) Jacob Manle
(b) Father Sudrez
(c) Louis Pasteur
(d) Joseph Lister

223. Scientist who developed aseptic technique and isolated bacteria in pure culture?

(a) Joseph Lister
(b) Metchnikoff
(c) Robert Koch
(d) Spencer

224. N_2 fixing bacteria in soil were discovered by:

(a) S.N. Winogradsky
(b) John Needham
(c) Waddington
(d) Walter Reed

225. In which of the following micro-organisms surface area/volume ratio is larges:

(a) Fungi
(b) Bacteria
(c) Yeast
(d) Algae

226. Teichoi acids are present in:

(a) Cell wall of G + bacteria
(b) Cell wall of G + bacteria
(c) Nucleoid of G + bacteria
(d) Plasma membrane of G + bacteria

227. The production of CAMP from ATP is catalysed by which of the following membrane bound enzyme:

(a) Adenylate cyclase
(b) Acetyl coenzyme – A
(c) ATP ase
(d) Catalases

228. L.T. plasmid in E. coli produces:

(a) Enterotoxin
(b) Bacteriocin
(c) Tetanalysin
(d) Tetanospasmin

229. Distinction between prokaryotic and eukaryotic cells was made on the basis of:

(a) Structure of cell wall
(b) Presence or absence of golgi bodies and Lysosomes
(c) Absence or presence of nuclear membrane
(d) Absence or presence of ribosomes

230. Organelles such a mitochondria and chloroplasts in eukaryotic cells evolved from prokaryotic organisms by the engulfment by primitive nucleated cells is known as:

(a) Cell theory
(b) Protoplast theory
(c) Endosymbiotic theory
(d) Theory of abiogenesis

231. The correct way to represent the scientific name for a bacterium is:

(a) *Escherichia coli*
(b) *Escherichia coli*
(c) *Escherichia coli*
(d) *Escherichia coli*

232. Structure essential for living cells:

(a) Ribosomes
(b) Capsule
(c) Cell wall
(d) Cytoplasmic membrane

233. The process of production of endospores from a vegetative cell is called as:

(a) Germination
(b) Sporogenesis
(c) Encystment
(d) Conidia formation

234. Spherical bacteria arranged in a grape bunch manner:

(a) Spirillum
(b) Cocci
(c) Bacilli
(d) Staphylococci

235. Resistance to antibiotics is carried on:

(a) R factor plasmid
(b) Flagellum
(c) Slime layer
(d) Sex pilli

236. Which of the following has chain on conidia growing on aerial hypahe?

(a) Actinomyces (b) Nocardia

(c) Streptomyces (d) Nitrobacter

237. In some bacteria, capsule surrounding the cell wall is made up of:

(a) Cellulose (b) Polysaccharides

(c) Polypeptide (d) Teichoic acids

238. Growth in bacteria reflects:

(a) Growth in number (b) Growth in weight

(c) Growth in volume (d) Changes in shape and size

239. Which of the following bacteria oxidases ferrous compounds into ferric ones:

(a) *Leptothrix orcharacae* (b) *Pseuformonas*

(c) *Rhizobium* (d) *Lactobacillus*

240. Which of the following bacteria oxidizes molecular hydrogen into water in soil:

(a) *Penicillium notatum* (b) *Rhizobium noduliformes*

(c) *Bacillus pentotrophus* (d) None of the above

241. Extra chromosomal genetic materials is:

(a) Nucleoid (b) Nucleus

(c) Nucleolus (d) Plasmid

242. Structure involved in the transfer of genetic material during conjugation of bacteria:

(a) Cilia (b) Flagella

(c) Pilus (d) Cytoplasmic membrane

243. Infecting a bacterium:

(a) Bacteriophage (b) Viroid

(c) Viroid (d) Mycoplasma

244. Small, circular, self-replicating DNA found in *E. coli*:

(a) Chromosomes (b) Plastids

(c) Plasmids (d) Ribosomes

245. A segment of DNA that is transcribed, but removed from the DNA sequences:

(a) Introns (b) Exons

(c) Split genes (d) Jumping genes

246. Total genetic material in a cell constitutes:

(a) Genome (b) Gene Bank

(c) Gene (d) Ganome DNA

247. Which of the following is not a prokaryote?

(a) Bacterium (b) Amoeba

(c) Blue Green Algae (d) Archaeobacteria

248. Thermophiles grow in conditions of high:

(a) Temperature (b) Salt

(c) CO_2 (d) Food

249. Halophiles grow in conditions of high:

(a) Salt (b) Temperature

(c) Acidity (d) Oxygen

250. Photoautotrophic metabolism involves:

(1) Use of inorganic CO_2 as the only source of carbon

(2) Sun light as the source of energy

(3) Energy derived from the metabolism or organic components

(4) Organic carbon compounds as a source of energy

(5) Energy obtained from the oxidation of inorganic molecules

(a) 1,2 (b) 1

(c) 2,3,4 (d) 4,5

251. Bacteria multiplying in the dead sea are mostly likely to be:

(a) Aerobic (b) Anaerobic

(c) Thermophiles (d) Holophiles

252. An organism that can grow either in the presence or in the absence of mol. O_2 is called a (n):

(a) Obligate aerobe (b) Facultative anaerobe

(c) Obligate anaerobe (d) Acidophiles

253. Bacteria which grow at a temperature around 10° are called as:

(a) Phychrophites (b) Thermophiles

(c) Mesophiles (d) Acidophiles

254. How many generation did the population go through, when a bacterial culture began with 2 cells and ended with 128 cells?

(a) 12 (b) 10

(c) 6 (d) 5

255. Bacterial broth culture stops growing and enters a stationary phase, because it:

(a) Depletes an essential nutrient
(b) Accumulates essential nutrients
(c) Becomes too crowded
(d) Produces plenty of oxygen

256. Which of the following processes does not result in recombination?

(a) Conjugation
(b) Transformation
(c) Transduction
(d) Translation

257. Which of the following can now be produced in microorganisms?

1. Insulin
2. Somatostatin
3. Vaccine for malaria
4.. Blood clotting factor

(a) 1,2,4,5
(b) 1,2,3,4,5
(c) 1,2,4
(d) 1,2,5

258. Which of the following can now be produced in microorganisms?

(a) E.H. Haeckel
(b) Carolus Linnaeus
(c) Dobzhansky
(d) R.H. Whittaker

259. Yeast is a:

(a) Bacteria
(b) Virus
(c) Fungi
(d) Algae

260. Scientist who for the first time demonstrated the principle of immunization:

(a) Louis Pasteur
(b) Edward Jenner
(c) John Tyndall
(d) Robert Koch

261. Name of the scientist who observed micro-organisms for the first time:

(a) Griffith
(b) A.V. Leeuwenhoek
(c) Lazaro Spallanzani
(d) Robert Hooke

262. Germ theory of disease was first proposed by:

(a) Jacob Manle
(b) Father Sudrez
(c) Louis Pasteur
(d) Joseph Lister

263. Scientist who developed aseptic technique and isolated bacteria in pure culture:

(a) Joseph lister
(b) Metchnikoff
(c) Robert Koch
(d) Spencer

264. N_2 fixing bacteria in soil were discovered by:

(a) S.N. Winogradsky
(b) John Needham
(c) Waddington
(d) Walter Reed

265. In which of the following micro-organisms surface area/volume ratio is larges:

(a) Fungi
(b) Bacteria
(c) Yeast
(d) Algae

266. Teichoi acids are present in:

(a) Cell wall of G+bacteria
(b) Cell wall of G + bacteria
(c) Nucleoid of G+bacteria
(d) Plasma membrane of G+ bacteria

267. The production of CAMP from ATP is catalysed by which of the following membrane bound enzyme:

(a) Adenylate Cyclase
(b) Acetyl Coenzyme – A
(c) ATP ase
(d) Catalases

268. LT Plasmid in E. coli produces:

(a) Enterotoxin
(b) Bacteriocin
(c) Tetanalysin
(d) Tetanospasmin

269. Distinction between prokaryotic and eukaryotic cells was made on the basis of:

(a) Structure of cell wall
(b) Presence or absence of Golgi bodies and Iysosomes
(c) Absence or presence of nuclear membrane
(d) Absence or presence of ribosomes

270. Organelles such as mitochondria and chloroplasts in eukaryotic cells evolved from prokaryotic organisms by the engulfment by primitive nucleated cells is known as:

(a) Cell theory
(b) Protoplast theory
(c) Endosymbiotic theory
(d) Theory of abiogenesis

271. The correct way to represent the scientific name for a bacterium is:

(a) *Escherichia coli*
(b) *Escherichia coli*
(c) *Escherichia coli*
(d) *Escherichia coli*

272. Structure essential for living cells:

(a) Ribosomes
(b) Capsule
(c) Cell wall
(d) Cytoplasmic membrane

273. The process of production of endospores from a vegetative cell is called as:

(a) Germination
(b) Sporogenesis
(c) Encystment
(d) Conidia formation

274. Spherical bacteria arranged in a grape bunch manner:

(a) *Spirillum* (b) *Cocci*

(c) *Bacilli* (d) *Staphylococci*

275. Resistance to antibiotics is carried on:

(a) R factor plasmid (b) Flagellum

(c) Slime layer (d) Sex pilli

276. Which of the following has chain on conidia growing on aerial hypahe?

(a) Actinomyces (b) Nocardia

(c) Streptomyces (d) Nitrobacter

277. In some bacteria, capsule surrounding the cell wall is made up of:

(a) Cellulose (b) Polysaccharides

(c) Polypeptide (d) Teichoic acids

278. Growth in bacteria reflects:

(a) Growth in number (b) Growth in weight

(c) Growth in volume (d) Changes in shape and size

279. Which of the following bacteria oxidizes ferrous compounds into ferric ones:

(a) *Leptothrix ochracae* (b) *Pseuformonas*

(c) *Rhizobium* (d) *Lactobacillus*

280. Which of the following bacteria oxidizes molecular hydrogen into water in soil:

(a) *Penicillium notatum* (b) *Rhizobium noduliformes*

(c) *Bacillus pentotrophus* (d) None of the above

Answer Sheet

1.	(b)	31.	(d)	61.	(a)	91.	(a)	121.	(b)
2.	(b)	32.	(b)	62.	(b)	92.	(b)	122.	(b)
3.	(a)	33.	(b)	63.	(c)	93.	(b)	123.	(c)
4.	(b)	34.	(c)	64.	(a)	94.	(c)	124.	(c)
5.	(a)	35.	(b)	65.	(a)	95.	(b)	125.	(b)
6.	(b)	36.	(b)	66.	(d)	96.	(a)	126.	(a)
7.	(d)	37.	(a)	67.	(c)	97.	(b)	127.	(c)
8.	(a)	38.	(b)	68.	(c)	98.	(d)	128.	(b)
9.	(b)	39.	(d)	69.	(a)	99.	(c)	129.	(a)
10.	(b)	40.	(a)	70.	(c)	100.	(d)	130.	(b)
11.	(b)	41.	(b)	71.	(c)	101.	(d)	131.	(a)
12.	(b)	42.	(b)	72.	(a)	102.	(c)	132.	(d)
13.	(b)	43.	(b)	73.	(c)	103.	(a)	133.	(a)
14.	(c)	44.	(d)	74.	(a)	104.	(d)	134.	(d)
15.	(c)	45.	(a)	75.	(c)	105.	(d)	135.	(b)
16.	(c)	46.	(b)	76.	(a)	106.	(a)	136.	(b)
17.	(a)	47.	(b)	77.	(b)	107.	(a)	137.	(b)
18.	(b)	48.	(b)	78.	(d)	108.	(c)	138.	(d)
19.	(d)	49.	(c)	79.	(c)	109.	(b)	139.	(d)
20.	(b)	50.	(b)	80.	(a)	110.	(d)	140.	(a)
21.	(b)	51.	(b)	81.	(b)	111.	(c)	141.	(c)
22.	(a)	52.	(b)	82.	(c)	112.	(d)	142.	(b)
23.	(c)	53.	(c)	83.	(c)	113.	(d)	143.	(c)
24.	(b)	54.	(b)	84.	(d)	114.	(a)	144.	(c)
25.	(d)	55.	(c)	85.	(c)	115.	(b)	145.	(d)
26.	(b)	56.	(c)	86.	(d)	116.	(d)	146.	(d)
27.	(a)	57.	(b)	87.	(d)	117.	(b)	147.	(c)
28.	(c)	58.	(a)	88.	(b)	118.	(d)	148.	(c)
29.	(c)	59.	(a)	89.	(c)	119.	(c)	149.	(d)
30.	(b)	60.	(c)	90.	(b)	120.	(d)	150.	(b)

151.	(c)	177.	(d)	203.	(a)	229.	(c)	255.	(d)
152.	(a)	178.	(c)	204.	(c)	230.	(c)	256.	(d)
153.	(c)	179.	(b)	205.	(a)	231.	(c)	257.	(d)
154.	(d)	180.	(a)	206.	(a)	232.	(d)	258.	(d)
155.	(b)	181.	(c)	207.	(b)	233.	(b)	259.	(c)
156.	(a)	182.	(b)	208.	(a)	234.	(d)	260.	(a)
157.	(d)	183.	(a)	209.	(a)	235.	(a)	261.	(b)
158.	(c)	184.	(c)	210.	(a)	236.	(c)	262.	(a)
159.	(c)	185.	(b)	211.	(d)	237.	(b)	263.	(a)
160.	(a)	186.	(d)	212.	(b)	238.	(a)	264.	(a)
161.	(b)	187.	(a)	213.	(a)	239.	(a)	265.	(b)
162.	(b)	188.	(b)	214.	(c)	240.	(c)	266.	(a)
163.	(c)	189.	(a)	215.	(d)	241.	(d)	267.	(a)
164.	(b)	190.	(a)	216.	(d)	242.	(c)	268.	(a)
165.	(a)	191.	(a)	217.	(a)	243.	(a)	269.	(c)
166.	(c)	192.	(a)	218.	(d)	244.	(c)	270.	(c)
167.	(a)	193.	(b)	219.	(c)	245.	(a)	271.	(c)
168.	(d)	194.	(a)	220.	(a)	246.	(a)	272.	(d)
169.	(b)	195.	(d)	221.	(b)	247.	(b)	273.	(b)
170.	(b)	196.	(a)	222.	(a)	248.	(a)	274.	(d)
171.	(a)	197.	(b)	223.	(a)	249.	(a)	275.	(a)
172.	(c)	198.	(a)	224.	(a)	250.	(a)	276.	(c)
173.	(a)	199.	(d)	225.	(b)	251.	(d)	277.	(d)
174.	(a)	200.	(b)	226.	(a)	252.	(b)	278.	(a)
175.	(a)	201.	(d)	227.	(a)	253.	(a)	279.	(a)
176.	(d)	202.	(c)	228.	(a)	254.	(c)	280.	(c)

Chapter 14

Wildlife and Conservation

1. **The Latin name for black buck is:**
   (a) *Antilope cervicarpa* (b) *Moschiala memima*
   (c) *Rusa unicolor* (d) Axis axis
2. **The floating national park is situated in the state of:**
   (a) Kerala (b) Assam
   (c) Manipur (d) Tamil Nadu
3. **Wildlife protection act was enacted during:**
   (a) 1975 (b) 1980
   (c) 1972 (d) 1995
4. **The project tiger was launched in the year:**
   (a) 1972 (b) 1995
   (c) 1973 (d) 1980
5. **Corbet National Park is located in:**
   (a) Madhya Pradesh (b) Uttarakhand
   (c) Uttar Pradesh (d) Himachal Pradesh
6. **The latin name for lion tailed macaque is:**
   (a) *Macaca silenus* (b) *Macaca aatoides*
   (c) *Macaca nemtoetrina* (d) *Macaca assamensis*

7. **Wildlife protection act of 1972 was amended during the year:**
   (a) 1988 (b) 1990
   (c) 1991 (d) 1980

8. **The Keoladeo Ghana National Park is situated in the state of:**
   (a) Rajasthan (b) Odisha
   (c) Assam (d) Uttar Pradesh

9. **Ponking call is attributable to:**
   (a) Sambar (b) Chital
   (c) Wolf (d) Tiger

10. **In Wildlife Protection Act 1972, Nilgiri Tahr comes under:**
   (a) Schedule I part II (b) Schedule I part I
   (c) Schedule II part II (d) Schedule III

11. **The census method followed for Asiatic lion is:**
   (a) Pugmark (b) Call
   (c) Waterhole (d) Transect

12. **Wild Asiatic Water Buffalo is distributed in the state of:**
   (a) Assam (b) Himachal Pradesh
   (c) Karnataka (d) Kerala

13. **Kanha National Park is established in the year:**
   (a) 1948 (b) 1950
   (c) 1955 (d) 1961

14. **Kanha National Park is located in the state:**
   (a) Karnataka (b) Uttar Pradesh
   (c) Madhya Pradesh (d) Assam

15. **Which of the following is the largest bird in India?**
   (a) Hornbill (b) Great Indian Bustard
   (c) Sarus crane (d) Monal pheasant

16. **Andaman teal (*Anas gibberifrons algularis*) is a:**
   (a) Mammal (b) Reptile
   (c) Amphibian (d) Bird

17. **Indian has seen the extinction of one of the finest cats namely:**
   (a) *Panther pardus* (b) *Axis tigris c*
   (c) *Actnonyx jubatus venaticus* (d) *Neofelis nebulosa*

18. *Chelonia mydas* is commonly called as:

(a) Olive Ridley turtle (b) Green turtle

(c) Travancore tortoise (d) Starred tortoise

19. Ophiophagus Hannah is commonly called as:

(a) Indian cobra (b) King cobra

(c) Rat snake (d) Python

20. Common langur (*Presbytis entellus*) is the animal coming under:

(a) Schedule I (b) Schedule II

(c) Schedule III (d) Schedule IV

21. Which one of the following animals is not a vermin?

(a) Common crow (b) Fruit bat

(c) Field rat (d) Wild cat

22. Who wrote the autobiographical book, My Life: My Trees?

(a) Joy adamson (b) Richard St. Barbe Baker

(c) Konard Lorenz (d) K.M. Tewari

23. In which biogeographical realm does most part of the Asian continent fall?

(a) Nearctic realm (b) Neotropical realm

(c) Palaearctic realm (d) Oriental realm

24. Which tree yields the heaviest wood in the World?

(a) Balsa (*Ochroma pyramidale*)

(b) Black Iron Wood (*Olea laurifolia*)

(c) Durian (*Durio zibethinus*)

(d) Mango (*Mangifera indica*)

25. Which is the only animal in the world to have four horns?

(a) Elk (*Alces alces*)

(b) Chausingha (*Tetracerus quadricornis*)

(c) Black Buck (*Antilope cervicapra*)

(d) Bharal (*Pseudonis nayaur*)

26. Name the only country in the world where both the Lion (Panthera Leo) and the tiger (*Panthera tigris*) are found in the wild?

(a) Uganda (b) India

(c) Zaire (d) Pakistan

27. Which bird constructs the largest unit nest structure in the world?

(a) Red-billed Quelea (*Quelea quelea*)

(b) Sociable weaver (*Philetairus socius*)

(c) Mountain Drongo (*Chaetorhynchus Papuensis*)

(d) Ostrich (*Struthio camelus*)

28. Which is the longest reptile in the world?

(a) King Cobra (*Ophiophagus hannnah*)

(b) Reticulated Python (*Python reticulates*)

(c) Emerald Tree Boa (*Corallus caninus*)

(d) Black Mamba (*Dendroapsis polylepsis*)

29. Which is the world's largest private international nature conservation organization?

(a) Bombay Natural History Society

(b) Wold Wide Fund for nature

(c) Audubon Society

(d) Ranthambhore foundation

30. The largest in terms of area (673 sq km), this bird sanctuary is also a Ramsar site:

(a) Neelapattu

(b) Kolleru

(c) Rollapadu

(d) Rudrasagar lake

31. The first man made bird sanctuary is:

(a) Kaladeo Ghana

(b) Vedanthangal

(c) Ranganthitoo

(d) Bharatpur

32. India has _______ number of tiger reserves:

(a) 50

(b) 23

(c) 14

(d) 20

33. What is Ictheologists?

(a) Study of fish

(b) Study of Birds

(c) Study of Animals

(d) Study of Mammals

34. India's oldest national park is:

(a) Kaziranga

(b) Corbett

(c) Kanha

(d) Kaladeo Ghana

35. To the indiscriminate slaughter to which wild animals were subjected to prior to the 1920s' the idea of the _______ was born.

(a) Game sanctuary

(b) National park

(c) Forest reserve

(d) Wild life sanctuary

36. The Rajaji Sanctuary is located in the state of:

(a) Tamil Nadu
(b) Karnataka
(c) Andhra Pradesh
(d) Uttarakhand

37. The Vettankudi Bird Sanctuary in Tamil Nadu is situated in the district of:

(a) Tirunelveli
(b) Chengalpattu
(c) Ramanathapuram
(d) Thanjavur

38. Which one of the following is incorrectly matched?

(a) Asiantic Lion – *Panthera leopersica*
(b) Leopard – *Panthera bengalensis*
(c) Tiger – *Panthera tigris*
(d) Snow leopard – *Panthera uncia*

39. The Indian Board for Wildlife was first constituted in the year:

(a) 1972
(b) 1982
(c) 1958
(d) 1952

40. The scientific name of Nilgiri tahr is:

(a) *Asinus hemionus*
(b) *Felis sylvestris*
(c) *Canis lupus*
(d) *Hemitrogus hylocrius*

41. One horned rhinoceros is found in:

(a) Kanha National Park
(b) Periyar tiger reserve
(c) Kaziranga national park
(d) Corbet national park

42. How many Biosphere reserve exist in India?

(a) 18
(b) 23
(c) 12
(d) 17

43. Which of the following areas have the highest number of endemic species of Amphibians?

(a) Eastern Ghats
(b) Terais
(c) Western Ghats
(d) Central Indian Highlands

44. The state animals of Tamil Nadu is:

(a) Sambar
(b) Nilgiri thar
(c) Tiger
(d) Lion-tailed Macaque

45. How many project tiger areas are there in India now?

(a) 18
(b) 39
(c) 19
(d) 30

46. **Special game is included in which of the following schedules of the wildlife protection act of 1972?**
    (a) II (b) III
    (c) IV (d) V

47. **The simplest behavior of animal is:**
    (a) Kineses (b) Taxes
    (c) Reflexes (d) None of the above

48. **Kelakad Mumdanthurai Tiger reserve is located in:**
    (a) Andhra Pradesh (b) Kerala
    (c) Karnataka (d) Tamil Nadu

49. **Nilgiris Biosphere reserve is found in:**
    (a) Tamil Nadu (b) Kearala
    (c) Karnataka (d) All of the above

50. **The Tamil Nadu Animals and Birds Sacrifices Prohibition Act was enacted in the year:**
    (a) 1950 (b) 1960
    (c) 1970 (d) 1980

51. **The staffing of wild animals is:**
    (a) Animal article (b) Animal trophy
    (c) Taxidermy (d) None of the above

52. **The animal coming under Schedule I part II of the wildlife protection act, 1972 is:**
    (a) Agra monitor lizard (b) Blackbuck
    (c) Caracal (d) Dugong

53. **Ptyas mucosus is commonly called as:**
    (a) Indian cobra (b) King cobra
    (c) Rat snake (d) Python

54. **Indian cobra is scientifically called as:**
    (a) *Ophiophagus hannah* (b) *Naja naja*
    (c) *Python molurus* (d) *Ptyas mucosus*

55. **Digitigrade locomotion is used by:**
    (a) Dogs (b) Cats
    (c) Wolves (d) All of the above

56. Unguligrade locomotion is found in:

(a) Cow (b) Deer

(c) Horse (d) All of the above

57. The dropping of carnivores are called as:

(a) Dung (b) Pellet

(c) Bolus (d) Scat

58. Olive Ridley turtle is scientifically called as:

(a) *Chelonian mydas* (b) *Lepidochelys divacea*

(c) *Geochelone travancoria* (d) *Geochelone elegans*

59. Green turtle is scientifically called as:

(a) *Chelonian mydas* (b) *Geochelone elegans*

(c) *Geochelone travancoria* (d) *Lepidochelys divacea*

60. Geochelone elegans is called as:

(a) Starred tortoise (b) Travancore tortoise

(c) Green turtle (d) Olive Ridley turtle

61. Animal adaption to climbing is:

(a) Scansorial (b) Fossorial

(c) Cursorial (d) Volant

62. Animal adaption to flight is:

(a) Scansorial (b) Fossorial

(c) Cursorial (d) Volant

63. The animal coming under perissodactyla includes:

(a) Rhinoceros (b) Horse

(c) Zebra (d) All of the above

64. The study of the combination of animals behavior and ecology is:

(a) Bioecology (b) Ethology

(c) Autecology (d) Synecology

65. The competitive exclusion principle in animal is called as:

(a) Liebig's law minimum (b) Shelford's law of tolerance

(c) Gause's principle (d) None of the above

66. _______ National Park in Kerala celebrates its 25th Year of declaration as a national Park in 2010?
    (a) Silent Valley National Park
    (b) Periyar National Park
    (c) Eravikulam National Park
    (d) Pambadum Shoal National Park

67. Which animals are captured in large numbers for pet trade?
    (a) Parrots
    (b) Love birds
    (c) Corals
    (d) All of the above

68. What is wild life symbol of WWF?
    (a) Red panda
    (b) Tiger
    (c) Elephant
    (d) Leopard

69. When does the wild life week is celebrated in India?
    (a) First week of June
    (b) First week of October
    (c) First week of March
    (d) First week of August

70. Name the only Indian state where all three species of Indian crocodiles are found wild:
    (a) Uttar Pradesh
    (b) Bihar
    (c) Odisha
    (d) Andhra Pradesh

71. Cheetah is found in:
    (a) Gir forest
    (b) West Bengal
    (c) Deccan Plains
    (d) Extinct

72. The first marine national park is in:
    (a) Tamil Nadu
    (b) Gujarat
    (c) Bihar
    (d) West Bengal

73. Rhino usually produces one calf after a gestation period of _______ months:
    (a) 12 months
    (b) 15 months
    (c) 18 months
    (d) 24 months

74. Asian elephants have a life span of _______ years.
    (a) 20-30 years
    (b) 30-40 years
    (c) 50-60 years
    (d) 60-70 years

Answer Sheet

1.	(a)	16.	(d)	31.	(a)	46.	(a)	61.	(a)
2.	(b)	17.	(c)	32.	(b)	47.	(c)	62.	(d)
3.	(c)	18.	(b)	33.	(a)	48.	(d)	63.	(d)
4.	(c)	19.	(b)	34.	(c)	49.	(d)	64.	(b)
5.	(b)	20.	(b)	35.	(a)	50.	(a)	65.	(c)
6.	(a)	21.	(d)	36.	(d)	51.	(c)	66.	(a)
7.	(c)	22.	(b)	37.	(c)	52.	(a)	67.	(d)
8.	(a)	23.	(c)	38.	(a)	53.	(c)	68.	(a)
9.	(a)	24.	(b)	39.	(d)	54.	(b)	69.	(b)
10.	(b)	25.	(b)	40.	(d)	55.	(d)	70.	(c)
11.	(c)	26.	(b)	41.	(c)	56.	(d)	71.	(d)
12.	(a)	27.	(b)	42.	(c)	57.	(d)	72.	(b)
13.	(c)	28.	(b)	43.	(c)	58.	(b)	73.	(c)
14.	(c)	29.	(b)	44.	(b)	59.	(a)	74.	(d)
15.	(c)	30.	(b)	45.	(b)	60.	(a)		

Chapter 15

Forest Economics and Marketing

1. **Economic laws are not determine as:**
   (a) Physical laws
   (b) Moral laws
   (c) Government Laws
   (d) All of the above
2. **Economics is an _______:**
   (a) Art
   (b) Science
   (c) Art and Science
   (d) None of the above
3. **Economics is primarily a study of _______:**
   (a) Man
   (b) Wealth
   (c) Wants
   (d) None of the above
4. **_______ is th best measure to find worthiness of investment:**
   (a) Payback period
   (b) Benefit cost ratio
   (c) Net present value
   (d) Internal rate of return
5. **Principle of substitution is used in:**
   (a) Product min
   (b) Selection of enterprise
   (c) Selection of practice
   (d) None of the above
6. **The least cost principle is sued in:**
   (a) What to produce
   (b) How much to produce
   (c) How to produce
   (d) When to produce

7. **When total product means constant, the marginal product will be:**
   (a) Zero (b) Negative
   (c) Positive (d) Constant

8. **When Average Product < Marginal Product, Average Product is:**
   (a) Remains constant (b) At maximum
   (c) Increasing (d) Decreasing

9. **A production function will unitary elasticity indicates:**
   (a) Increasing returns (b) Decreasing returns
   (c) Nothing (d) Constant returns

10. **In the relational region of production function, choice of indicator is:**
    (a) Factor factor price ratio (b) Factor product prices
    (c) Product product price ratio (d) Factor product quantity

11. **The relationship between total product and total cost is:**
    (a) Directly proportional (b) No relationship
    (c) Inversely related (d) None of the above

12. **When marginal cost is at lowest, marginal product will be:**
    (a) Decreasing (b) Maximum
    (c) Increasing (d) Minimum

13. **In balanced Zone:**
    (a) Marginal Cost = Average cost (b) Marginal Cost > Average cost
    (c) Marginal Cost < Average cost (d) None of the above

14. **If Marginal Rate of product substitution between two products is negative, then two goods are:**
    (a) Substituting (b) Complementing
    (c) Supplementing (d) No relationship

15. **In case of diminishing substitution the curve will be:**
    (a) Convex to origin (b) Straight line
    (c) Concave to origin (d) Vertical or Horizontal

16. **Optimum choice of enterprises or combination of enterprises is made based on:**
    (a) Principle of least cost (b) Principle of variable proportions
    (c) Principle of opportunity cost (d) Principle of cost

17. In LP, t he availability and input output coefficients as well as prices are known with certainty then assumption is:

(a) Linearity
(b) Additively
(c) Finiteness
(d) Single value expectations

18. In maximum returns interaction, the outgoing column will be located with the help of:

(a) Highest positive value
(b) Lowest positive value
(c) Lowest negative value
(d) Highest negative value

19. Marketable surplus is given by:

(a) Marketable surplus = Total production + Total Requirement
(b) Marketable surplus = Total production + Total Requirement/2
(c) Marketable surplus = Total production - Total Requirement
(d) Marketable surplus = Total production -Total Requirement

20. DMI headquarters is located at:

(a) Hyderabad
(b) Faisalabad
(c) Ghaziabad
(d) Faridabad

21. World Standard Day is celebrated on:

(a) 14th January
(b) 14th November
(c) 14th October
(d) 14th December

22. Co-operative Credit Societies Act was passed in:

(a) 1902
(b) 1903
(c) 1904
(d) 1912

23. First Land Mortgage bank in 1920 established at:

(a) Karnataka
(b) Bihar
(c) West Bengal
(d) Punjab

24. Nationalization of banks was carried out on:

(a) 19th June, 1969
(b) 19th July, 1969
(c) 19th August, 1969
(d) 19th October, 1969

25. In case of amortized even repayment plan, the interest rate:

(a) Increases
(b) Declines
(c) Remains constant
(d) None of the above

26. If liabilities > assets, then net worth is placed at:

(a) Liabilities side
(b) Assets side
(c) Income side
(d) None of the above

27. Book value of durable assets refers to:

(a) Market price of assets (b) Purchase price of assets

(c) Production of assets (d) Value at cost

28. Break even analysis is carried out by:

(a) Break even analysis = price per unit/fixed cost –variable cost

(b) Break even analysis =fixed cost/price per unit – variable cost

(c) Break even analysis = Total cost/fixed cost – variable cost

(d) Break even analysis = price per unit – variable cost/fixed cost

29. Marginal value products of scarce resources is known by:

(a) Linear programming (b) Production function

(c) Planning (d) Both (a) and (b)

30. Tangible benefit of project is:

(a) Income distribution (b) Standard living

(c) National prosperity (d) Higher yield level

31. Shadow prices of resources are used in:

(a) Economic analysis (b) Financial analysis

(c) Econometric analysis (d) Technical analysis

32. Benefit Cost ratio is used to evaluate mostly:

(a) Construction of dam (b) Construction of building

(c) Construction of house (d) Construction of cattle shed

33. Elasticity cannot be explained by:

(a) Regression (b) GLS

(c) OLS (d) Correlation

34. 'Economics is what economist do' given by:

(a) Alfred Marshall (b) Lord Robbins

(c) JM Keynes (d) Jacob veneer

35. Marginal utility of money cannot be:

(a) Positive (b) Negative

(c) Zero (d) Both (a) and (b)

36. Indifference curves for perfect substitutes will be:

(a) Convex to origin (b) Concave to origin

(c) Perpendicular to origin (d) Straight line

37. If income consumption curve is bending towards Y axis then commodity X is:

(a) Inferior
(b) Luxury
(c) Normal
(d) None of the above

38. For Giffen goods price consumption curve will be:

(a) Backward going
(b) Horizontal
(c) Downward sloping
(d) Upward sloping

39. Quantity purchased will vary inversely with price when income effect is for normal goods:

(a) Zero
(b) Positive
(c) Negative
(d) None of the above

40. Consumer is neither worse off nor better off than before in:

(a) Hicks substitution effect
(b) Marshall substitution effect
(c) Shitsky substitution effect
(d) None of the above

41. When with a change in price the total outlay on commodity remains constant, it is a case of:

(a) Perfect elasticity
(b) Unit elasticity
(c) Perfect in elasticity
(d) Zero elasticity

42. Change in quantity demand refers to:

(a) Upward shift of demand
(b) Downshift of demand
(c) Movement on same curve
(d) None of the above

43. When both a price of substitute and price of a component of commodity X rise, demand for X:

(a) Falls
(b) Rises
(c) No change
(d) All of the above

44. Cross elasticity of demand between petrol and automobiles is:

(a) Negative
(b) Zero
(c) High
(d) Infinite

45. For inelastic goods, fall in the price leads to:

(a) Increase in total elasticity
(b) No change in total elasticity
(c) Decrease in total elasticity
(d) None of the above

46. The engel curve has all goods.

(a) Positive slope
(b) Constant
(c) Negative slope
(d) All of the above

47. The difference between value in use and value in exchange is:

(a) Consumer surplus (b) Marketing surplus

(c) Farm production (d) None of the above

48. Marginal Rate of product Substitution on the isoclines will be:

(a) Positive (b) Zero

(c) Negative (d) Constant

49. Marginal Cost = Marginal Returns = Average Cost = Average Returns shos the long run, equilibrium positive of the:

(a) Monopolist firm (b) Oligopolistic firm

(c) Competitive firm (d) None of the above

50. The concept of 'reserve army of labor' is due to:

(a) David Richardo (b) JS Mill

(c) Adam Smith (d) Karl Marx

51. PQli OF Morris D. Morris does not include:

(a) Infant mortality (b) Life expectancy

(c) Literacy (d) Standard of living

52. ICOR is in less developed countries:

(a) Low (b) High

(c) Constant (d) Zero

53. Intangible transactions are recorded in:

(a) Service account (b) Goods account

(c) Unilateral transfers (d) Tangible account

54. Snake in the tunnel system is followed in:

(a) USA (b) SE Asia

(c) African nations (d) European Union

55. Keynes is related to which inflation theory:

(a) Cost push (b) Demand pull

(c) Structural theory (d) None of the above

56. Aggregate supply of labour in case of classical is:

(a) Horizontal line (b) Vertical line

(c) Both horizontal and vertical (d) Intermediate

57. The difference between national income and domestic product is accounted by:

(a) Net factor income from abroad
(b) Budgetary surplus/deficit
(c) Export surplus/deficit
(d) Inflow of capital

58. Gross National Product deflator is:

(a) Nominal Gross National Product/Nominal Gross Domestic Product
(b) Nominal Gross National Product/Real Gross Domestic Product
(c) Nominal Gross National Product/Real Gross Domestic Product
(d) Real Gross National Product/Nominal Gross Domestic Product

59. The demand for capital in essence is a:

(a) Direct demand
(b) Derived demand
(c) Reciprocal demand
(d) Complementary demand

60. Capital formation in agriculture over the years:

(a) Decreasing
(b) Increasing
(c) Both (a) and (b)
(d) Constant

61. Private investment is to public investment in agriculture:

(a) Less than
(b) More than
(c) Equal
(d) Can't say

62. Project planning and budgeting are used to known _______:

(a) Commercial aspects
(b) Financial aspects
(c) Economic aspects
(d) Technical aspects of project

63. Sensitivity analysis is used in:

(a) Preparation of project
(b) Appraisal
(c) Implementation
(d) Forecasting/evaluation

64. 'economic a' is written by:

(a) Marshal
(b) Aristotle
(c) Smith
(d) Mill

65. Nature of economic law is:

(a) Normative
(b) Destructive
(c) Exactness
(d) Positive

66. Law of substitution is given by:

(a) Marshal
(b) Smith
(c) Gossen
(d) Pigon

67. Concept of 'crystallized labor' is given by:

(a) Marshal
(b) Keynes
(c) Mill
(d) Karl Marx

68. When Marginal Physical Product is zero:

(a) Marginal cost is Maximum
(b) Marginal cost become vertical
(c) Marginal cost is Minimum
(d) Marginal cost becomes horizontal

69. When the demand curve is relatively flatter, then demand is:

(a) Relatively inelastic
(b) Relatively elastic
(c) Unitary elastic
(d) All of the above

70. If the demand is elastic, for monopoly:

(a) Higher prices, more profit
(b) Lower price, more profits
(c) Higher price, less profit
(d) Lower price, lower profit

71. The 'Drain Theory' about poverty in India is associated with:

(a) Baldev Ram Miridha
(b) VKRV Rao
(c) Dada Bhai Naoroji
(d) C Bose

72. Benefit cost Ratio is used to evaluate:

(a) Social project
(b) Public project
(c) Both (a) and (b)
(d) None of the above

73. 'Giffen' goods refers to:

(a) Stock of goods
(b) Lack of goods
(c) Inferior goods
(d) Superior goods

74. Perfect elasticity is shown by:

(a) Vertical straight line
(b) Horizontal straight line
(c) Slopping curve from left to right
(d) Slopping curve from right to left

75. Production is a function of:

(a) Profit
(b) Factors
(c) Price
(d) Cost

76. Price theory is a part of:

(a) Micro-economics
(b) Macro-economics
(c) Farm management
(d) Marketing

77. The overhead cost is also known as:

(a) Variable cost (b) Total cost

(c) Marginal cost (d) Fixed cost

78. The ultimate aim of farm management:

(a) Increase gross income (b) Reduce total cost

(c) Optimize factor utilization (d) None of the above

79. Government of India signed World Trade Organization agreement in:

(a) 1981 (b) 1994

(c) 1995 (d) 1996

80. The cause of inflation is:

(a) Increase in money supply

(b) Fall in production

(c) Increase in money supply and fall in production

(d) Decrease in money supply and fall in production

81. The Central Agmark Lab is located at:

(a) Mumbai (b) Nagpur

(c) Bangalore (d) Kolkata

82. IBRD is synonymous to:

(a) World Bank (b) Asian Development Bank

(c) International Monetary Fund (d) World Trade Organization

83. Which of the following is not a farming system?

(a) State farming (b) Collective farming

(c) Rainfed farming (d) Co-operative farming

84. Livestock insurance is included in:

(a) Variable cost (b) Special cost

(c) Fixed cost (d) Marginal cost

85. Farm management is an:

(a) Intra farm science (b) Inter farm science

(c) Inter regional farm science (d) International farm science

86. One of the important fixed cost on a farm is:

(a) Seed cost (b) Land rent

(c) Irrigation cost (d) Feed cost

87. Principle of Equip marginal return is applied when:

(a) The resources are unlimited

(b) The price of resources is high

(c) The price o resources vary

(d) The resources are limited

88. Who pro founded the population theory first?

(a) Adam smith (b) Ricardo

(c) Malthus (d) Marshall

89. When Marginal Physical Product is Zero, Total Physical Product is at its maximum because:

(a) Further addition to Marginal Physical Product is negative

(b) Further addition to Marginal Physical Product is positive

(c) Marginal cost = marginal returns

(d) Elasticity of production is more than one

90. Regulated market ensures:

(a) Procurement price (b) Support price

(c) Remunerative price (d) Fair price

91. Farming system approach ensures the best use of:

(a) Land

(b) Land+Labour

(c) Land+Labour+capital

(d) Land+Labour+capital+management

92. Internal economy refers to:

(a) Transport (b) Banks

(c) Communication (d) Specialization

93. Input-Output relationship is called:

(a) Cost function (b) Production function

(c) Price function (d) None of the above

94. Value Added Tax is a/an:

(a) Direct tax (b) Indirect tax

(c) Both (a) and (b) (d) Cannot be classified

95. Inflation is measured by:

(a) WPI (b) CPI

(c) Both (a) and (b) (d) None of the above

96. Which one of the following refers to 'system of farming'?

(a) Economic (b) Operational

(c) Organizational (d) Cultural

97. An application of the accounting principles to the business of farming:

(a) Farm planning (b) Farm budgeting

(c) Farm accounting (d) Marketing

98. Beef and hides are examples of:

(a) Independent enterprise (b) Supplementary enterprise

(c) Joint enterprise (d) Competitive enterprise

99. Indian farming is published by:

(a) Ministry of agriculture Government of India

(b) ICAR

(c) National seed corporation of India

(d) None of the above

100. A consumer spends his income according to the law of:

(a) Diminishing marginal utility (b) Increasing cost

(c) Equi-marginal utility (d) Least cost

101. A group of technical units are known as:

(a) Farm firm (b) Plant

(c) Economic unit (d) All of the above

102. A market structure with large number of firms selling homogenous product is known as:

(a) Monopoly market (b) Monopolistic competition

(c) Perfect competition (d) None of the above

103. A single convenient unit in production for which technical co-efficient are calculated is known as:

(a) Units of accounting (b) Technical units

(c) Marginal units (d) None of the above

104. A table is a:

(a) Material good (b) Non-material good

(c) Free good (d) None of the above

105. About 15 per cent farmers had long holding in between:

(a) 2-4 ha (b) 0-1ha

(c) 1-2ha (d) None of the above

106. Air is:

(a) Economic goods
(b) Free goods
(c) Transferable goods
(d) All of the above

107. An income increase the expenditure on light, fuel and rend generally:

(a) Increases
(b) Decreases
(c) Remains the same
(d) None of the above

108. At equilibrium, the slope of the indifference curve is:

(a) Equal the slope of the budget line
(b) Greater than the slope of the budget line
(c) Smaller than the slope of the budget line
(d) None of the above

109. At present, the farming system of India has become:

(a) Total fixed cost/100
(b) Total fixed cost/total variable cost
(c) Total fixed cost/output
(d) Total fixed cost/input

110. Average product is equal to marginal product when:

(a) Average product is maximum
(b) Average product is minimum
(c) Marginal product is zero
(d) None of the above

111. Who was the first economist to have evolve the technique of indifference curve Analysis:

(a) David Richardo
(b) Bohm Bawerk
(c) F.Y. Edgeworth
(d) A.C. Pigou

112. Break-even points is determined by:

(a) Marginal cost = marginal returns
(b) Marginal cost =Average variable cost
(c) Marginal cost =average cost = average returns
(d) Total returns = total cost

113. Building machinery and implements are examples:

(a) Variable resources
(b) Fixed resources
(c) Flow resources
(d) None of the above

114. By preparing chairs the woods creates:

(a) Form utility (b) Time utility

(c) Place utility (d) Knowledge utility

115. Central co-operative Bank operates normally at 116:

(a) District level (b) Village level

(c) National level (d) None of the above

116. Chairman of planning commission is:

(a) Prime minister (b) President

(c) Appointed by prime minister (d) Appointed by president

117. Co-operatives for tribes are called as:

(a) Camp co-operative societies (b) Multi-purpose societies

(c) Lead Bank (d) Sativa

118. Creation of utility By way of marketing is known as:

(a) Consumption (b) Production

(c) Distribution (d) Exchange

119. Criteria for goods market information is:

(a) Comprehensive (b) Accuracy

(c) Relevance (d) All of the above

120. Crop farming and milk production is an example of:

(a) Competitive enterprise (b) Complementary enterprise

(c) Supplementary enterprise (d) None of the above

121. Decisions regarding irrigation, conservation and reclamation programs are comes under:

(a) Demand change due to change in price

(b) Demand changes not due to the change of price but due to income

(c) Both (a) and (b)

(d) None of the above

122. Demand for necessary goods is:

(a) Elastic (b) Inelastic

(c) Perfectly elastic (d) None of the above

123. The marginal productivity theory is based on the assumption of:

(a) Imperfect competition (b) Monopolistic competition

(c) Perfect competition (d) Elastic competition

124. Expenditure on fertilizers is called:

(a) Fixed cost (b) Variable cost

(c) Marginal cost (d) Total cost

125. Extension is demand means:

(a) Chang in demand due to change in price

(b) Change in demand due to change in income

(c) Both (a) and (b)

(d) None of the above

126. Extent of land forest in India is:

(a) 21 per cent (b) 20 per cent

(c) 27 per cent (d) 16 per cent

127. Earn budget are used to:

(a) Compare economic profitability of different production activities within among farms

(b) Indicate whether proposed changes will be profitable under a given set of factors

(c) Explore conditions under which certain farm practices profitable or unprofitable

(d) All of the above

128. Farm management concern with obtaining the Maximum yield per hectare and considers per unit cost therefore it is:

(a) Practical science (b) Integrating science

(c) Profitability oriented (d) Broader field

129. Farm management deals with:

(a) Judicious decisions

(b) Use of scarce resources

(c) Profit maximization and family satisfaction

(d) All of the above

130. Who propounded the theory of optimum population first:

(a) Sigwick (b) Malthus

(c) Carr-saunders (d) Kuczyansky

131. Farming planning means:

(a) Farm budgeting (b) Cropping sequence

(c) Types of enterprise (d) None of the above

132. For purchasing of seeds, fertilizers, *etc.*, the loan distributed is called:

(a) Long term loan
(b) Short term loan
(c) Medium term loan
(d) All of the above

133. For purchasing the pump sets, tractor *etc.*, the loan is distributed for:

(a) Long term
(b) Short term
(c) Medium term
(d) All of the above

134. Free goods have a great:

(a) Value in use
(b) Value in exchange
(c) Transferability
(d) Both (b) and (c)

135. Oligopoly is a market situation which is characterized by the existence of:

(a) One firm
(b) Two firms
(c) Several firms
(d) Few firms

136. Human wants are generally:

(a) Static
(b) Recurring
(c) Non-recurring
(d) Limited

137. Monophony imply the existence of a:

(a) Single buyer
(b) Single seller
(c) Single seller Vs Single buyer
(d) All of the above

138. If substitution ratio is equal to price ratio than cost will be:

(a) Minimum
(b) Increasing
(c) Maximum
(d) Decreasing

139. If the income elasticity or demand is greater than one, the commodity is:

(a) A necessity
(b) A Luxury
(c) An inferior good
(d) A essentially

140. If the quantity of a commodity demanded does not change when its price, the demand curve is:

(a) Negatively sloped
(b) Positively slope
(c) Vertically sloped
(d) None of the above

141. If the quantity of commodity demand remains unchanged as its price, the coefficient of price elasticity of demand is:

(a) Greater than one
(b) Zero
(c) Smaller than one
(d) Equal to one

142. If two items are complementary than demand of one item will increase for in piece of another item:

(a) Increase
(b) Decrease
(c) No effect
(d) Not known

143. In classical production function when average product is increasing:

(a) Marginal product is less than average product
(b) Marginal product is equal to average product
(c) Marginal product is greater than average product
(d) None of the above

144. In economics the term labour means:

(a) Only manual labour
(b) Only mental work
(c) Manual and mental work
(d) None of the above

145. Who wrote the book, the theory of Monopolistic competition:

(a) Joan Robinson
(b) E.H. Chamberlin
(c) A.C. Pigou
(d) Frederic Benham

146. In perfect competition, there are:

(a) Large number of buyers and few sellers in the market
(b) Large number of sellers and little number of buyers in the market
(c) Number of buyers and sellers are large and nearly equal
(d) All of the above

147. In zamindari system the land revenue was charged:

(a) Thorough Landlords
(b) Directly form the farmers
(c) Through village panchayats
(d) Through either of these

148. Income tax is a:

(a) Direct tax
(b) Indirect tax
(c) Both (a) and (b)
(d) None of the above

149. Initial utility is the utility of the:

(a) First unit
(b) Last unit
(c) Marginal unit
(d) Third unit

150. Man's wants are:

(a) Limited
(b) Unlimited
(c) Countable
(d) Scarce

151. Marginal cost is equal to:

(a) The average cost of the total output declared

(b) The difference between average cost and the cost of the last unit produced

(c) The cost of the last but one unit produced

(d) The extra cost of an additional unit produced

152. Per hectare calculated cost is known as:

(a) Cost of cultivation (b) Cost of production

(c) Cost of farming (d) Total cost

153. Patents and copy rights are:

(a) Material goods (b) Non-material goods

(c) Transferable goods (d) Scarce goods

154. Net return per hectare is:

(a) Overall efficiency measure (b) Partial efficiency measure

(c) Specific efficiency measure (d) None of the above

155. Market demand curved is faced by:

(a) Monopolist (b) Competitive firm

(c) Joint firm (d) Oligopolistic firm

156. Marshall defines economics as:

(a) The study of mankind in the ordinary business of life

(b) The study of human behavior as a relationship between the ends scarce means which have alternative uses

(c) An enquiry into the nature and causes of the wealth of nations

(d) Fundamentally a science of scarcity

157. Marshall definition of economics has been criticized by:

(a) Richardo (b) Robbins

(c) Ruskin (d) Cartyle

158. Maximum profit will be at the point where:

(a) Marginal cost = marginal returns

(b) Marginal cost is more than marginal returns

(c) Marginal cost is less than marginal returns

(d) None of the above

159. Per centage population of world lives in India is:

(a) 16 (b) 18

(c) 20 (d) 19

160. Regional rural banks are financed by:

(a) Reserve Bank of India (b) Government of India

(c) Nationalized Banks (d) All of the above

161. SFDA are formed on the recommendation of All India Rural Review Committee headed by:

(a) M.L. Danthawala (b) C. Swaminathan

(c) K.N. Raj (d) Ashok Mehta

162. Perfect elasticity is shown by:

(a) Vertical straight line (b) Horizontal straight line

(c) Sloping curve right to left (d) Sloping curve left to right

163. The biggest nationalized Bank is:

(a) Union Bank of India (b) Bank of Baroda

(c) Central Bank of India (d) Bank of India

164. The Cannons of taxation was given by:

(a) Adam Smith (b) Robbins

(c) Pigou (d) Marshall

165. The classical theory of interest is also called as:

(a) The real theory of interest (b) Monetary theory of interest

(c) Lon able funds theory (d) Liquidity preference theory

166. To whom do you attribute the concept of consumers surplus:

(a) AC Pigou (b) Alfred Marshall

(c) HH Gossen (d) Barbura Wotton

167. The law which is very useful in determining the best use of limited resources is:

(a) Law of opportunity cost (b) Law of diminishing return

(c) Law of substitution (d) None of the above

168. The main cause of inflation is in money supply and decrease in production:

(a) Decrease (b) Increase

(c) Remains same (d) None of the above

169. The owner of land is state farming is:

(a) Farmer (b) Capitalist

(c) Government (d) All of the above

170. Who was first economist to have propounded theory of rend?

(a) Joan Robinson
(b) RGD Allen
(c) Edward Chamberlin
(d) Alfred Marshall

171. The principle of taxation was given by:

(a) Marshall
(b) Adam Smith
(c) Hicks
(d) None of the above

172. The process of determination of wages, rent, interest and profit is called:

(a) Consumption
(b) Production
(c) Distribution
(d) None of the above

173. The statement "in the process of production and distribution of gods, Income is generated and distributed amongst the factors of production" was given by:

(a) JB Say
(b) Classical economist
(c) Keynes
(d) Marshall

174. Which of the following definitions of economics including the economic concept of Scales of preferences:

(a) Wealth definition
(b) Welfare definition
(c) Scarcity definition
(d) Isolation definition

175. The total fixed curve is:

(a) Increase with the level of output
(b) Remains constant at all the output levels
(c) Decreases as output increases
(d) Both (b) and (c)

176. The want satisfying power of commodity is called:

(a) Demand
(b) Utility
(c) Consumption
(d) Production

177. Total cost is equal to:

(a) Fixed cost + variable cost
(b) Fixed cost + variable cost/100
(c) Fixed cost – variable cost
(d) None of the above

178. Utility in economic means:

(a) Usefulness
(b) Pleasure
(c) Want-satisfying power of a commodity
(d) Satisfaction

179. Utility of the last unit is called:

(a) Marginal utility
(b) Initial utility
(c) Total utility
(d) Zero utility

180. When marginal product is zero, the total product is:

(a) Maximum
(b) Minimum
(c) Zero
(d) None of the above

181. When total product is highest, marginal product will be:

(a) Maximum
(b) Minimum
(c) Zero
(d) Increasing

182. When variable cost is zero, the total cost will be?

(a) Equal to variable cost
(b) Equal to fixed cost
(c) Equal to average variable cost
(d) None of the above

183. Zamindari system was introduced by:

(a) British
(b) Moguls
(c) Marathas
(d) All of the above

184. World Trade Organization (WTO) was established in:

(a) 1992
(b) 1998
(c) 1995
(d) 1997

185. The Government established Tribal Cooperative Marketing Development Federation of India Limited (TRIFE(d) in:

(a) 2001
(b) 1997
(c) 1987
(d) 2004

186. Reserve Bank of India was set up in:

(a) 1934
(b) 1935
(c) 1904
(d) 1949

187. National Bank for Agriculture and Rural Development (NABARD) Act was passed in 1979:

(a) 1980
(b) 1981
(c) 1982
(d) 1983

188. Export-import Bank of India was set up on 01 January:

(a) 1980
(b) 1982
(c) 1981
(d) 1983

189. National Bank for Agriculture and Rural Development came into presence on:

(a) July 12, 1982
(b) July 22, 1984
(c) October 15, 1986
(d) August 10, 1983

Answer Sheet

1.	(a)	31.	(a)	61.	(b)	91.	(d)	121.	(d)
2.	(c)	32.	(a)	62.	(b)	92.	(d)	122.	(b)
3.	(a)	33.	(d)	63.	(d)	93.	(b)	123.	(d)
4.	(b)	34.	(d)	64.	(b)	94.	(b)	124.	(b)
5.	(a)	35.	(d)	65.	(d)	95.	(c)	125.	(a)
6.	(c)	36.	(d)	66.	(a)	96.	(b)	126.	(a)
7.	(a)	37.	(a)	67.	(d)	97.	(c)	127.	(a)
8.	(c)	38.	(a)	68.	(d)	98.	(b)	128.	(b)
9.	(d)	39.	(b)	69.	(b)	99.	(c)	129.	(d)
10.	(b)	40.	(a)	70.	(b)	100.	(c)	130.	(a)
11.	(c)	41.	(b)	71.	(c)	101.	(b)	131.	(a)
12.	(b)	42.	(c)	72.	(c)	102.	(c)	132.	(b)
13.	(b)	43.	(a)	73.	(c)	103.	(b)	133.	(c)
14.	(a)	44.	(a)	74.	(b)	104.	(a)	134.	(b)
15.	(a)	45.	(c)	75.	(b)	105.	(a)	135.	(d)
16.	(c)	46.	(a)	76.	(a)	106.	(b)	136.	(b)
17.	(d)	47.	(a)	77.	(d)	107.	(c)	137.	(a)
18.	(d)	48.	(d)	78.	(a)	108.	(a)	138.	(a)
19.	(c)	49.	(c)	79.	(b)	109	(c)	139.	(b)
20.	(d)	50.	(d)	80.	(c)	110.	(a)	140.	(c)
21.	(c)	51.	(d)	81.	(b)	111.	(c)	141.	(d)
22.	(c)	52.	(a)	82.	(a)	112.	(d)	142.	(a)
23.	(d)	53.	(a)	83.	(b)	113.	(b)	143.	(c)
24.	(b)	54.	(d)	84.	(c)	114.	(a)	144.	(c)
25.	(c)	55.	(b)	85.	(a)	115.	(a)	145.	(b)
26.	(b)	56.	(b)	86.	(b)	116.	(a)	146.	(c)
27.	(d)	57.	(a)	87.	(d)	117.	(a)	147	(a)
28.	(b)	58.	(b)	88.	(c)	118.	(b)	148	(a)
29.	(d)	59.	(b)	89.	(a)	119.	(d)	149.	(a)
30.	(d)	60.	(a)	90.	(d)	120.	(b)	150.	(b)

151.	**(d)**	**159.**	**(a)**	**167.**	**(a)**	**175.**	**(b)**	**183.**	**(a)**
152.	**(a)**	**160.**	**(d)**	**168.**	**(b)**	**176.**	**(b)**	**184.**	**(c)**
153.	**(b)**	**161.**	**(c)**	**169.**	**(c)**	**177.**	**(a)**	**185.**	**(c)**
154.	**(a)**	**162.**	**(b)**	**170.**	**(d)**	**178.**	**(c)**	**186.**	**(b)**
155.	**(b)**	**163.**	**(c)**	**171.**	**(b)**	**179.**	**(a)**	**187.**	**(c)**
156.	**(a)**	**164.**	**(a)**	**172.**	**(c)**	**180.**	**(c)**	**188.**	**(b)**
157.	**(b)**	**165.**	**(a)**	**173.**	**(a)**	**181.**	**(c)**	**189.**	**(a)**
158.	**(a)**	**166.**	**(b)**	**174.**	**(c)**	**182.**	**(b)**		

Chapter 16

Forest Entomology

1. **The eco-friendly insecticide is:**
   (a) Malathion
   (b) Thiodan
   (c) Nimbicidine
   (d) Permethrin

2. **Pheromone trap attracts:**
   (a) Female bugs
   (b) Caterpillars
   (c) Female moths
   (d) Male moths

3. **Which of the following is not a bio-pesticide?**
   (a) Bioneem
   (b) Biolep
   (c) Dipel
   (d) Carbaryl

4. **Which one of the following is a carbamate pesticide?**
   (a) Aldicarb
   (b) Heptachlor
   (c) Malathion
   (d) Endosulfan

5. **Dimethoate is:**
   (a) Contact insecticide
   (b) Systemic insecticide
   (c) Fumigant
   (d) Stomach poison

6. **Eri silkworm is reared on:**
   (a) Mulberry
   (b) Castor
   (c) Coconut
   (d) Banana

7. **The worker bee is:**
   (a) Drone (b) Sterile male
   (c) Sterile female (d) None of the above

8. **One of the major insect enemy of honey bee is:**
   (a) *Vesa cincta* (b) *Galleria mellonella*
   (c) *Acherontia styx* (d) *Formicaria rufa*

9. **The major predator of lac insect is:**
   (a) *Mantis religiosa* (b) *Eublemma amabilis*
   (c) *Coccinella septumpanata* (d) *Chrysoperla carnea*

10. **The most widely used bacterium in insect control is:**
    (a) *Fusarium* sp. (b) *Bacillus papillae*
    (c) *Bacillus cereus* (d) *Bacillus thuringiensis*

11. **Use of resistant varieties in the IPM of insect is an example of:**
    (a) Legal control (b) Cultural control
    (c) Chemical control (d) Biological control

12. **Which one of the following is the example of insect hormone?**
    (a) Lindane (b) Ecdysone
    (c) Chiti (d) Allethrin

13. **Certain secretions attract the opposite sex of the same species which are termed as:**
    (a) Hormones (b) Kairomones
    (c) Synomones (d) Pheromones

14. **The dwelling place of termites is also known as:**
    (a) Canal (b) Tunnel
    (c) Mound (d) Cave

15. **The insecticide which is being opposed by environmentalist is:**
    (a) DDVP (b) Chlorpyriphos
    (c) Sulphur (d) Endosulphan

16. **Following is the acaricide widely recommended for mite management:**
    (a) Kelthane (b) Rogor
    (c) DDVP (d) Monocrotophos

17. **Following is the pesticide which is of plant origin:**
    (a) Spinoside (b) Bulldoc
    (c) NSKE (d) Allethrin

18. Following granular insecticide is having long residual effect:

(a) Phorate (b) Carbofuran

(c) Aldicarb (d) Quinalphos

19. Following insecticide belongs to organophosphate group:

(a) Sulphur (b) Carbofuran

(c) Malathion (d) Allethirn

20. The insecticide which is used in mosquito mats is:

(a) Endosulphan (b) Diazinon

(c) Malathion (d) Allethrin

21. Following anticoagulant is used for rat control:

(a) Zinc-phosphide (b) Al2Po4

(c) Hydrogen cyanide (d) Warfarin

22. Aphids and white flies are effectively controlled by:

(a) Fumigants (b) Stomach poisons

(c) Contact insecticides (d) Systematic insecticide

23. IPM relies heavily on:

(a) Economic threshold level (b) Total loss of crops

(c) Resistant pest (d) Susceptible pest

24. Project Directorate of Biological Control is located at:

(a) Bangalore (b) New Delhi

(c) Faridabad (d) Jaipur

25. The following is the most notorious polyphagous pest:

(a) *Chilo incertulas* (b) *Mythimna separate*

(c) *Helicoverpa armigera* (d) *Gangara thyrsis*

26. Anamorphosis is encountered the following insect order:

(a) Protura (b) Dictyoptera

(c) Ephemereptera (d) Coleopteran

27. Sting in worker honey bee is a modified part of:

(a) Antenna (b) Ovipositor

(c) Styli (d) Cerci

28. Following is an example for secondarily wingless insect:

(a) Blister beetle (b) Head louse

(c) Collembolan (d) Silver fish

29. _______ is an important major pest species in the complex of pest attacking a crop:

(a) Sporadic pests (b) Regular pest
(c) Key pest (d) Endemic pest

30. Flooding of the fields will help in control of:

(a) White grubs (b) Termites
(c) Cut worms (d) All of these

31. Destructive Insects Pest Act (DIP(a) was passed on:

(a) 1914 (b) 1920
(c) 19374 (d) 1968

32. Directorate of Plant Protection Quarantine and Storage (DPPQS) was established in 1946 with head quarters:

(a) Ghaziabad (b) Faizabad
(c) Faridabad (d) Faislabad

33. Biocontrol agent for lantana weed (*Lantana camera*) is:

(a) *Ophiomyia lantanae* (b) *Neochetina eichorniae*
(c) *Crytobagus singularis* (d) *Zygogramma bicolorata*

34. *Bacillus thuringiensis* was discovered by:

(a) S. Ishiwata (b) Berliner
(c) Flacher (d) Frisch

35. Commercially granulosis virus (GV) used for the control of:

(a) *Helicoverpa armigera* (b) *Spodoptera litura*
(c) *Chilo infuscatellus* (d) None of the above

36. Which of the following repellent is used against termites?

(a) Oil of citronella (b) Deet
(c) Pentachloro phenol (d) Bordeaux mixture

37. The term "third generation pesticides" is given for:

(a) Juvenile hormone (b) Moulting hormone
(c) Brain hormone (d) Anti Juvenile hormone

38. The non-preference plant resistance to insects is also known as:

(a) Antibiosis (b) Antixenosis
(c) Tolerance (d) None of the above

39. Father of host plant resistance is:

(a) Kogan and Ortman
(b) RH Painter
(c) Zeidler
(d) Paul Muller

40. Which of the following pairs is not correct?

(a) Endosulfan Thiodan
(b) Chlorpyrifos Dursban
(c) Dimethoate Rogar
(d) Phorate Nuvacron

41. Which one of the following pairs is not correctly matched?

(a) Phosphomidon – Dimecron
(b) Aldicarb – Temik
(c) Carbofuran – Furadan
(d) Carbaryl –Celphos

42. *Bacillus thuringiensis* is available in market as:

(a) Vertilec
(b) Dipel, Delfin
(c) Larvocel, Boverin
(d) Biomax

43. Atropine is a/an:

(a) Repellent
(b) Antifeedant
(c) Antidote
(d) Pheromone

44. Central Insecticides Laboratory (CIL) is located at:

(a) Gurgaon
(b) Faridabad
(c) Ghaziabad
(d) Kanpur

45. Which of the following pair is not correct?

(a) Tasar silk worm –*Antheraea mylitta*
(b) Munga silk worm – *Anthera assame*
(c) Eri silk worm- Sanja *Cynthia ricini*
(d) Mulberry silk worm – *Antheraea* spp.

46. Scientific name of the Dammeer bee is:

(a) *Apis dorsata*
(b) *Melipona irridipennis*
(c) *Apis mellifera*
(d) *Apis indica*

47. The process of leaving off the colony by queen is known as:

(a) Absconding
(b) Swarming
(c) Supersedure
(d) Queen excluder

48. The first abdominal segment when fused with metathorax in bees ins known as:

(a) Propodeum
(b) Mandibles
(c) Basitarsus
(d) Mesopodeum

49. Safest chemical for honey bee is:

(a) Malathion
(b) Methyl parathion
(c) Endosulfan
(d) None of the above

50. Bee venom is used for _______ and released from acid glands:

(a) Curing pains at joints
(b) Repellent against greater wax moth
(c) Pollinization
(d) None of the above

51. In honey bees, the queen substance is secreted by:

(a) Mandibular glands
(b) Acid glands
(c) Pharyngeal glands
(d) Wax glands

52. Rangeeni strain is related to:

(a) Bee
(b) Silk worm
(c) Lac insect
(d) None of the above

53. In India more honey is obtained from which bee?

(a) Little bee
(b) Indian bee
(c) Italian bee
(d) Rock bee

54. Which pesticide is suitable for controlling the white grub?

(a) Malathion
(b) Chlorpyriphos
(c) Phorate
(d) Aldicarb

55. Insects that attack young twigs, stems or buds are _______.

(a) Bark borers
(b) Defoliators
(c) Sap suckers
(d) Tip feeders

56. Caterpillars are an example of _______ who may destroy all or part of t he leaf.

(a) Bark borers
(b) Defoliators
(c) Sap suckers
(d) Tip feeders

57. The study of forest insects is called _______.

(a) Forest insectology
(b) Forest entomology
(c) Forest pestology
(d) Forest ornithology

58. Aphids damage forest trees by _______.

(a) Chewing
(b) Sucking
(c) Defoliating
(d) All of the above

59. Most insect damage is done while the insects are in the _______ stage.

(a) Egg (b) Larva

(c) Pupa (d) Adult

60. Soft cambium and newly formed wood is damaged by these enemies which leave tunnels as a sign of damage.

(a) Bark borers (b) Defoliators

(c) Sap suckers (d) Tip feeders

61. Aphids and scales are examples of _______.

(a) Bark borers (b) Defoliators

(c) Sap suckers (d) Tip feeders

62. Insects that produce abnormal growth on twigs, limbs, trunks or leaves are _______:

(a) Gall forming insects (b) Root feeders

(c) Seed eaters (d) Wood borers

63. Which one of the following insect pests lays down a silky trail on foliage or bark?

(a) Aphid (b) Gypsy moth

(c) Spider mite (d) Tent caterpillar

64. Grubs and wire worms are examples of _______:

(a) Defoliators (b) Gall makers

(c) Root feeders (d) Wood borers

65. The term biological control refers to _______:

(a) Pest control chemicals containing mostly carbon

(b) Pest control using natural agents

(c) Pest control chemicals that are created in a lab by biologists

(d) Synthetic pest chemicals

66. Caterpillars are an example of _______.

(a) Bark borers (b) Defoliators

(c) Sap suckers (d) Tip feeders

67. Which of the following is a biological pest management strategy?

(a) Coat uninfest wood with paint

(b) Harvest mature trees or stands

(c) Release parasites or predators

(d) Use pesticides

68. Convulsions, secretions from nose and ears, loss or consciousness and inability to breathe are symptoms of _______.

(a) Heat stroke
(b) Hypothermia
(c) Severe pesticide poisoning
(d) Smoke inhalation

69. The control of a damaging insect population by its natural enemies is an example of _______:

(a) Mechanical control
(b) Chemical control
(c) Management control
(d) Natural control

70. The most destructive group of forest insects are the _______:

(a) Borers
(b) Leaf feeders
(c) Plant sucking insects and mites
(d) Gall insects and mites

71. The severity of damage to forest trees from leaf feeders depends on _______.

(a) The season when the foliage damage takes place
(b) The extent of the damage
(c) The frequency of the damage
(d) All of the above

72. Severe defoliation that occurs _______ is most harmful to trees:

(a) Late in the season
(b) In September and October
(c) As the spring foliage is unfolding
(d) In July and August

73. The gypsy moth over winters in the _______ stage.

(a) Adult
(b) Pupae
(c) Egg
(d) Larva

74. The damage done by the gypsy moth is when it is in the _______ stage.

(a) Adult
(b) Pupae
(c) Egg
(d) Larva

75. The male gypsy moth locates the female because she _______:

(a) Secretes a powerful scent called a pheromone
(b) Is bright orange in color
(c) Is about 5 inches long and quite colorful
(d) Emits a shrill mating call

76. Chemicals used to control only forest insects are called _______.

(a) Forest pesticides
(b) Forest herbicides
(c) Forest insecticides
(d) Forest fungicides

77. A notorious weed, *Lantana camara* is controlled by:

(a) Lantana bug
(b) Shoot fly
(c) Cochnial insect
(d) None of the above

78. Directorate of plant protection, quarantine and storage started in:

(a) 1940 (b) 1946

(c) 1950 (d) 1552

79. Fossil records indicate that insect evolved during:

(a) Cambrian period (b) Devonian period

(c) Silurian period (d) Precambrian period

80. Economic threshold level is always:

(a) Lower than economic injury level (EIL)

(b) Equal to EIL

(c) More than EIL

(d) Reached when severe damage is done

81. Entomological Society of India started in:

(a) 1934 (b) 1936

(c) 1938 (d) 1940

82. Factor on which penetration of insecticide through insect cuticle does not depend:

(a) Thickness of the cuticle

(b) Chemical nature of the cuticle

(c) Components of the cuticle]

(d) Nature of the carriers and solvents

83. Feeding in insects can be checked by the use of:

(a) Hormone (b) Pheromone

(c) Feeding deterrent (d) Antibiosis

84. In India, lac is mainly produced in:

(a) Tamil Nadu (b) Bihar

(c) Kerala (d) Rajasthan

85. India is exporting shellac to:

(a) China (b) Japan

(c) USA (d) Pakistan

86. India's place in the production of mulberry silk in the world is:

(a) First (b) Second

(c) Third (d) Fourth

87. Insecticide having antagonistic effect with NPV:

(a) DDT (b) Aldrin
(c) Endosulfan (d) Malathion

88. Insecticides approved for aerial spray:

(a) Lindane 20 EC (b) Malathion 50 EC
(c) Phos amidon 85 WSC (d) All of the above

89. Most of the soil fumigants are applied with:

(a) Applicator (b) Injector
(c) Generator (d) Dispenser

90. Light traps are used for the collection of:

(a) Positively phototropic (b) Negatively phototropic
(c) Geotropic (d) Hydrotropic

91. Malathion is a _______ insecticide:

(a) Slow fumigant (b) Contact plus systemic
(c) Contact plus stomach (d) Systemic

92. Most of the insects possess:

(a) One pair of wings (b) Two pair of wings
(c) Three pair of wings (d) None of the above

93. Number of insects known:

(a) About 50000 (b) About 1 million
(c) About 2 million (d) About 3 million

94. Pheromones are also known as:

(a) Antibiotics (b) Ectohormones
(c) Hormones (d) Attractants

95. Pheromones are also known as:

(a) Thoracic glands (b) Glandular epidermal
(c) Neuro-secretory cells (d) All of the above

96. Phorate is _______ insecticide:

(a) Systemic (b) Contact plus systemic
(c) Contact (d) Contact plus stomach

97. Termite control can be also done by:

(a) Tilling (b) Clean cultivation
(c) Irrigation (d) Trap cropping

98. The author of "Indian insect life":

(a) Hussain (b) Fletcher

(c) Lefroy (d) Mani

99. The central insecticide Act came into force from:

(a) 1950 (b) 1966

(c) 1971 (d) 1975

100. The damaging stage of cut worm is:

(a) Caterpillar (b) Nymph

(c) Adult (d) All of the above

101. The density of pest population at which control measures should be applied:

(a) GEL (b) PBL

(c) EIL (d) ETL

102. The first Government entomologist of India:

(a) Niceville (b) Stebbing

(c) Pruthi (d) Ramakrishna Ayyar

103. The first discovered protective insecticide was:

(a) B.H.C. (b) D.D.T.

(c) Diazinon (d) None of the above

104. The first Indian imperial Entomologist:

(a) Hem Singh Pruthi (b) Afzal Hussain

(c) Ram Chandra Rao (d) C.S. Mishra

105. The grasshopper possess:

(a) Natatortial legs (b) Foragial legs

(c) Saltatorial legs (d) Ambulatorial legs

106. The grinding of food in grasshopper is done by:

(a) Mandibles (b) Crop

(c) Gizzard (d) Midgut

107. The Indian lac Research Institute is located at:

(a) Patna (b) Cuttack

(c) Gaya (d) Ranchi

108. The latest philosophy in insect pest is control pest control is:

(a) Eradication (b) Pest management

(c) Control (d) Integrated control

109. The main characteristic feature of an insect is:
(a) Presence of wings (b) Three pairs of legs
(c) Three body regions (d) Pair of antennae

110. The host tree for the cultivation of lac insect is:
(a) Papal (b) Palas
(c) Khejri (d) Babul

111. The main host tree of eri–silk worm is:
(a) Mulberry (b) Sal
(c) Castor (d) Ber

112. The nervous system of insect is derive from:
(a) Ectoderm (b) Mesoderm
(c) Dendoderm (d) Endoderm and mesoderm

113. The paralysis of honey bees is caused by:
(a) Bacteria (b) Virus
(c) Fungus (d) Mycoplasma

114. The population of fruit flies takes placed in:
(a) Soil (b) Rind of fruit
(c) Stubbles in the field (d) Pulp of fruit

115. The prothoracicotropic hormone is secreted by:
(a) Prothoracic glands (b) Neurosecretory cells
(c) Corpora allata (d) Corpora cardiac

116. The queen in the honey colony feeds on:
(a) Royal jelly (b) Nector
(c) Queen substance (d) Honey

117. The ratio of zinc phosphide in poison bait for rats is:
(a) 1 per cent (b) 2 per cent
(c) 3 per cent (d) 4 per cent

118. The queen termite produces _______ egg per second:
(a) 40 (b) 50
(c) 60 (d) 70

119. The reproductive system of insect is derived from:
(a) Ectoderm (b) Endoderm
(c) Ectoderm and mesoderm (d) Ectoderm and Endoderm

120. Use of resistant varieties constitutes methods of:

(a) Genetic control (b) Self control

(c) Biotic control (d) Natural control

121. Vertical resistance is also called as:

(a) Oligogenic (b) Specific

(c) Monogenic (d) All of the above

122. Where the Directorate of plant protection, Quarnatine and storage is situated?

(a) New Delhi (b) Faridabad

(c) Hissar (d) Gurgaon

123. Which animal acts as natural tiller of soil?

(a) Termite (b) Earthworm

(c) Snake (d) Millipede

124. Which insecticides belongs to synthetic pyrethroid group?

(a) Pyrethrum (b) Cypermethrin

(c) Rotenone (d) Sabadilla

125. Which insecticides has contact and fumigant action group?

(a) Phosphamidon (b) Diazinon

(c) Lindane (d) Endosulfan

126. Which insecticide has systemic action?

(a) Endosulfan (b) Chlorpyriphos

(c) Methyldemeton (d) Diazinon

127. Which is the best suiting species for bee culture?

(a) *Apis dorsata* (b) *Apis florae*

(c) *Apis indica* (d) *Milipona* sp.

128. Which are the following are systemeic fungicides?

(a) Vitravax (b) Benlate

(c) Brestonal (d) All of the above

129. Which of the following is called "water scorpion"?

(a) Belostoma (b) Nepa

(c) Corixa (d) Notonecta

130. Which of the following is larval parasitoid?

(a) Xanthopimpla (b) Tetrastichus

(c) Apanteles (d) None of the above

131. Which of the following is pupal parasitoid?

(a) Elasmus (b) Tetrasticus

(c) Xanthopimpla (d) All of the above

132. Which of the following is systemic insecticide?

(a) Parathion (b) Malathion

(c) Cabaryl (d) Carbofuran

133. Which of the following organism is not storage pest?

(a) Rodents (b) Nematodes

(c) Bacteria (d) Fungi

134. Which one is used for inoculation of host tree for lac cultivation?

(a) Seed lac (b) Broad lac

(c) Phunki lac (d) Eri lac

135. Which one of the following insects has not much changed from its fossil records?

(a) Butter fly (b) Praying mantis

(c) Cockroach (d) Grasshopper

136. Which state is famous for sericulture?

(a) Kerala (b) Karnataka

(c) Andhra Pradesh (d) Uttar Pradesh

137. Which type of special reproduction is found in aphid?

(a) Polyembryony (b) Pedogenesis

(c) Parthenogenesis (d) Viviparity

138. Which is often referred to as the 'Father of animal classification'?

(a) Aristotle (b) Carlous Linnnaeus

(c) Wallace (d) Hugo de Varies

139. Honey bee queen produces _______ thousand eggs per day:

(a) 0.5-1.5 (b) 2.0-3.0

(c) 4.0-4.5 (d) 5.0-5.5

140. Inorganic salts are toxic to insects by:

(a) Contact (b) Digestion

(c) Both (a) and (b) (d) Inhalation

141. Insects feeding on plants of several genera within a family are called:

(a) Phytophagous (b) Polyphagous

(c) Oilgophagous (d) Monophagous

142. Integrated Pest Management (IPM) aims at:

(a) Keeping pest population below injurious levels

(b) Increasing natural enemies of the pest

(c) Strengthening the host

(d) Billing the post

143. It has been estimated that insects had existed on this earth for at least _______. Million years:

(a) 100 (b) 150

(c) 200 (d) 250

144. Lac research institute is situated at:

(a) Bangalore (b) Behrampur

(c) Lucknow (d) Namkum

145. Name the insecticide which is harmless to the honey bees:

(a) BHC (b) Carbaryl

(c) Endosulfan (d) Permethrin

146. *Odontotermes obesus* is the scientific name of:

(a) Cutworm (b) Termite

(c) Shoot berer (d) Top Borer

147. Originally the phytopahgous insects were:

(a) Monophagous (b) Oligophagous

(c) Polyphagous (d) All of the above

148. Rodentiside are used to kill:

(a) Rats and mites (b) Insects

(c) Weed (d) None of the above

149. Who coined the term 'Pest Management'?

(a) Geier (b) Bartlett

(c) Dr. Perry Adkisson (d) Dr. Ray F. Smith

150. Which insecticide is more persistent in the environment?

(a) Parathion (b) Aldrin

(c) DDT (d) BHC

151. Marine borer: (Pick the odd man out):

(a) Limnoria (b) Teredo

(c) Banksias (d) Cerambicids

152. Sal borer is:

(a) *Hoplocerambyx spinicornis*
(b) *Sthenias grisator*
(c) *Hyblea peura*
(d) *Atteiva fabriciella*

153. Sisham defoliator is:

(a) *Hyblea peura*
(b) *Plecopteera reflexa*
(c) *Calosomma beesoni*
(d) *Hypsipyla robusta*

154. Hyblaea puera is a defoliator on:

(a) Teak
(b) Deodar
(c) Sal
(d) Toon

155. What is another term for complete metamorphosis?

(a) *Ametabolous*
(b) *Hemimetabolous*
(c) *Holometabolous*
(d) *Paurometabolous*

156. Which insect develops using complete metamorphosis?

(a) Caddisfly
(b) Katydid
(c) Mayfly
(d) Water strider

157. Attack by scales and aphids would result in:

(a) Death by girdling
(b) Defoliation
(c) Forked and crooked trees
(d) Stunted growth

Answer Sheet

1.	(c)	29.	(c)	57.	(b)	85.	(a)	113.	(b)
2.	(d)	30.	(d)	58.	(b)	86.	(c)	114.	(a)
3.	(d)	31.	(a)	59.	(b)	87.	(d)	115.	(b)
4.	(a)	32.	(c)	60.	(a)	88.	(d)	116.	(a)
5.	(b)	33.	(a)	61.	(c)	89.	(b)	117.	(b)
6.	(b)	34.	(a)	62.	(a)	90.	(a)	118.	(d)
7.	(c)	35.	(c)	63.	(d)	91.	(c)	119.	(b)
8.	(b)	36.	(c)	64.	(c)	92.	(b)	120.	(a)
9.	(b)	37.	(a)	65.	(b)	93.	(b)	121.	(d)
10.	(d)	38.	(b)	66.	(b)	94.	(b)	122.	(b)
11.	(b)	39.	(b)	67.	(c)	95.	(b)	123.	(b)
12.	(c)	40.	(d)	68.	(c)	96.	(b)	124.	(b)
13.	(d)	41.	(d)	69.	(d)	97.	(c)	125.	(b)
14.	(c)	42.	(b)	70.	(a)	98.	(d)	126.	(b)
15.	(d)	43.	(c)	71.	(d)	99.	(c)	127.	(c)
16.	(a)	44.	(b)	72.	(c)	100.	(a)	128.	(d)
17.	(c)	45.	(d)	73.	(c)	101.	(d)	129.	(b)
18.	(c)	46.	(b)	74.	(c)	102.	(a)	130.	(c)
19.	(c)	47.	(b)	75.	(a)	103.	(b)	131.	(d)
20.	(d)	48.	(a)	76.	(c)	104.	(a)	132.	(d)
21.	(d)	49.	(c)	77.	(b)	105.	(c)	133.	(b)
22.	(d)	50.	(a)	78.	(a)	106.	(a)	134.	(b)
23.	(a)	51.	(a)	79.	(b)	107.	(d)	135.	(c)
24.	(a)	52.	(c)	80.	(a)	108.	(b)	136.	(b)
25.	(c)	53.	(d)	81.	(c)	109.	(b)	137.	(c)
26.	(a)	54.	(c)	82.	(b)	110.	(b)	138.	(b)
27.	(b)	55.	(d)	83.	(c)	111.	(c)	139.	(b)
28.	(b)	56.	(b)	84.	(b)	112.	(a)	140.	(b)

141.	(c)	145.	(c)	149.	(a)	153.	(b)	157.	(d)
142.	(a)	146.	(b)	150.	(c)	154.	(a)		
143.	(c)	147.	(c)	151.	(d)	155.	(c)		
144.	(d)	148.	(a)	152.	(a)	156.	(a)		

Chapter 17

Forest Pathology

1. **Sandal spike disease of sandal (*Santalum album*) is caused by:**
   (a) Bacteria (b) Fungi
   (c) MLO (d) Virus
2. **Fungi imperfecti includes in:**
   (a) Deuteromycotina (b) Basidiomycotina
   (c) Ascomycotina (d) Oomycetes
3. **Rust includes in:**
   (a) Deuteromycotina (b) Basidiomycotina
   (c) Ascomycotina (d) Oomycetes
4. **Smut includes in:**
   (a) Deuteromycotina (b) Basidiomycotina
   (c) Ascomycotina (d) Oomycetes
5. **Powdery mildew include in:**
   (a) Deuteromycotina (b) Basidiomycotina
   (c) Ascomycotina (d) Oomycetes
6. **Downy mildew include in:**
   (a) Oomycetes (b) Chytridiomycetes
   (c) Zygomycetes (d) Hypochytridiomycetes

7. **Father of Indian Mycology:**
   (a) EJ Butler (b) KC Mehta
   (c) Mundakur (d) RS Singh

8. **Micrografting is used to produce plants free from:**
   (a) Virus (b) RLO's
   (c) MLO's (d) Bacteria

9. **Bacterial diseases are controlled by use of chemicals:**
   (a) Kelthane (b) Fungicide
   (c) Antibiotics (d) Viricides

10. **Application of potash increases:**
   (a) Resistance to water logging (b) Frost resistance in plants
   (c) Disease resistance in plants (d) None of the above

11. **Decomposition of organic matter in submerged soil is carried out by:**
   (a) Bacteria (b) Actinomycetes
   (c) Fungi (d) Earthworm

12. **Select the pair which is not correctly matched:**
   (a) Longidorus – Needle nematode
   (b) Pratylenchus – Lesion nematode
   (c) Ditylenchus – Bulb nematode
   (d) Tylenchulus – Cyst nematode

13. **The root-knot nematode generally have interaction with:**
   (a) *Pseudomonas solanacearum* (b) *Xanthomonas citri*
   (c) *Bacillus subtilis* (d) *Streptomyces scabies*

14. **Select the systemic fungicide was first time discovered by Von Schmeling and Kulka?**
   (a) Pyrimidines (b) Triazoles
   (c) Metalaxyl (d) All of the above

15. **Major genes contributes disease resistance in the form of:**
   (a) Vertical resistance (b) Horizontal resistance
   (c) Both (a) and (b) (d) None of the above

16. **Most widely used fungicide for smut fungi is:**
   (a) Vitavax (b) Plantvax
   (c) Dithane M-45 (d) Dithane Z-78

17. **Race specific resistance called:**
    (a) Vertical resistance
    (b) Horizontal resistance
    (c) Adult plant resistance
    (d) Apparent resistance

18. **When plant showed the partial resistance against all the races of pathogen then it is a type of ?**
    (a) Vertical resistance
    (b) Horizontal resistance
    (c) Induced resistance
    (d) Non-host resistance

19. **Nuclear polyhedrosis virus (NPV) is the most effective for control of:**
    (a) *Chilo partellus*
    (b) *Pectionophora gossypiella*
    (c) *Helicoverpa armigera*
    (d) *Diacrasia oblique*

20. **Solar treatment is used for:**
    (a) Loose smut
    (b) Stem rust
    (c) Powdery mildew
    (d) Blast

21. **A fungus disease which may cause death or damage to newly emerging seedlings is called _______.**
    (a) Ground rot
    (b) Damping off
    (c) Mulching
    (d) Powdery mildew

22. **The study of diseases of forest trees is called:**
    (a) Forest dendrology
    (b) Forest entomology
    (c) Forest pathology
    (d) Forest technology

23. **For which of the following pests would a fungicide be used?**
    (a) Aphids
    (b) Damping off
    (c) Scale
    (d) Sunscald

24. **Pathogens are:**
    (a) Disease causing organisms
    (b) Wood boring insects
    (c) Used in biological control
    (d) Both (a) and (c)

25. **Other than wildfires, which of the following forest enemies are the most damaging?**
    (a) Diseases and insects
    (b) Domestic livestock
    (c) Environment
    (d) Wildlife

26. **2,4-D is a(n):**
    (a) Fungicide
    (b) Herbicide
    (c) Insecticide
    (d) Pesticide

27. Fungicides are chemicals used to control _______.

(a) Rust diseases
(b) Fungus diseases
(c) Powdery mildew
(d) All of the above

28. The disease fire blight is caused by _______.

(a) Fungus
(b) Virus
(c) Bacteria
(d) All of the above

29. Salt damage to a tree is identified by _______.

(a) Bent or broken stems
(b) Tip curl
(c) Foliage necrosis
(d) Twisting of needles or leaves

30. A symptom of winter injury in conifers may be _______:

(a) Blackened and dead roots
(b) Foliage turns red and dies
(c) Small spots on the needles
(d) Stem cankers

31. Which of the following is symptomatic of needle cast diseases?

(a) Brown needles, die back and top kill
(b) Needles bent at the end of the fascicle
(c) Orange discoloration at the base of the needles
(d) Spots on the needles, death of the affected needles

32. For which of the following would a fungicide be used?

(a) Aphids
(b) Damping off
(c) Scale
(d) Sunscald

33. The introduction of a pathogen that attacks certain insects is called a _______.

(a) Biological control
(b) Chemical control
(c) Management control
(d) Mechanical control

34. Forest pathology is study of _______.

(a) Forest ecosystems
(b) Forest insects
(c) Tree diseases
(d) Tree identification

35. Abnormal tissue growth on trees that is cussed by insects or mites is called _______:

(a) Conk
(b) Fungus
(c) Gall
(d) Pitch

36. A canker would best be described as a:

(a) Dead tree top
(b) Globule swelling
(c) Sunken area
(d) Tumor

37. The best way to control forest insects or diseases is to _______.

(a) Spray as soon as a disease or insect is noticed

(b) Use integrated pest management

(c) Harvest all trees near the infestation immediately

(d) All of the above

38. The vector for spike disease in sandal is:

(a) Jassid (b) Psyllid

(c) Green plant hopper (d) White fly

39. Root rot in sal is caused by:

(a) Pseudomonas (b) Polyporus

(c) Ganoderma (d) Pythium

40. Corticium salmonicolor is a disease which attacks:

(a) *Casuarinas* (b) *Acacia*

(c) *Eucalyptus* (d) Tamarind

41. An organism on or within which another organism feeds and develops is called:

(a) Parasites (b) Epiphytes

(c) Host (d) None of the above

42. Stemwilt in Casuarina is caused by:

(a) *Pseudomonas* (b) *Trichosporium*

(c) *Coricium* (d) *Ganoderma*

43. Fungal roots are otherwise called as:

(a) *Frankia* (b) *Mycorrhiza*

(c) *Rhizobium* (d) *Azospirillum*

44. Which of the following is considered a parasitic cause of disease?

(a) Drought (b) Fungi

(c) Ice storms (d) Salt

45. _______.,before 1910 initiated exhaustive study of Indian fungi and disease caused by them.

(a) E.J. Butler (b) JC Mosemam

(c) RE Nelson (d) ALivey

46. A bacterium is devoid of:

(a) Chlorophyll (b) Well defined nucleus

(c) Cell wall (d) Both (a) and (b)

47. A virus particle with both nucleic acid and protein coat is known as:

(a) Virion
(b) Viriod
(c) Capsid
(d) All above

48. Agar agar is produced from:

(a) *Nostoc*
(b) *Macrocrocystis*
(c) *Gelidium*
(d) *Oscillatoria*

49. Rust is a:

(a) Obligate parasite
(b) Saprophyte parasite
(c) Facultative parasite
(d) Predator

50. Alexander Fleming (1929) discovered:

(a) Streptomycetease
(b) Penicillin
(c) Tyrocidine
(d) Giseofulvin

51. Association tinsel flagellum is characteristics of:

(a) Lichen
(b) Mycorrhiza
(c) Ascolichen
(d) Basidiolichen

52. B.B. Mundkur wrote:

(a) Fungi and disease in plants
(b) Fungi and plant diseases
(c) The fungi
(d) All of the above

53. Bacilli are:

(a) Spiral shaped
(b) Spherical shaped
(c) Rod shaped
(d) None of the above

54. Bacterial cell surrounded by flagella is called:

(a) Monotrichous
(b) Lophotrichous
(c) Amphitricous
(d) Peritrichous

55. Plant pathology is _______ word:

(a) English
(b) Latin
(c) Greek
(d) Mandarin

56. Bordeaux mixture was discovered by P.A. Millardet of France during the year 1982 following his chance observation of farmer's practice for protection against:

(a) *Plasmopara viticola* on grapevine
(b) *Uncinula nector* on grapevine
(c) *Podosphaera leucotricha* on apple
(d) *Veturia inequalis* on apple

57. Carborundum power is:

(a) Silicon carbide
(b) Aluminium oxide
(c) Diatomaceous earth
(d) Calcium carbonate

58. Cocci are:

(a) Spiral shaped
(b) Spherical shaped
(c) Rod shaped
(d) None of the above

59. Crystallization of virsus infected plant described as:

(a) Mayer
(b) Stanely
(c) Iwanowski
(d) Bejerinck

60. Cysts are formed by;:

(a) Meloidogyne
(b) Aphelenchoides
(c) Heterodera
(d) None of the above

61. Death of tissue in virus infected plant described as:

(a) Motting
(b) Mosaic
(c) Necrosis
(d) Curling

62. Dithane M-45 is a:

(a) Bactericide
(b) Fungus
(c) Fungicide
(d) Nematicide

63. First plant parasitic bacteria was reported by:

(a) TJ Burrill
(b) Needham
(c) Louis Pasteur
(d) Leeuwenhoeck

64. First plant parasitic nematode was discovered by:

(a) Meloidogyne
(b) Heterodera
(c) Anguina
(d) Globodera

65. First plant parasitic nematode was reported by:

(a) Needham
(b) Spallanzzani
(c) Cob
(d) Mayer

66. Fungi which can grow on living host plant are called:

(a) Obligate saprophyte
(b) Obligate parasite
(c) Facultative parasite
(d) Saprophytes

67. Fusarium belong to order:

(a) Melaconiales
(b) Moniliales
(c) Sporobolomycetales
(d) Cryptococcales

68. Gelatin plate technique was discovered by:

(a) Burrill (b) Fanny Hesse
(c) Robert Koch (d) Louis Pasteur

69. How much quantity of nematicide is used to control the nematodes in one hectare are?

(a) 10-15 kg (b) 20-25 kg
(c) 50-60 kg (d) 100-200 kg

70. In fermentation industry, pasteurization is done at temperature:

(a) 90.7°C (b) 100°C
(c) 62.8°C (d) 80.0°C

71. Indian Journal of Mycology and Plant Pathology is published from:

(a) IARI, New Delhi
(b) Botany Department, Lucknow University
(c) RCA, Udaipur
(d) Botany Department, Madras University

72. Indian Phytopathological Society was established by:

(a) JF Dastur (b) EJ Butler
(c) BB Mundkar (d) KC Mehta

73. Lichen is an association between:

(a) Algae and Fungus (b) Bacterium and Fungus
(c) Fungus and Virus (d) Fungus and Nematode

74. Most effective fungicide against powdery mildew disease:

(a) Dithane (b) Captan
(c) Karathane (d) Bavistin

75. Most of the plant parasitic nematodes belong to order:

(a) Tylenchida (b) Dorylaimida
(c) Aphelenchida (d) Rhabditida

76. Mycology is the study of:

(a) Bacteria (b) Algae
(c) Virus (d) Fungi

77. Mycoplasmsa is sensitive to:

(a) Penicillin (b) Streptomycin
(c) Bavistin (d) Tetracycline

78. Nematodes are also known as:

(a) Tapeworms
(b) Trematodes
(c) Earthworms
(d) Round worms

79. Once the pathogen has established in an area, steps taken to remove or destroy is known as:

(a) Exclusion
(b) Eradication
(c) Protection
(d) Management

80. Phycomycetes belong to:

(a) Fungi
(b) Bacteria
(c) Yeast
(d) None of above

81. Phytophthera belongs to family:

(a) Peronophythoraceae
(b) Pythiaceae
(c) Pareonosporaceae
(d) Albuginaceae

82. Plant parasitic nematodes posses:

(a) Two-part oesophagus
(b) Three-part oesophagus
(c) Four-part oesophagus
(d) None of the above

83. Plant pathogenic viroids contain;:

(a) Externally seed-borne disease
(b) Externally seed-borne and soil-borne disease
(c) Air-borne disease
(d) Internally seed-borne disease

84. Powdery mildew disease is caused by:

(a) Bacterium
(b) Fungus
(c) Virus
(d) Nematode

85. Powdery mildew fungi belong to the family:

(a) Erysiphaceae
(b) hypocereaceae
(c) Hypomycetaceae
(d) Nectriaceae

86. Powdery mildews are controlled by spray of:

(a) Copper fungicides
(b) Mercury fungicides
(c) Sulphur fungicides
(d) Miscellaneous fungicides

87. Rhizopus belongs to class:

(a) Oomycetes
(b) Zygomycetes
(c) Trichomycetes
(d) Ascomycetes

88. Root rot of papaya could be controlled by application of:

(a) Bordeaux mixture (b) Dithane M-45
(c) Benelate (d) Calaxin

89. The founder of mycology was:

(a) Linnaeus (b) Anton de bary
(c) PA Micheli (d) Saccardo

90. The organisms lacking a well defined nucleus are termed:

(a) Fungi (b) Prokaryote
(c) Eukaryote (d) Protozoa

91. The plant mycoplasmal diseases are transmitted by:

(a) Aphids (b) Leaf hoppers
(c) Mites (d) All of these

92. Virus is transmitted by nematode:

(a) *Globodera* (b) *Tylenchorhynchus*
(c) *Anguina* (d) *Xiphinema*

93. Which fungus produces aflatoxin?

(a) *Penicillium notatum* (b) *Aspergillus flavus*
(c) *Albugo candida* (d) *Trichoderma viride*

94. Which is citrus nematode?

(a) *Hoploimus* (b) *Pratylenchus*
(c) *Rotylenchulus* (d) *Tylenchus*

95. Which is sulphur bacteria?

(a) *Nitrosomonas* (b) *Cladethrix*
(c) *Thiothrix* (d) *Bacillus*

96. Which is the root knot nematode?

(a) *Angunia* (b) *Heterodera*
(c) *Meloidogyne* (d) *Tylenchorhynchus*

97. Which of the following nematicide is a carbamate?

(a) DBCP (b) DDT
(c) Phorate (d) Aldicarb

98. Which of the nutrient is supplied by mycorrhiza to the plant?

(a) Nitrogen (b) Phosphorus
(c) Sulphur (d) Iron

99. Which of them is the gram negative bacterium?

(a) *Xanthomonas* (b) *Pseudomonas*

(c) *Erwinia* (d) *Corynebacterium*

100. Which of them is a poisonous mushroom?

(a) *Agaricus* (b) *Volvariella*

(c) *Pleaurotus* (d) *Amanita*

101. Which one of the following is known as downy mildew fungus?

(a) *Erysiphe* (b) *Sclerospora*

(c) *Taphrina* (d) *Urocystic*

102. Which one the following tissues is most suitable for tissue culture to produce virus-free plants of potato?

(a) Leaf tissue (b) Pollen

(c) Embryo tissue (d) Meristem tip tissue

103. Which shows the symbiotic relationship?

(a) *Rhizobium* (b) *Azotobacter*

(c) *Clostridium* (d) *Nitrosomonas*

104. Who for the first time discovered antibiotics?

(a) Alexander flemming (b) Stanely

(c) Anton de bary (d) Louis Pasteur

105. Who gave the term pasteurization?

(a) Anton de bary (b) Louis Pasteur

(c) KC Mehta (d) Leeuwenhoeck

106. Who is known as the father of nematology?

(a) NA Cobb (b) G Thorne

(c) FGW Jones (d) De-man

107. With regard to nutrient Erysiphe is:

(a) Ectoparasite (b) Endoparasite

(c) None of the above (d) Predator

108. _______ proposed a classification of fungi which is now widely accepted.

(a) Walker (1957) (b) Stakman and Harrar (1957)

(c) Whitaker (1969) (d) Ainworth (1973)

Answer Sheet

1.	(c)	23.	(b)	45.	(a)	67.	(b)	89.	(b)
2.	(a)	24.	(a)	46.	(b)	68.	(c)	90.	(b)
3.	(b)	25.	(a)	47.	(c)	69.	(b)	91.	(b)
4.	(b)	26.	(b)	48.	(c)	70.	(c)	92.	(d)
5.	(c)	27.	(d)	49.	(a)	71.	(c)	93.	(b)
6.	(a)	28.	(c)	50.	(b)	72.	(c)	94.	(d)
7.	(a)	29.	(b)	51.	(b)	73.	(a)	95.	(b)
8.	(a)	30.	(b)	52.	(b)	74.	(c)	96.	(c)
9.	(c)	31.	(d)	53.	(c)	75.	(a)	97.	(d)
10.	(c)	32.	(b)	54.	(d)	76.	(d)	98.	(b)
11.	(c)	33.	(a)	55.	(c)	77.	(d)	99.	(d)
12.	(d)	34.	(c)	56.	(a)	78.	(a)	100.	(b)
13.	(a)	35.	(c)	57.	(a)	79.	(b)	101.	(b)
14.	(d)	36.	(b)	58.	(a)	80.	(a)	102.	(a)
15.	(a)	37.	(b)	59.	(b)	81.	(b)	103.	(a)
16.	(a)	38.	(a)	60.	(a)	82.	(a)	104.	(a)
17.	(a)	39.	(b)	61.	(b)	83.	(a)	105.	(b)
18.	(a)	40.	(c)	62.	(c)	84.	(b)	106.	(a)
19.	(c)	41.	(c)	63.	(a)	85.	(a)	107.	(a)
20.	(a)	42.	(b)	64.	(c)	86.	(c)	108.	(d)
21.	(b)	43.	(b)	65.	(a)	87.	(b)		
22.	(c)	44.	(b)	66.	(b)	88.	(a)		

Chapter 18

Forest Policy and Law

1. **First Indian Forest Act was drafted in the year:**
   (a) 1865 (b) 1848
   (c) 1894 (d) 1927
2. **The last forest policy was formulated during the year:**
   (a) 1998 (b) 1972
   (c) 1988 (d) 1952
3. **Forest Conservation Act was enacted during:**
   (a) 1880 (b) 1980
   (c) 1988 (d) 1998
4. **The first Inspector general of forest was:**
   (a) Sir Dietrich Brandis (b) Howard
   (c) Chaturvedi (d) Alfred moller
5. **The National Commission on Agriculture submitted its report during:**
   (a) 1970 (b) 1976
   (c) 1967 (d) 1980
6. **The first Independent National Forest Policy was enunciated during:**
   (a) 1950 (b) 1952
   (c) 1954 (d) 1956

7. **Forest and wildlife were brought under concurrent list as per:**
   (a) 40th amendment (b) 42nd amendment
   (c) 44th amendment (d) 48th amendment

8. **The Head Quarters of Inspector General of Forest:**
   (a) Dehradun (b) Missouri
   (c) Delhi (d) Mumbai

9. **The first Forest policy was enunciated during:**
   (a) 1864 (b) 1874
   (c) 1884 (d) 1894

10. **Declaration of national parks is covered in which section of wildlife protection act, 1972?**
    (a) Section 25 (b) Section 35
    (c) Section 36 (d) Section 37

11. **Conservation of ecological fragile ecosystem and preservation of biological diversity in terms of flora and fauna was emphasized in:**
    (a) Forest policy, 1894 (b) Forest policy, 1952
    (c) Forest policy, 1988 (d) None of the above

12. **Wildlife protection act was enacted during:**
    (a) 1975 (b) 1980
    (c) 1972 (d) 1995

13. **The project tiger was launched in the year:**
    (a) 1972 (b) 1995
    (c) 1973 (d) 1980

14. **India's First Forest Policy was enunciated during:**
    (a) 1894 (b) 1927
    (c) 1952 (d) 1988

15. **Wildlife protection act of 1972 was amended during the year:**
    (a) 1988 (b) 1990
    (c) 1991 (d) 1980

16. **In wildlife protection act 1972, Nilgiri Tahr comes under:**
    (a) Schedule I part II (b) Schedule I part I
    (c) Schedule II part II (d) Schedule III

17. World Forest Day is celebrated on:

(a) 15th June (b) 1st July

(c) 1st October (d) 21st March

18. ICFRE was established during the year:

(a) 1986 (b) 1980

(c) 1975 (d) 1990

19. Forest ranger under which section has got the power for evicting any person un authorisedly occupying any land in reserve forest.

(a) 68A (b) 78A

(c) 71A (d) 61A

20. The Environment (Protection) Act is attributed to the year:

(a) 1972 (b) 1980

(c) 1986 (d) 1988

21. The Tamil Nadu Rosewood trees (Conservation Act) was enacted in the year:

(a) 1974 (b) 1984

(c) 1994 (d) 1998

22. The Tamil Nadu Animals and Birds sacrifices prohibition act was enacted in the year:

(a) 1950 (b) 1960

(c) 1970 (d) 1980

23. The Tamil Nadu Hills Areas (preservation of trees) 1955 includes:

(a) Ootacamund municipal areas (b) Coonoor municipal area

(c) Kodikanal municipal area (d) All of the above

24. The Indian board of wildlife was constitutes in the year:

(a) 1952 (b) 1972

(c) 1966 (d) 1994

25. The national board of wildlife has:

(a) The prime Minister as chairperson

(b) The minister in charge of forests and wildlife as chairman

(c) The director, ICFRE as a chairman

(d) None of the above

26. The chief wildlife warden may permit any person to enter in a sanctuary for:

(a) Study of wildlife (b) Photography
(c) Tourism (d) All of the above

27. The measures of immunization against communicable diseases of livestock are taken within:

(a) 1 km of sanctuary (b) 2 km of sanctuary
(c) 5 km of sanctuary (d) None of the above

28. The staffing of wild animals is:

(a) Animal article (b) Animal trophy
(c) Taxidermy (d) None of the above

29. Trade or commerce in wild animals, animal articles are tropics dealt in which section of wildlife protection act, 1972?

(a) Section 38 (b) Section 39
(c) Section 40 (d) Section 42

30. The animal coming under schedule I part II of the wildlife protection act, 1972 is:

(a) Agar Monitor Lizard (b) Black Buck
(c) Caracal (d) Dugong

31. The Tamil Nadu Animals and Birds in urban Areas (control and regulation) act was enacted in the year:

(a) 1984 (b) 1987
(c) 1994 (d) 1997

32. The animals covered in the Tamil Nadu animals and Birds in Urban Areas (control and regulation) act, 1997 is:

(a) Elephant (b) Tige
(c) Lion (d) None of the above

33. The Indian Board of Wildlife (IBWL) was set up by the government of India in the year:

(a) 1950 (b) 1951
(c) 1952 (d) 1962

34. Bombay National History Society was started in the year:

(a) 1880 (b) 1883
(c) 1948 (d) 1951

35. Dhebar Commission studied the problem of tribals during the year:

(a) 1948-49 (b) 1960-61

(c) 1971-72 (d) 1980-82

36. National Forest Policy, 1988 envisages that _______ per cent of the total geographical area of India should be under forest:

(a) 22 (b) 33

(c) 40 (d) 40

37. GATT (General Agreement on Trade and Tariff) was set up in the year:

(a) 1972 (b) 1979

(c) 1984 (d) 1990

38. India's Patent Act, 1970 does not permit patenting of:

(a) Herb (b) Shrub

(c) Medicinal plants (d) All of the above

39. Sub chapter 6.2 of the Indian Forest Act, 1927 deals with the:

(a) Protected forests

(b) Penalties and procedures

(c) Duty on timber and other forest produce

(d) Reserved forests

40. It shall be the duty of every citizen of India to protect and improve the natural environment including forests, lakes, rivers and wildlife and to have compassion for living creatures. In which article of the constitution of India does it appear?

(a) Article 48A (b) Article 49A

(c) Article 51A (d) Article 52A

41. Common langur is the animal coming under:

(a) Schedule I (b) Schedule II

(c) Schedule III (d) Schedule IV

42. National parks and wildlife sanctuary come under which section of wildlife protection act, 1972?

(a) u/s 15 (b) u/s 25

(c) u/s 35 (d) u/s 45

43. The Tamil Nadu sandal wood transit rules were made in:

(a) 1966 (b) 1967

(c) 1968 (d) 1969

44. Madras forest act is attributable to the year:

(a) 1878 (b) 1882

(c) 1865 (d) 1871

45. Biggame is included in which of the following schedules of the wildlife protection act of 1972?

(a) II (b) III

(c) IV (d) V

46. Man and biosphere (MAB) programme was started during:

(a) 19714 (b) 1980

(c) 1982 (d) 1990

47. The "Charter of Indian Forest" was based on the report given by:

(a) Mc Clelland (b) Voelcker

(c) Brandis (d) Jack Westoby

48. The challans of forest offences are prepared under section:

(a) 12 (b) 14

(c) 26 (d) 36

49. Forest law is:

(a) A general law (b) A special law

(c) An international law (d) A constitutional law

50. The forests are brought on concurrent list by 42nd amendment to the constitution in the year:

(a) 1972 (b) 1976

(c) 1980 (d) 1988

Answer Sheet

1.	(a)	11.	(c)	21.	(c)	31.	(d)	41.	(b)
2.	(c)	12.	(c)	22.	(a)	32.	(d)	42.	(c)
3.	(b)	13.	(c)	23.	(d)	33.	(c)	43.	(b)
4.	(a)	14.	(a)	24.	(a)	34.	(b)	44.	(b)
5.	(b)	15.	(c)	25.	(a)	35.	(b)	45.	(b)
6.	(b)	16.	(b)	26.	(d)	36.	(b)	46.	(a)
7.	(b)	17.	(d)	27.	(c)	37.	(c)	47.	(a)
8.	(c)	18.	(a)	28.	(c)	38.	(d)	48.	(c)
9.	(d)	19.	(a)	29.	(b)	39.	(d)	49.	(d)
10.	(b)	20.	(c)	30.	(a)	40.	(c)	50.	(b)